The Mathematics and Topology of Fullerenes

CARBON MATERIALS: CHEMISTRY AND PHYSICS

A comprehensive book series which encompasses the complete coverage of carbon materials and carbon-rich molecules from elemental carbon dust in the interstellar medium, to the most specialized industrial applications of the elemental carbon and derivatives. A great emphasis is placed on the most advanced and promising applications ranging from electronics to medicinal chemistry. The aim is to offer the reader a book series which not only consists of self-sufficient reference works, but one which stimulates further research and enthusiasm.

Series Editors

Dr. Prof. Franco Cataldo
Via Casilina 1626/A
00133 Rome
Italy

Professor Paolo Milani
University of Milan
Department of Physics
Via Celoria, 26
20133, Milan, Italy

VOLUME 4:
THE MATHEMATICS AND TOPOLOGY OF FULLERENES

Volume Editors

Dr. Franco Cataldo
Prof. of Chemistry
Dept. of Materials Science
Tor Vergata University
Rome, Italy

Prof. Ante Graovac
Faculty of Science, University of Split
Nikole Tesle 12
HR-21000 Split, Croatia

NMR Center
The "Ruđer Bošković" Institute
Bijenička c. 54
HR-10002 Zagreb, Croatia

IMC, University of Dubrovnik
Branitelja Dubrovnika 29
HR-20000 Dubrovnik, Croatia

Dr. Ottorino Ori
Correspondent Member
Actinium Chemical Research
Rome, Italy

For further volumes:
http://www.springer.com/series/7825

Franco Cataldo · Ante Graovac · Ottorino Ori
Editors

The Mathematics and Topology of Fullerenes

Springer

Editors
Dr. Franco Cataldo
Prof. of Chemistry
Dept. of Materials Science
Tor Vergata University
Rome, Italy
franco.cataldo@fastwebnet.it

Dr. Ottorino Ori
Correspondent Member
Actinium Chemical Research
Rome, Italy
ottorino.ori@alice.it

Prof. Ante Graovac
Faculty of Science
University of Split
Nikole Tesle 12
HR-21000 Split
Croatia

NMR Center
The "Ruđer Bošković" Institute
Bijenička c. 54
HR-10002 Zagreb
Croatia

and

IMC
University of Dubrovnik
Branitelja Dubrovnika 29
HR-20000 Dubrovnik
Croatia
ante.graovac@irb.hr

ISSN 1875-0745
e-ISSN 1875-0737
ISBN 978-94-007-0220-2
e-ISBN 978-94-007-0221-9
DOI 10.1007/978-94-007-0221-9
Springer Dordrecht Heidelberg London New York

Library of Congress Control Number: 2011921407

Printed on acid-free paper

Springer is part of Springer Science+Business Media (www.springer.com)

to the patient love of Stefania,
Biserka, Floriana

Foreword

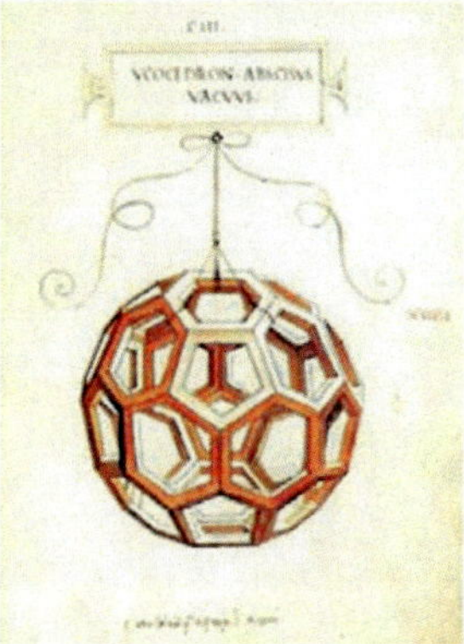

A Quintessential Aspect of the Human Condition – The Appreciation of Beauty in All Its Forms

It is a pleasure to write the foreword for this book probing relationship between mathematics and chemistry as well as computing as I am very conscious of the fact that mathematics made some extremely important and elegant contributions to the story of C_{60} Buckminsterfullerene. Indeed in some ways the C_{60} story epitomizes many aspects of the ways mathematics and science as well as art, architecture and engineering interplay to mix-and-match and inspire sensations of beauty which are at one and the same time complex in deeper aspects and yet simple to appreciate visually. It is not only children that appreciate the elegant beauty of highly symmetric structures such as the one epitomized in the magnificent drawing above by Leonardo. More complex appreciation is engendered in the minds of people who have some familiarity with mathematics which enables them to appreciate the elegance of the general equation governing such structures polyhedral structures:

$$12 = 3 \times n_3 + 2 \times n_4 + 1 \times n_5 + 0 \times n_6 - 1 \times n_7 - 2 \times n_8 \ldots$$

In this equation n_m is the number of m-sided polygons in a closed network with trivalent interconnections; when limited to hexagons and pentagons (i.e., only n_5 and n_6 may be non zero) the equation indicates that the network must contain 12 pentagons but there is no limitation on the number of hexagons. Note however that in this case $n_6 \neq 1$. The juxtaposition of Leonardo's drawing with this generalized form of Euler's Equation encapsulates almost the complete spectrum of what it means to be human. The motivation for such a selection of articles as this one reflects perfectly the way in which the patterns of the physical world, often buried deeply in mathematical concepts can be revealed by the combined drives of human curiosity allied with the desire for the cathartic experience of recognizing such patterns for the first time.

Tallahassee, Florida Harold Kroto

Preface

This book has been conceived during the 24th MATH/CHEM/COMP MCC Conference in Dubrovnik, placed along the beautiful and inspiring coast of Croatia in June 2009, and involves some of the most authoritative experts in this elegant field of nanoscience, placed at the border between mathematics and chemistry.

Since their discovery, fullerenes molecules are intimately connected to mathematics. Geometry, topology, number theory and other mathematical instruments greatly assist researchers to classify fullerenes structures and to predict their unique physical and chemical properties. The name itself of these hollow molecules of pure carbon is rooted in geometry, homage to Richard Buckminster Fuller and to his geodesic dome, the inspiring structure of the "buckminsterfullerene" C_{60} discovered by Robert Curl, Harold Kroto and Richard Smalley in 1985. This epochal, and in somehow epical, discovery has been awarder by 1996 Nobel Prize for Chemistry.

The amplitude of the topics and the level of the contributions are prominent characters of this scientific book that will help researchers in studying and understanding fullerenes properties. For this, we are fully indebted with all authors for their competent and patient works and with Prof. Harold Kroto for his introduction to this monograph that, in many pages, shows that beauty is present in science. We would also like to thank Springer for the opportunity to publish this book and Springer people who allowed all our efforts to become a real text.

Dubrovnik, Croatia

Ante Graovac
Ottorino Ori
Franco Cataldo

Contents

Contributors

Ali Reza Ashrafi Faculty of Science, Department of Mathematics and Computer Science, University of Kashan, Kashan 87317-51167, Iran, ashrafi@kashanu.ac.ir

Giorgio Benedek Donostia International Physics Centre (DIPC), 20018 Donostia-San Sebastián, Spain; Dipartimento di Scienza dei Materiali, Universitá di Milano-Bicocca, 20125 Milano, Italy, giorgio.benedek@mater.unimib.it

Marco Bernasconi Dipartimento di Scienza dei Materiali, Universitá di Milano-Bicocca, 20125 Milano, Italy, marco.bernasconi@unimib.it

Franco Cataldo Actinium Chemical Research, Via Casilina 1626/A, 00133, Rome, Italy, franco.cataldo@fastwebnet.it

Eugenio Cinquanta Dipartimento di Scienza dei Materiali, Universitá di Milano-Bicocca, 20125 Milano, Italy, eugenio.cinquanta@mater.unimib.it

Luca D'Alessio Dipartimento di Scienza dei Materiali, Universitá di Milano-Bicocca, 20125 Milano, Italy; Department of Physics, Boston University, Boston, MA 02215, USA, dalessio@buphy.bu.edu

Marzio De Corato Dipartimento di Scienza dei Materiali, Universitá di Milano-Bicocca, 20125 Milano, Italy, m.decorato@campus.unimib.it

Michel Deza JAIST, Nomi, Ishikawa-ken, 923-1292, Japan, Michel.Deza@ens.fr

Mircea V. Diudea Faculty of Chemistry and Chemical Engineering, Babes-Bolyai University, 400028 Cluj, Romania, diudea@chem.ubbcluj.ro

Helena Dodziuk Institute of Physical Chemistry, Polish Academy of Sciences, Kasprzaka 44/52, 01-224 Warsaw, Poland, dodziuk@ichf.edu.pl

Mathieu Dutour Sikirić Ruđer Bošković Institute, Bijenička 54, 10000 Zagreb, Croatia, mdsikir@irb.hr

Ernesto Estrada Department of Mathematics and Statistics, Department of Physics, Institute of Complex Systems, University of Strathclyde, Glasgow G1 1XQ, UK, ernesto.estrada@strath.ac.uk

Patrick W. Fowler Department of Chemistry, University of Sheffield, Sheffield S3 7HF, UK, p.w.fowler@sheffield.ac.uk

Modjtaba Ghorbani Department of Mathematics, Faculty of Science, Shahid Rajaee Teacher Training University, Tehran, 16785-136, I.R. Iran, mghorbani@srttu.edu

Ante Graovac Faculty of Science, University of Split, Nikole Tesle 12, HR-21000 Split, Croatia; NMR Center, The "Ruđer Bošković" Institute, Bijenička c. 54, HR-10002 Zagreb, Croatia; IMC, University of Dubrovnik, Branitelja Dubrovnika 29, HR-20000 Dubrovnik, Croatia, ante.graovac@irb.hr

Naomichi Hatano Institute of Industrial Science, University of Tokyo, Meguro, Tokyo 153-8505, Japan, hatano@iis.u-tokyo.ac.jp

Haruo Hosoya Ochanomizu University (Emeritus), Bunkyo-ku, Tokyo 112-8610, Japan, hosoya@is.ocha.ac.jp

Ali Iranmanesh Department of Mathematics, Tarbiat Modares University, P.O. Box: 14115-137, Tehran, Iran, iranmanesh@modares.ac.ir

E.C. Kirby Resource Use Institute, Pitlochry, Scotland PH16 5DS, UK, graphtheory@edwardkirby.eu

István László Department of Theoretical Physics, Institute of Physics, Budapest University of Technology and Economics, H-1521 Budapest, Hungary, laszlo@eik.bme.hu

R.B. Mallion School of Physical Sciences, University of Kent, Canterbury, CT2 7NH, England, UK, R.B.Mallion@kent.ac.uk

Adelio R. Matamala Facultad de Ciencias Químicas, Departamento de Físico-Química, Universidad de Concepción, Concepción, Chile, amatamal@udec.cl

Ottorino Ori Actinium Chemical Research, Via Casilina 1626/A, 00133 Rome, Italy, ottorino.ori@alice.it

P. Pollak The King's School, Canterbury, Kent CT1 2ES, UK

Milan Randić National Institute of Chemistry, SI-10000 Ljubljana, Slovenia, mrandic@msn.com

Tamás Réti Szechenyi István University, Egyetem tér 1, H-9026 Győr, Hungary, reti.tamas@bgk.bmf.hu

Aniela E. Vizitiu Faculty of Chemistry and Chemical Engineering, "Babes-Bolyai" University, Arany Janos Str. 11, 400084 Cluj, Romania, anielavizitiu@yahoo.com

Damir Vukičević Department of Mathematics, University of Split, Nikole Tesle 12, HR-21000 Split, Croatia, vukicevi@pmfst.hr

Chapter 1
Omega Polynomials of Fullerenes and Nanotubes

Ali Reza Ashrafi, Modjtaba Ghorbani, Mircea V. Diudea, and Ante Graovac

Abstract A counting polynomial $C(G,x)$ is a sequence description of a topological property so that the exponents express the extent of its partitions while the coefficients are related to the occurrence of these partitions. Basic definitions and properties of the Omega polynomial $\Omega(G,x)$ and Sadhana polynomial $Sd(G,x)$ are presented. These polynomials for some infinite classes of fullerenes and nanotubes are also computed.

1.1 Introduction

A graph can be described by: a connection table, a sequence of numbers, a single derived number (called sometimes a topological index), a matrix, or a polynomial (Diudea et al. 2001).

A finite sequence of some graph-theoretical categories/properties, such as the distance degree sequence or the sequence of numbers of the k-independent edge sets, can be described by so-called counting polynomials:

$$P(G,x) = \sum\nolimits_k p(G,k) \cdot x^k \tag{1.1}$$

where $p(G,x)$ is the frequency of occurrence of the property partitions of G, of length k, and x is simply a parameter to hold k.

Counting polynomials were introduced, in the Mathematical Chemistry literature, by Hosoya with his Z-counting (independent edge sets) and the distance degree polynomials, initially called **Wiener** and later **Hosoya polynomials** (Hosoya 1971, 1988). Their roots and coefficients are used for the characterization of topological nature of hydrocarbons.

Hosoya also proposed the **sextet polynomial** (Hosoya 1990; Ohkami et al. 1981; Ohkami and Hosoya 1983) to count the resonant rings in a benzenoid molecule.

A.R. Ashrafi (✉)
Faculty of Science, Department of Mathematics and Computer Science, University of Kashan, Kashan 87317-51167, Iran
e-mail: ashrafi@kashanu.ac.ir

F. Cataldo et al. (eds.), *The Mathematics and Topology of Fullerenes*,
Carbon Materials: Chemistry and Physics 4, DOI 10.1007/978-94-007-0221-9_1,

The sextet polynomial is important in connection to the Clar aromatic sextets (Clar 1964, 1972) expected to stabilize the aromatic molecules.

The **independence polynomial** (Gutman and Hosoya 1990; Gutman 1991a, b, 1992; Stevanović 1998) counts the number of distinct k-element independent vertex sets of G. Other related graph polynomials are the *king*, *color* and *star or clique polynomials*. (Motoyama and Hosoya 1977; Balasubramanian and Ramaraj 1985; Farrell 1978, 1989, 1994; Farrell and De Matas 1988a, b, c; Hoede 1994; Stevanović 1997). More about polynomials the reader can find in (Diudea et al. 2001).

Vertex contributions to a polynomial $P(G, x)$, based on distance counting, can be written as:

$$P(i,x) = (1/2)\sum\nolimits_k p(i,k) \cdot x^k \tag{1.2}$$

where $p(i,k)$ is the contribution of vertex i to the partition $p(G,k)$ of the global molecular property $P = P(G)$. Note that $p(i,k)$'s are just the entries in Layer **LM** or **Shell SM** matrices, more exactly 1/2 the value because each vertex contribution is counted twice, (Diudea et al. 2003).

Usually, the vertex contribution varies from one atom to another, so that the polynomial for the whole graph is obtained by summing all vertex contributions:

$$P(G,x) = \sum\nolimits_i P(i,x) \tag{1.3}$$

In a vertex transitive graph, the vertex contribution is simply multiplied by N:

$$P(G,x) = N \cdot P(i,x) \tag{1.4}$$

Hence, $P(G)$ is easily obtained as the polynomial value in $x = 1$:

$$P(G) = P(G,x)|_{x=1} \tag{1.5}$$

A **distance-extended property** $D\_P(G)$ can be calculated by the *first derivative* of the polynomial in $x = 1$ (Konstantinova and Diudea 2000; Diudea 2002a, b)

$$D\_P(G) = P'(G,x) = \sum\nolimits_k k \cdot p(G,k) \cdot x^{k-1}|_{x=1} \tag{1.6}$$

In Sagan et al. (1996), the authors presented a treatment apparently independent of Hosoya's. Perhaps the most interesting property of $H(G,x)$ is the first derivative, evaluated at $x = 1$, which equals the Wiener index: $H'(G,1) = W(G)$. One of us (ARA) continued the line of the mentioned paper of Sagan et al. to introduce the notion of **PI polynomial** of a molecular graph G as:

$$PI(G,x) = \sum\nolimits_{(u,v)=e\in E(G)} x^{N(u,v)} \tag{1.7}$$

where $N(u, v) = n_{eu}(e|G) + n_{ev}(e|G)$ and $n_{eu}(e|G)$ is the number of edges lying closer to u than v (*i.e.*, the **non-equidistant** edges) while the number of edges

equidistant to the edge $e = uv \in E(G)$ is given by: $N(e) = |E(G)| - N(u, v)$, where $E(G)$ denotes the set of all edges of the graph G. In (Ashrafi et al. 2006) the authors have shown that this new polynomial has the same basic properties as the Wiener polynomial. Thus, its first derivative gives the PI index, which can also be calculated by subtracting the total number of equidistant edges in G from the square of the edge set cardinality:

$$PI(G) = PI'(G, 1) = (|E|)^2 - \sum_e N(e) \tag{1.8}$$

relation also find in (John et al. 2002) to calculate the PI index.

The interested readers can consult some recent papers (Ashrafi and Mirzargar 2008c; Ashrafi and Shabani 2009a; Ashrafi et al. 2006, 2007, 2008a, b; 2009b, c, d; 2010; Ghorbani et al. 2009; Manoochehrian et al. 2007). There are four important softwares for such calculations. These are TopoCluj, Omega 1.1, HyperChem and GAP (Diudea et al. 2002; Cigher and Diudea 2005; HyperChem package Release 7.5 for Windows 2002; The GAP Team 2005). Our notations are standard and taken from (Gutman and Polansky 1986; Trinajstić 1992).

The basic definitions and properties of the Omega polynomial $\Omega(G, x)$ are presented in the second section. In the third section, the Omega polynomials of some well-known graphs are computed. A forth section will present our latest results in computing Omega polynomials of some infinite classes of fullerenes. Conclusions and references will close this chapter.

1.2 Omega Polynomial

The **Omega polynomial** is a counting polynomial introduced by one of the present authors MVD (Diudea 2006, 2008, 2009). In recent years, several papers on methods for computing Omega polynomial of molecular graphs have been published (Diudea et al. 2008, 2011; Vizitiu et al. 2007).

Let G be a connected bipartite graph with the vertex set $V = V(G)$ and edge set $E = E(G)$, without loops. Two edges $e = ab$ and $f = xy$ of G are called **co-distant** (briefly: $e\ co\ f$) if for $k = 0, 1, 2, \ldots$, there exist the relations: $d(a, x) = d(b, y) = k$ and $d(a, y) = d(b, x) = k + 1$ or vice versa. For some edges of a connected graph G there are the following relations satisfied:

$$e\,co\,f \tag{1.9}$$

$$e\,co\,f \Leftrightarrow f\,co\,e \tag{1.10}$$

$$e\,co\,f \text{ and } f\,co\,g \Rightarrow e\,co\,g \tag{1.11}$$

though, the relation (1.11) is not always valid.

Let $C(e) := \{e' \in E(G); e'\,co\,e\}$ denote the set of all edges of G which are *co*-distant to the edge e. If all the elements of $C(e)$ satisfy the relations (1.9), (1.10), and (1.11) then $C(e)$ is called an **orthogonal cut** *oc* of the graph G. The graph G is called a ***co*-graph** if and only if the edge set $E(G)$ is the union of disjoint orthogonal cuts: $C_1 \cup C_2 \cup ... \cup C_k = E$ and $C_i \cap C_j = \emptyset$ for $i \neq j$, $i, j = 1, 2, ..., k$.

We now assume that G has a plane representation F. If S is the set of all faces forming the interior regions then every edge appears in at most two members of S. Suppose T denotes the outside edges of G. Start with an edge e of G. If there is not an edge e_1 different from e with the property that $e\, co\, e_1$ and $\{e, e_1\}$ lie in the same face of G then we define $H = \{e_1\}$ and choose another edge f of G. Otherwise, there exists the edge e_1 such that $e\, co\, e_1$. Continue this process by e_1 to construct the sequence $e\, co\, e_1\, co\, e_2\, co\, \ldots\, co\, e_r$. If $e \in T$ then define $H = \{e, e_1, e_2, \ldots, e_r\}$. If not, there exists an edge f_1 of G different from e_1 such that $f_1\, co\, e$ and $\{e, f_1\}$ lie in the same face of G. By this algorithm a sequence $H = \{f_t, \ldots, f_1, e, e_1, e_2, \ldots, e_r\}$ is constructed. H is called a **quasi-orthogonal cut** or a **qoc strip** or also an opposite edge strip **ops,** because the co-relation, applied inside a polyhedral face accounts for opposite edges, which are topologically parallel to each other. It is an easy fact that the relation ops is not necessarily transitive. In the case that G is bipartite, then every member of S have an even number of edges and so $f_t, e_r \in T$. Notice that a*n* ops strip starts and ends either out of G (at an edge with endpoints of degree lower than 3, if G is an open lattice,) or in the same starting polygon (if G is a closed lattice). Any *ocs* set is an *ops* strip but the reverse is not always true.

Suppose $E_1, E_2, \ldots, E_r$ are *qoc/ops* strips of a connected planar bipartite graph G. We claim that $X = \{E_1, E_2, \ldots, E_r\}$ is a partition of $E = E(G)$. To do this we assume that $e \in E$ is an arbitrary edge of G. Using a similar argument as those given above one can find a sequence $f_t\; co\; f_{t-1}\; co \ldots co f_1\; co\, e\; co\; e_1 \ldots co\, e_r$. Therefore there exists j, $1 \leq j \leq r$, such that $\{f_t, \ldots, f_1, e, e_1, \ldots, e_r\} \subseteq E_j$. This implies that $e \in E_j$ and so $E = E_1 \cup E_2 \cup \ldots \cup E_r$. To complete our claim, we must prove $E_i \cap E_j = \emptyset$ for $1 \leq i \neq j \leq r$. Suppose $E_i = \{e_1, e_2, \ldots, e_n\}$, $E_j = \{f_1, f_2, \ldots, f_m\}$ and $e \in E_i \cap E_j$. Then there are r, s, $1 \leq r \leq n$, $1 \leq s \leq m$ such that $e = e_r = f_s$. But every edge appears in at most two members of S, so by using an inductive argument $E_i = E_j$. Therefore, X is a partition of E.

The *Omega* $\Omega(G, x)$ polynomial, accounting for *ops* strips in G is defined as:

$$\Omega(G, x) = \sum\nolimits_s m(G, s) \cdot x^s \tag{1.12}$$

with $m(G, c)$ being the number of strips of length s. The summation runs up to the maximum length of *ops* in G.

If G is bipartite, then an *ops* starts and ends out of G and so $\Omega(G, 1) = r/2$, in which r is the number of edges in out of G. On the other hand, one can easily seen that $\Omega'(G, 1) = \sum_s m \cdot s = e = |E(G)|$. Two single number descriptors are derived from $\Omega(G, x)$ as:

$$CI(G) = (\Omega'(1))^2 - (\Omega'(1) + \Omega''(1)) \tag{1.13}$$

$$I_\Omega(G) = (1/\Omega'(G, 1)) \cdot \sum\nolimits_d (\Omega^d(G, 1))^{1/d} \tag{1.14}$$

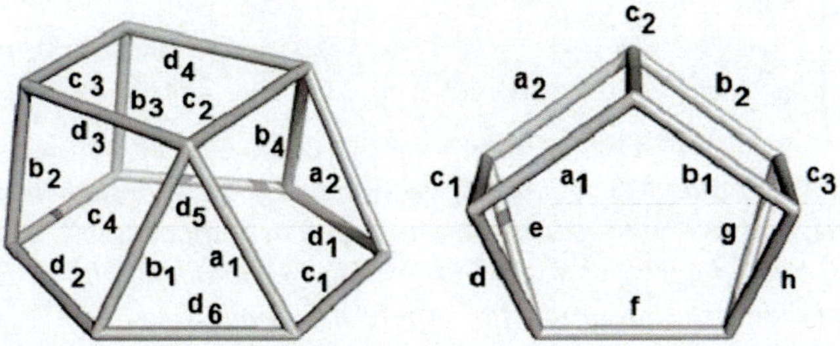

Fig. 1.1 Two graphs G_1 (*left*) and G_2 (*right*)

In case of I_Ω, summation runs over all possible derivatives d in the corresponding polynomial. When one or more edges do not belong to a counted strip, such edges are added as "strips of length 1".

It is easily seen that, for a single *ops*, one calculates the polynomial: $\Omega(G,x) = x^s$ and $CI(G) = s^2 - (s + s(s-1)) = 0$. There exist graphs for which *CI* equals *PI*. In fact, the two indices *CI* and *PI* will show identical values if the edge equidistance evaluation in the graph involves only the locally parallel edges. This is occurred for example in partial cubes. In this case, we have:

$$CI(G) = \left(\sum\nolimits_s m \cdot s\right)^2 - \left[\sum\nolimits_s m \cdot s + \sum\nolimits_s m \cdot s \cdot (s-1)\right] = e^2 - \sum\nolimits_s m \cdot s^2 = PI(G) \tag{1.15}$$

This counting polynomial is useful in topological description of benzenoid structures as well as in counting some single number descriptors, *i.e.*, topological indices. The *ops* strips could give account for the helicity of polyhex nanotubes and nanotori. The Omega 1.1 software program includes the *ops* strips procedure.

In the end of this section a simple counterexample for Eqs. (1.9), (1.10), and (1.11) is given in Fig. 1.1. In the graph G_1; {a} and {c} are *oc* sets; {b} and {d} do not have all elements co-distant to each other, so that {b} and {d} are *qoc* strips but all are *ops* strips. In the graph G_2; {a} and {b} and {c} are *oc* strips; {f} and {c_2} are equidistant but {f} and {c_1 or c_3} do not obey the symmetry relation (1.8) (and do not belong to one face) thus {f} does not belong to the strip {c}. Therefore, $\Omega(G_1,x) = x^2 + 2x^4 + x^6$ and $\Omega(G_2,x) = 5x + 2x^2 + x^3$.

1.3 Examples

In this section the Omega polynomial of some well-known graphs are computed. A general formula for computing Omega polynomial of the graph product is presented; in this way, it is possible to compute the Omega polynomials of nanotubes and nanotori covered by C_4. We begin by some well-known graphs.

Example 1 *Suppose T_n, C_n and K_n denote the an arbitrary tree, cycle and complete graph on n vertices, respectively. Then by simple calculations, one can see that*

$$\Omega(K_n,x)=\begin{cases}\frac{n}{2}\left(x^{\frac{n}{2}}+x^{\frac{n}{2}-1}\right) & 2|n\\ nx^{\frac{n-1}{2}} & 2\nmid n\end{cases},\quad \Omega(C_n,x)=\begin{cases}\frac{n}{2}x^2 & 2|n\\ nx & 2\nmid n\end{cases}\quad \text{and } \Omega(T,x)=(n-1)x.$$

The Cartesian product $G \times H$ of graphs G and H is a graph such that $V(G\times H) = V(G) \times V(H)$, and any two vertices (a,b) and (u,v) are adjacent in $G \times H$ if and only if $a = u$ and b is adjacent with v, or $b = v$ and a is adjacent with u. The following properties of the Cartesian product of graphs are crucial:

(a) $|V(G \times H| = |V(G)|\,|V(H)|$ *and* $|E(G \times H)| = |E(G)|\,|V(H)| + |V(G)|\,|E(H)|$;
(b) *$G \times H$ is connected if and only if G and H are connected;*
(c) *If (a, x) and (b, y) are vertices of $G \times H$ then* $d_{G\times H}((a,x),(b,y)) = d_G(a,b) + d_H(x,y)$;
(d) *The Cartesian product is associative.*

Theorem 1 *Let G and H be bipartite connected co-graphs. Then*

$$\Omega(G\times H,x)=\sum\nolimits_{s_1} m(G,s_1)\cdot x^{|V(H)|s_1}+\sum\nolimits_{s_2} m(H,s_2)\cdot x^{|V(G)|s_2}$$

Proof Suppose that for an edge $e = uv$ of an arbitrary graph L, $N_L(e) = |E| - (n_u(e) + n_v(e))$. Then by definition,

$$N_{G\times H}((a,x),(b,y))=\begin{cases}|V(G)|\,N(f) \text{ for } a=b \text{ and } xy=f\in E(H)\\ |V(H)|\,N(g) \text{ for } x=y \text{ and } ab=g\in E(G).\end{cases}$$

By the above paragraph and definition of the Omega polynomial, we have:

$$\Omega(G\times H,x)=\sum\nolimits_s m(G\times H,s)\cdot x^s=\sum\nolimits_{s_1} m(G,s_1)\cdot x^{|V(H)|\,s_1}+\sum\nolimits_{s_2} m(H,s_2)\cdot x^{|V(G)|\,s_2}$$

which completes the proof. ■

Corollary 1 *Let* $G_1, G_2, \ldots G_n$ *be bipartite connected co-graphs. Then we have*:

$$\Omega(G_1\times G_2\times\cdots\times G_n,x)=\sum_{i=1}^{n}\sum\nolimits_{s_i} m(G_i,s_i)\cdot x^{\prod_{j=1,\,j\neq i}^{n}|V(G_j)|s_i}.$$

Proof Use induction on n. By Theorem 1, the result is valid for $n = 2$. Let $n \geq 3$ and assume the theorem holds for $n - 1$. Set $G = G_1 \times \ldots \times G_{n-1}$. Then we have

$$\Omega(G\times G_n,x)=\sum\nolimits_s m(G,s)\cdot x^{|V(G_n)|s}+\sum\nolimits_{s_n} m(G_n,s_n)\cdot x^{|V(G)|s_n}$$

$$= \sum_{i=1}^{n-1} \sum\nolimits_{s_i} m(G_i, s_i) \cdot x^{\prod_{j=1, j\neq i}^{n} |V(G_j)| s_i} + \sum\nolimits_{s_n} m(G_n, s_n) \cdot x^{|V(G)| s_n}$$

$$= \sum_{i=1}^{n} \sum\nolimits_{s_i} m(G_i, s_i) \cdot x^{\prod_{j=1, j\neq i}^{n} |V(G_j)| s_i} \quad . \blacksquare$$

Example 2 *In this example the Omega polynomial of nanotubes and nanotori covered by C_4 are calculated. By definitions of Cartesian product of graphs and Omega polynomial, one can easily prove*:

$$\Omega(G \times H, x) = \sum\nolimits_{s_1} m(G, s_1) \times x^{|V(H)| s_1} + \sum\nolimits_{s_2} m(H, s_2) \times x^{|V(G)| s_2}. \quad (1.16)$$

Suppose R and S denote a C_4 tube and torus, respectively. Then by definition $R \cong P_n \times C_m$ and $S \cong C_k \times C_m$. Apply Theorem 1 to deduce that $\Omega(P_n \times P_m, x) = (n-1)x^m + (m-1)x^n$. On the other hand, we have:

$$\Omega(P_n \times C_m, x) = \begin{cases} (n-1)x^m + \dfrac{m}{2}x^{2n} & 2|\mathrm{m} \\ (n-1)x^m + mx^n & 2\nmid \mathrm{m} \end{cases},$$

$$\Omega(C_n \times C_m, x) = \begin{cases} nx^m + mx^n & 2\nmid \mathrm{m}, 2\nmid \mathrm{n} \\ nx^m + \frac{m}{2}x^{2n} & 2|\mathrm{m}, 2\nmid \mathrm{n} \\ \frac{n}{2}x^{2m} + mx^n & 2\nmid \mathrm{m}, 2|\mathrm{n} \\ \frac{n}{2}x^{2m} + \frac{m}{2}x^{2n} & 2|\mathrm{m}, 2|\mathrm{n} \end{cases}.$$

Note, these formulas calculate Omega strips including the tube cross-section. Usually, the Omega polynomial calculations are specified by the $Face_{max}$ or $Ring_{max}$ considered.

Example 3 *Consider the graph of a nanocone $C[a, n]$, Fig. 1.2.*

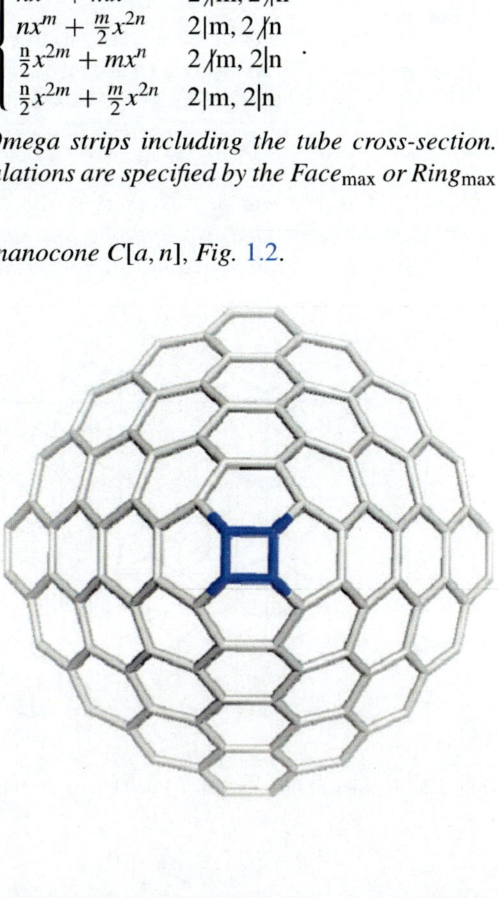

Fig. 1.2 The carbon nanocone C[4,4]

This graph has exactly $a(n+1)^2$ vertices, a being the apex polygon while n is the number of hexagon rows around the apex. Function of apex a-parity, there are different formulas, as shown in Table 1.1.

Table 1.1 Formulas to calculate omega polynomial in nanocones

Formulas
Case: a = even
$\Omega(C(a_{even}, n), x) = (a/2) \cdot x^{2(n+1)} + \sum_{k=2}^{n+1} a \cdot x^{n+k}$
$\Omega'(C(m_{even}, n), 1) = (m/2)(3n^2 + 5n + 2)$
$CI(C(a_{even}, n)) = (1/12)a(n+1)(27an^3 - 28n^2 + 63an^2 + 48an + 12a - 50n - 24)$
Examples
C(4, 3): $4x^5 + 4x^6 + 4x^7 + 2x^8$; CI = 7176
C(8, 4): $8x^6 + 8x^7 + 8x^8 + 8x^9 + 4x^{10}$; CI = 76160
Case: a = odd
$\Omega(C(a_{odd}, n), x) = \sum_{k=1}^{n+1} a \cdot x^{n+k}$
$\Omega'(C(a_{odd}, n), 1) = (a/2)(3n^2 + 5n + 2)$
$CI(C(a_{odd}, n)) = (1/12)a(n+1)(27an^3 - 28n^2 + 63an^2 + 48an + 12a - 38n - 12)$
$CI(C(a_{odd}, n)) = CI(C(a_{even}, n)) + a(n+1)^2$
Examples
C(3, 5): $3x^6 + 3x^7 + 3x^8 + 3x^9 + 3x^{10} + 3x^{11}$; CI = 22056
C(5, 3): $5x^4 + 5x^5 + 5x^6 + 5x^7$; CI = 11470
C(7, 4): $7x^5 + 7x^6 + 7x^7 + 7x^8 + 7x^9$; CI = 58240

Example 4 *In the end of this section, the Omega polynomial in Du (Med(6,6)) TiO_2 pattern, embedded as nanotube and nanotorus are computed, Figs. 1.3 and 1.4. The corresponding graphs of these structures are denoted by G and H,*

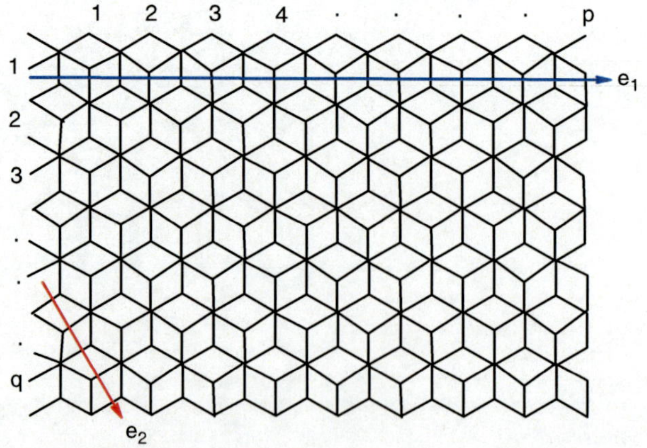

Fig. 1.3 The ops of the nanotube G = TU[p,q] in *Du*(*Med*(6,6)) TiO_2 pattern

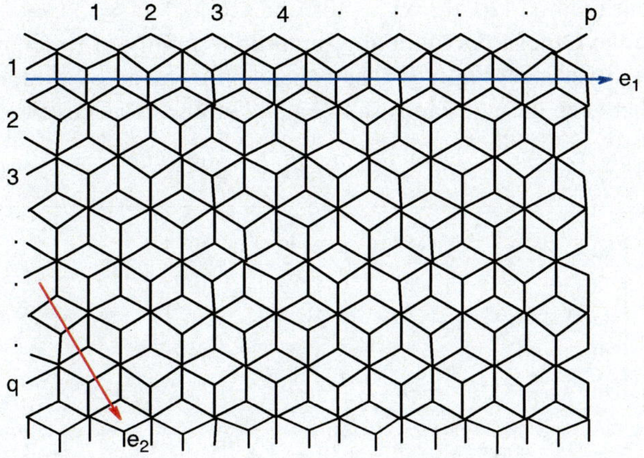

Fig. 1.4 The ops of the nanotorus T[p,q] in *Du*(*Med*(6,6)) TiO_2 pattern

respectively. From these figures, one can see that there are two different cases of ops strips. Suppose e_1 *and* e_2 *are representatives of the different cases. In the graph G,* $|s(e_1)| = 2p$ *and* $|s(e_2)| = 2q + 1$*. On the other hand, there are* $q(e_1)$ *and* $2p(e_2)$ *similar edges. This leads to the formula* $\Omega(G, x) = q \cdot x^{2p} + 2p \cdot x^{2q+1}$*. For the graph H,* $|s(e_1)| = 2p$ *and* $|s(e_2)| = 2pq$*. On the other hand, there are 2q similar edges for each of* e_1, e_2*, respectively. With the above considerations we have the following formula* $\Omega(H, x) = qx^{2p}+2x^{2pq}$.

1.4 Omega Polynomial of Fullerenes

In this section, the Omega polynomials of some classes of infinite fullerenes are investigated. Our method is simple. We first draw our molecular graph by HyperChem, then compute its adjacency and distance matrices by TopoCluj, then calculate the Omega polynomials by GAP.

Let's begin by small fullerenes C_{20} and C_{30} fullerenes, as depicted in Fig. 1.5.

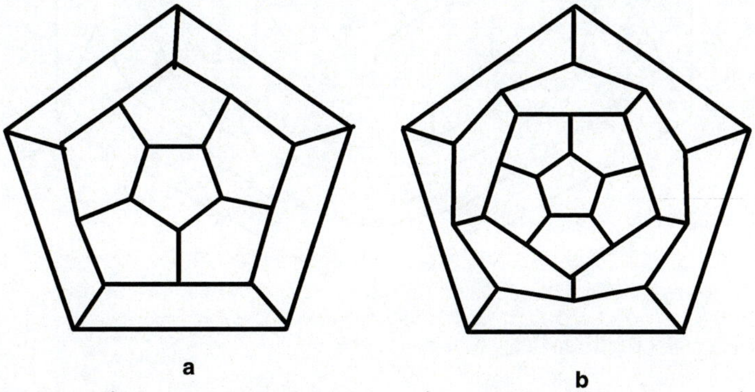

Fig. 1.5 Fullerene graphs: C_{20} (**a**) and C_{30} (**b**)

Then by our method $\Omega(C_{20}, x) = 30x$ and $\Omega(C_{30}, x) = 20x + 10x^2 + x^5$. We now compute the Omega polynomial of an infinite family of fullerene graphs with exactly 40*n* + 6 vertices, Fig. 1.6. The expansion in made by prolonging the tube TUH[20, *n*] between the two caps (of 44 and 42 atoms, respectively).

Theorem 2 *The Omega polynomial of fullerene graph* $G = C_{40n+6}(n \geq 2)$ *is computed as follows*:

$$\Omega(G,x) = \begin{cases} a(x) + 4x^{2n} + 4x^{2n+1} + 4x^{4n-1} + 2x^{4n} & 5|n \\ a(x) + 2x^{4n+3} + 8x^{2n-2} + 2x^{4n+4} + 2x^{4n+1} & 5|n-1 \\ a(x) + 8x^{2n} + 4x^{2n-1} + 2x^{4n} + 2x^{4n+2} & 5|n-2\ , \\ a(x) + 4x^{2n-2} + 4x^{2n+2} + 4x^{4n-1} + 2x^{4n+2} & 5|n-3 \\ a(x) + 4x^{2n-2} + 4x^{2n-1} + 4x^{2n} + 2x^{4n+3} + x^{8n+6} & 5|n-4 \end{cases}$$

where $a(x) = x + 9x^2 + 4x^3 + 2x^4 + (2n-3)x^{10}$.

Proof From Fig. 1.6, one can see that there are ten distinct cases of ops in G. We denote the corresponding edges by $e_1, e_2, \ldots, e_{10}$. By using calculations given in Table 1.2 and the Fig. 1.7, the proof is completed. ■

Next, we consider a class of fullerenes G_n with exactly 10*n* vertices, Fig. 1.8. Such a fullerene consists of two symmetric caps as halves of the dodecahedron and a distancing "zig-zag" tube TUH[10,n].

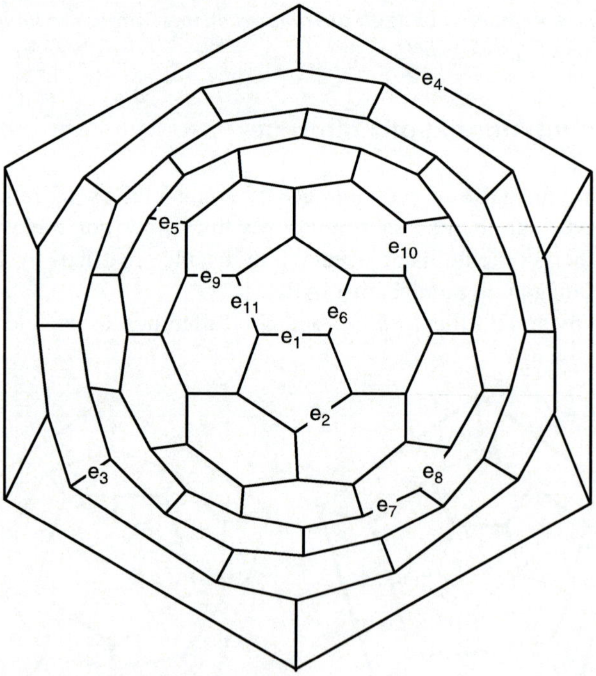

Fig. 1.6 The graph of fullerene C_{40n+6}, when $n = 2$

Table 1.2 The number of co-distant edges of e_i, $1 \le i \le 10$

No.	Number of co-distant edges	Type of edges
1	1	e_1
9	2	e_2
4	3	e_3
2	4	e_4
$2n-3$	10	e_5
2	$2n+1$ $\quad 5 \mid n$	e_6
	$4n+3$ $\quad 5 \mid n-1$	
	$2n$ $\quad 5 \mid n-4, n-2$	
	$2n+2$ $\quad 5 \mid n-3$	
2	$4n-1$ $\quad 5 \mid n-3$	e_7
4	$2n$ $\quad 5 \mid n, n-2$	
4	$2n-2$ $\quad 5 \mid n-1, n-4$	
4	$2n-2$ $\quad 5 \mid n-1, n-3$	e_8
4	$2n-1$ $\quad 5 \mid n-2, n-4$	
2	$4n-1$ $\quad 5 \mid n$	
1	$8n+6$ $\quad 5 \mid n-4$	e_9
2	$4n+2$ $\quad 5 \mid n-3$	
2	$4n+4$ $\quad 5 \mid n-1$	
2	$4n$ $\quad 5 \mid n, n-2$	
2	$4n-1$ $\quad 5 \mid n, n-3$	e_{10}
	$4n+1$ $\quad 5 \mid n-1$	
	$4n+2$ $\quad 5 \mid n-2$	
	$4n+3$ $\quad 5 \mid n-4$	
2	$2n+1$ $\quad 5 \mid n$	e_{11}
	$2n$ $\quad 5 \mid n-2, n-4$	
	$2n+2$ $\quad 5 \mid n-3$	

It is easily seen that there are six distinct cases of qoc strips as follows:

Theorem 3 *The Omega polynomial of $G_n (n \ge 2)$ is computed as follows*:

$$\Omega(G_n, x) = 20x + (n-2)x^5 + 10x^{n-1}$$

Proof We denote the representatives of edges regarding to co-distant relation by $e_1, e_2, \ldots, e_6$. Then $|s(e_1)| = |s(e_2)| = |s(e_3)| = |s(e_6)| = 1$, $|s(e_4)| = 5$ and $|s(e_5)| = n-1$. On the other hand there are five similar edges for each of edges e_1, e_2, e_3 and e_6, $n-2$ edges similar to e_4 and 10 edges similar to e_5. Therefore,

$$\Omega(G_n, x) = 20x + (n-2)x^5 + 10x^{n-1} \blacksquare$$

In Table 1.3, we list the Omega polynomial of F_n for $n \le 9$.

In what follows, new classes F_{10n} of fullerenes with $10n$ carbon atoms are considered. This series has two symmetric caps as [5:6$_5$] flower/circulene and a distancing “armchair” TUV[10,n].

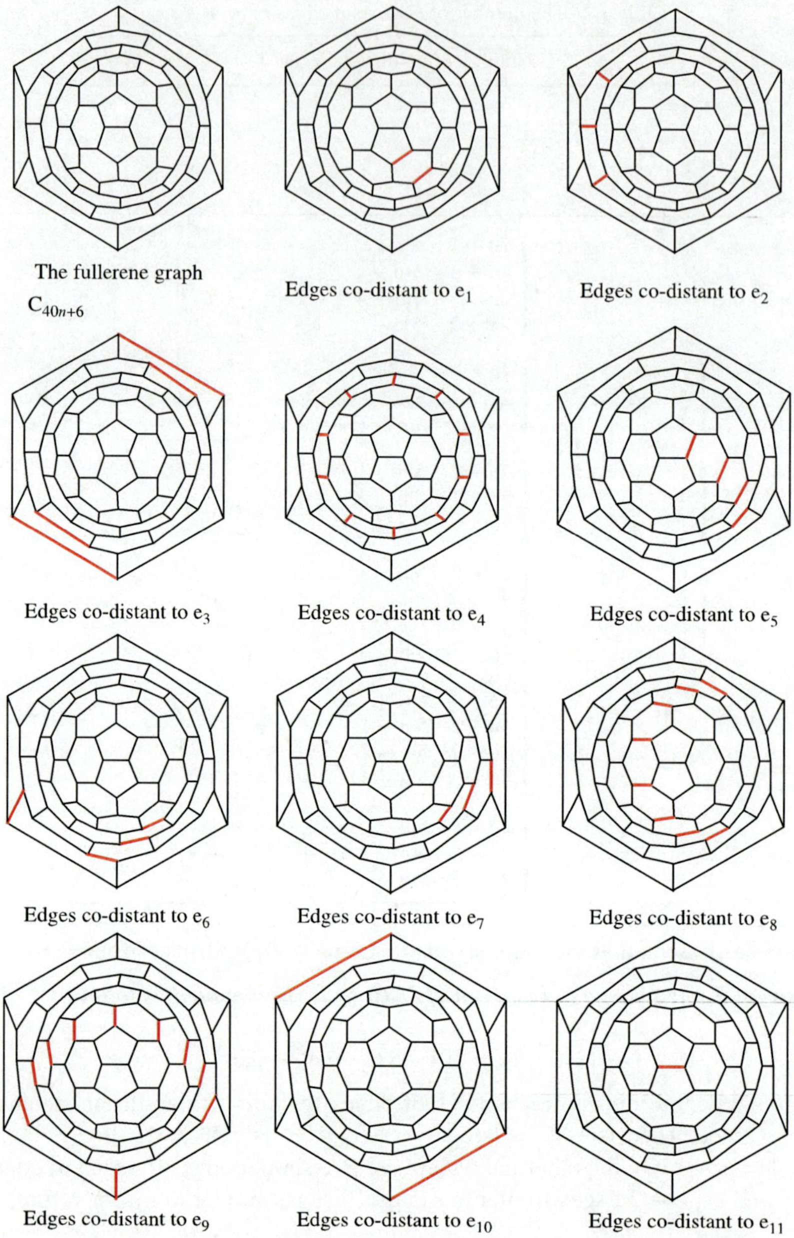

Fig. 1.7 Different cases of C_{40n+6} fullerene with co-distant edges

Fig. 1.8 The fullerene graph G_n, $n = 8$

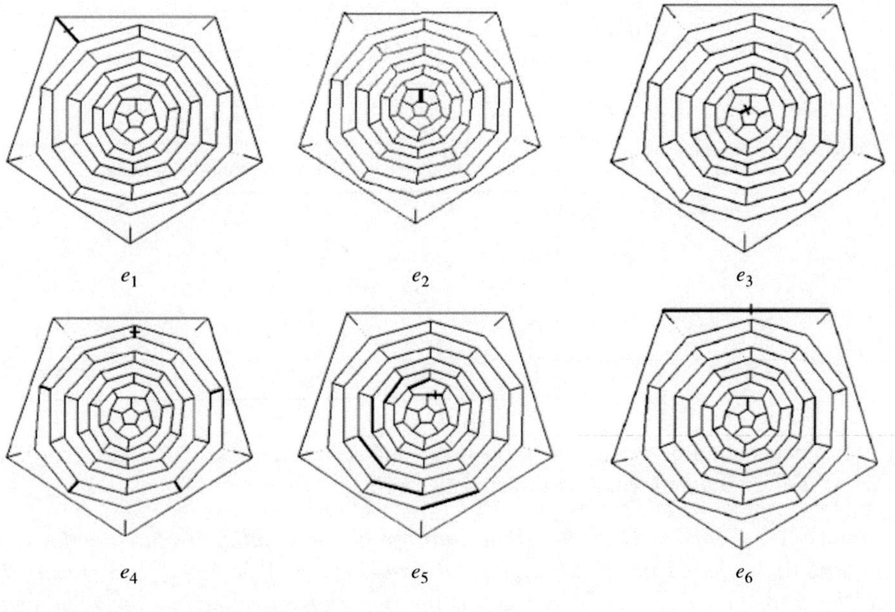

Fig. 1.9 The ops of edges $e_1, e_2, \ldots, e_6$ in G_n

Theorem 4 *The Omega polynomial of F_{10n} is computed as follows:*

$$\Omega(F_{10n}, x) = \begin{cases} 10x^3 + 10x^{\frac{n}{2}} + 10x^{n-3} & 2|n \\ 10x^3 + 5x^{\frac{n-3}{2}} + 5x^{\frac{n+3}{2}} + 10x^{n-3} & 2 \nmid n \end{cases}. \tag{1.17}$$

Table 1.3 The omega polynomial of F_{10n} for $n \leq 9$

Fullerenes	Omega polynomials
F_{20}	$30X^1$
F_{30}	$20X^1 + 1X^5 + 10X^2$
F_{40}	$20X^1 + 2X^5 + 10X^3$
F_{50}	$20X^1 + 3X^5 + 10X^4$
F_{60}	$20X^1 + 4X^5 + 10X^5$
F_{70}	$20X^1 + 5X^5 + 10X^6$
F_{80}	$20X^1 + 6X^5 + 10X^7$
F_{90}	$20X^1 + 7X^5 + 10X^8$

Proof To compute the Omega polynomial of F_{10n}, it is enough to calculate s(e) for every e ∈ E(G). In Tables 1.4 and 1.5, the number of co-distant edges of this fullerene, are computed.

Table 1.4 The number of co-distant edges, when 2|n

Type of edges	Number of co-distant edges	No
e_1	3	10
e_2	$n/2$	10
e_3	$n-3$	10

Table 1.5 The number of co-distant edges, when $2 \nmid n$

Type of edges	Number of co-distant edges	No
e_1	3	10
e_2	$\frac{n-3}{2}$	5
e_3	$\frac{n+3}{2}$	5
e_4	$n-3$	10

From calculations given in Tables 1.4 and 1.5 and Figs. 1.10 and 1.11 the Eq. (1.17) is obtained which completes the proof.

Theorem 5 *Suppose G is the molecular graph of a fullerene having the cap* [Cor:(5,6)$_6$], *where* Cor=[6:6$_6$] *and the distancer is now* TuV[12, n]. *It corresponds to the formulas:* $C_{12(2n+1)}$ *and* C_{24n}, *if the two caps are glued to the tube symmetrically (h-symmetry) and twisted (d-symmetry), respectively. Then the Omega polynomial of G is*

$$\Omega(G,x) = 12x^3 + 12x^{2n-2} + 6x^{n-1} + 3x^{2n+4}\ ,\ n \geq 2$$

$$\Omega(G,x) = 6x^{2n} + 12x^{2n-3} + 12x^3;\ n \geq 3$$

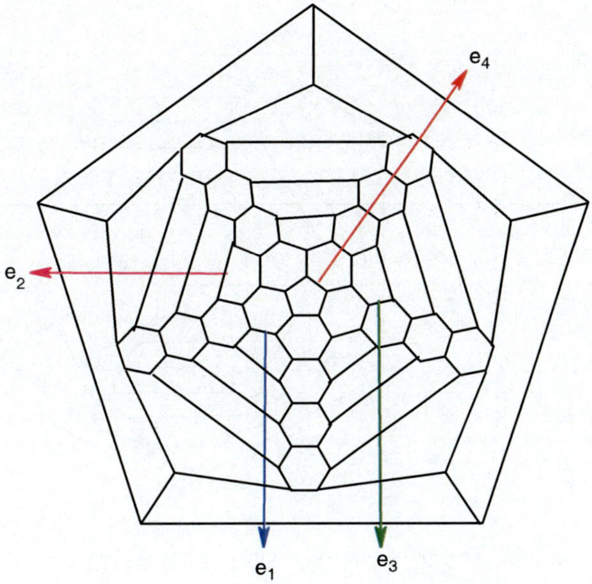

Fig. 1.10 The fullerene graph F_{10n} (n is odd)

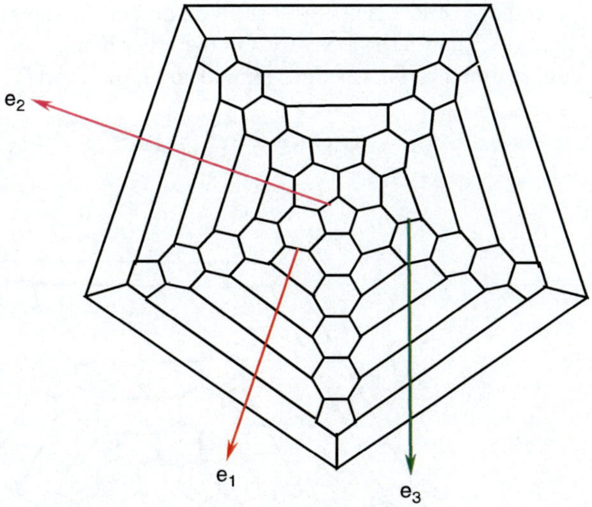

Fig. 1.11 The fullerene graph F_{10n} (n is even)

Proof It is easy to see that there are four different type of edges, f_1, f_2, f_3 and f_4, Fig. 1.12. One can see that $|s(e_1)| = 3$, $|s(e_2)| = 2n - 2$, $|s(e_3)| = 2n + 4$ and $|s(e_4)| = n - 1$. On the other hand, there are 12, 12, 3, and 6 similar edges for each of edges e_1, e_2, e_3, and e_4, respectively, then we have

$$\Omega(G, x) = 12x^3 + 12x^{2n-2} + 6x^{n-1} + 3x^{2n+4}\ ,\ n \geq 2$$

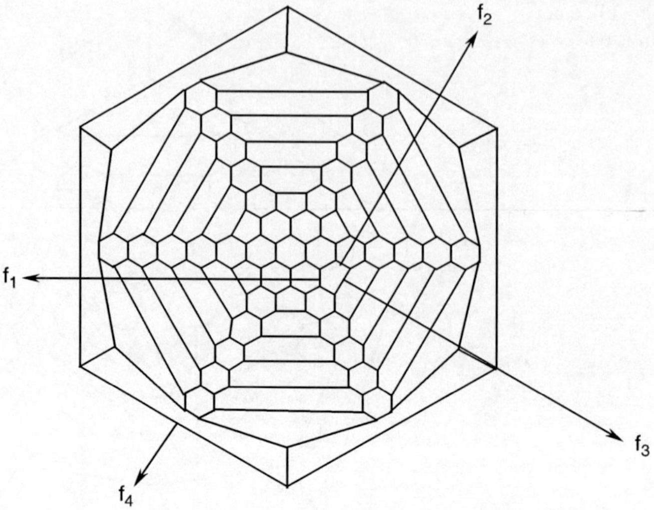

Fig. 1.12 The ops of edges in graph of fullerene $C_{12_{(2n+1)}}$

In the second case (Fig. 1.13) the number of edges co-distant to f_1, f_2, and f_3 are $2n$, $2n-3$ and 3 respectively. On the other hand, there are 6 edges similar to f_1, 12 edges similar to f_2, 12 edges similar to f_3, therefore,

$$\Omega(G,x) = 6x^{2n} + 12x^{2n-3} + 12x^3;\ n \geq 3$$

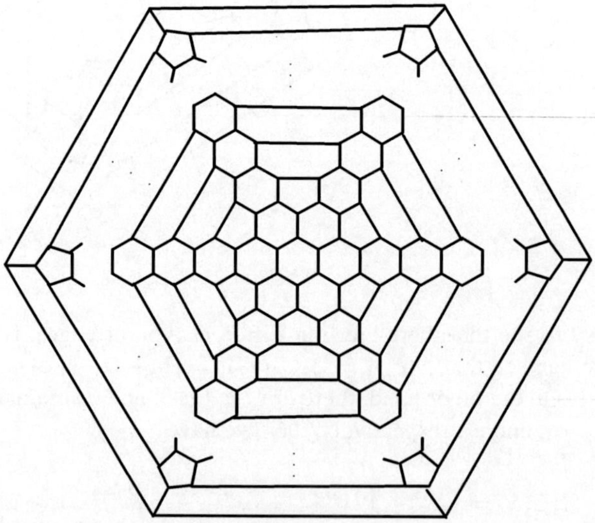

Fig. 1.13 The Schlegel graph of C_{24n} fullerene

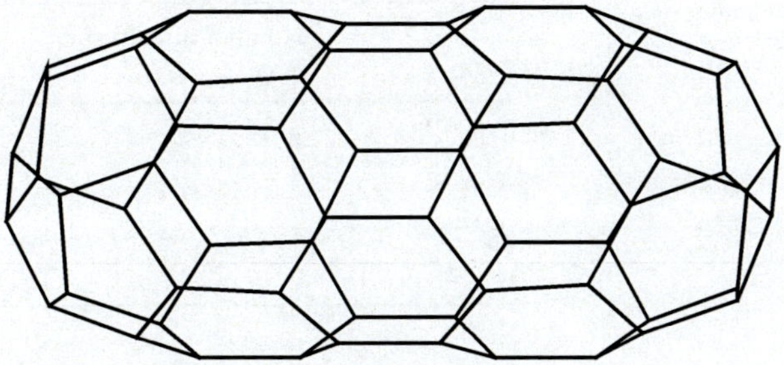

Fig. 1.14 The molecular graph of C_{12n+4} fullerene

Theorem 6 *The Omega polynomial of fullerene graph* C_{12n+4} *is as follows*:

$$\Omega(C_{12n+4},x) = 14x + 4x^2 + (n-2)x^6 + 4x^{n-1} + 8x^n; \quad n = 3,\ 5,\ 9,\ 11,\ 5\ldots$$

$$\Omega(C_{12n+4},x) = 14x + 4x^2 + (n-2)x^6 + 8x^{n-1} + 4x^{n+1}; \quad n = 7,\ 13,\ 19,\ 25,\ \ldots$$

Proof From Figs. 1.14 and 1.15, one can see: The cage is made by a symmetric cap and a TUH[12,*n*]. There are five distinct cases of ops. We denote the corresponding edges by $e_1, e_2, \ldots, e_5$. By Table 1.6, we can see that $|s(e_1) = 2|, |s(e_2)| = n-1, |s(e_3)| = n, |s(e_3)| = n, |s(e_4)| = 1$ and $|s(e_5)| = 6$. On the other hand, there are 4, 8, 4, 18 and $n-2$ similar edges for each of edges e_1, e_2, e_3, e_4 and e_5, respectively. So, we have

$$\Omega(C_{12n+4},x) = 14x + 4x^2 + (n-2)x^6 + 4x^{n-1} + 8x^n; \quad n = 3,\ 5,\ 9,\ 11,\ 15,\ldots$$

$$\Omega(C_{12n+4},x) = 14x + 4x^2 + (n-2)x^6 + 8x^{n-1} + 4x^{n+1}; \quad n = 7,\ 13,\ 19,\ 25,\ \ldots$$

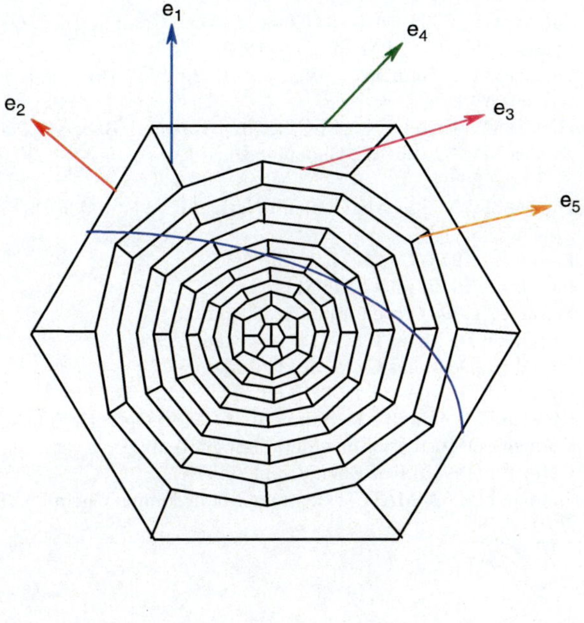

Fig. 1.15 The ops of edges $e_1, e_2, \ldots, e_5$ in C_{12n+4} fullerene

Table 1.6 The number of co-distant edges of e_i, $1 \le i \le 5$

No.	Number of co-distant edges	Type of edges
4	2	e_1
8	$n-1$	e_2
4	n	e_3
18	1	e_4
$n-2$	6	e_5

This completes the proof. ■

References

Ashrafi AR, Ghorbani M, Jalali M (2008b) Indian J Chem 47A:535

Ashrafi AR, Ghorbani M, Jalali M (2010) Fullerenes, nanotubes. Carbon Nanostruct 18:107

Ashrafi AR, Jalali M, Ghorbani M, Diudea MV (2008a) MATCH Commun Math Comput Chem 60:905

Ashrafi AR, Manoochehrian B, Yousefi-Azari H (2006) Util Math 71:97

Ashrafi AR, Manoochehrian B, Yousefi-Azari H (2007) Bull Iranian Math Soc 33:37

Ashrafi AR, Mirzargar M (2008c) MATCH Commun Math Comput Chem 60:897

Ashrafi AR, Shabani H (2009a) Optoelectron Adv Mater Rapid Comm 3:356

Ashrafi AR, Ghorbani M, Jalali M (2009b) J Theoret Comput Chem 8:451

Ashrafi AR, Ghorbani M, Jalali M (2009c) Optoelectron. Adv Mater Rapid Comm 3:823

Ashrafi AR, Shabani H (2009d) Optoelectron. Adv Mater Rapid Comm 3:1309

Balasubramanian K, Ramaraj R (1985) J Comput Chem 6:447

Cigher S, Diudea MV (2005) Omega 1.1, "Babes-Bolyai" University. Cluj, Romania

Clar E (1964) Polycyclic hydrocarbons. Academic, London

Clar E (1972) The aromatic sextet. Wiley, New York, NY

Diudea MV (2002a) Studia Univ "Babes-Bolyai" 47:131

Diudea MV (2002b) MATCH Commun Math Comput Chem 45:109

Diudea MV (2006) Carpathian J Math 22:43

Diudea MV (2008) MATCH Commun Math Comput Chem 60:935

Diudea MV (2009) J Math Chem 45:309

Diudea MV, Gutman I, Jäntschi L (2001) Molecular topology. Nova Sci. Huntington, New York, NY

Diudea MV, Ursu O, Nagy LCs (2002). Topocluj. Babes-Bolyai University, Cluj

Diudea MV, Vizitiu AE, Gholaminezhad F, Ashrafi AR (2008) MATCH Commun Math Comput Chem 60:945

Diudea MV, Vizitiu AE, Gholami-Nezhad F, Ashrafi AR (2011) MATCH Commun Math Comput Chem 65(1):173

Farrell EJ (1978) Canad Math Bull 2:35

Farrell EJ (1989) Int J Math Math Sci 12:77

Farrell EJ (1994) Proc Caribb Acad Sci 5:163

Farrell EJ, De Matas C (1988a) Ark Math 26:185

Farrell EJ, De Matas C (1988b) Util Math 33:33

Farrell EJ, De Matas C (1988c) Int J Math Math Sci 11:87

Ghorbani M, Ashrafi AR, Hemmasi M (2009) Optoelectron Adv Mater Rapid Comm 3:1306

Gutman I (1991a) Publ Inst Math (Beograd) 50:19

Gutman I (1991b) Rev Roum Chim 36:379

Gutman I (1992) MATCH Commun Math Comput Chem 28:139

Gutman I, Hosoya H (1990) Z Naturforsch 45a:645
Gutman I, Polansky OE (1986) Mathematical concepts in organic chemistry. Springer, New York, NY
Hoede C, Li XL (1994) Discr Math 125:219
Hosoya H (1971) Bull Chem Soc Jpn 44:2332
Hosoya H (1988) Discrete Appl Math 19:239
Hosoya H (1990) Top Curr Chem 153:255
Hyperchem package Release 7.5 for Windows (2002) Hypercube, Inc., Florida
John PE, Khadikar PV, Singh J (2002) J Math Chem 42:37
Konstantinova EV, Diudea MV (2000) Croat Chem Acta 73:383
Manoochehrian B, Yousefi-Azari H, Ashrafi AR (2007) MATCH Commun Math Comput Chem 57:653
Motoyama A, Hosoya H (1977) J Math Phys 18:1485
Ohkami N, Hosoya H (1983) Theoret Chim Acta 64:153
Ohkami N, Motoyama A, Yamaguchi T, Hosoya H (1981) Tetrahedron 37:1113
Sagan BE, Yeh Y-N, Zhang P (1996) Int J Quantum Chem 60:959
Stevanović D (1988) Graph Theory Notes New York 34:31
Stevanović D (1997) Publ Elektrotehn Fac (Beograd) Ser Mat 8:84
The GAP Team (1995) GAP, groups, algorithms and programming, lehrstuhl de für mathematik. RWTH, Aachen
Trinajstić N (1992) Chemical graph theory. CRC Press, Boca Raton, FL
Vizitiu AE, Cigher S, Diudea MV, Florescu MS (2007) MATCH Commun Math Comput Chem 57:457

Chapter 2
Wiener Index of Nanotubes, Toroidal Fullerenes and Nanostars

Ali Reza Ashrafi

Abstract A topological index for a molecular graph is a number correlated some physic-chemical properties of the molecule under consideration. The Wiener index is the first reported distance based topological index applicable in chemistry. It is defined as half sum of the distances between all the pairs of vertices in a molecular graph. In this survey article a brief account on the development of the Wiener in the fascinating field of nanoscience are discussed

2.1 Introduction

Mathematical chemistry is a branch of theoretical chemistry for discussion and prediction of the molecular structure using mathematical methods without necessarily referring to quantum mechanics (Gutman and Polansky 1986; Cyvin and Gutman 1988; Gutman 2006). ***Chemical graph theory*** is a branch of mathematical chemistry concerned with the study of chemical graphs. Chemical graphs are models of molecules in which atoms are represented by vertices and chemical bonds by edges of a graph. The basic idea of chemical graph theory is that physico-chemical properties of molecules can be studied by using the information encoded in their corresponding chemical graphs (Balaban 1976; Bonchev 1983; Trinajstić 1992). This theory had an important effect on the development of the chemical sciences. The pioneers of the chemical graph theory are Alexandru Balaban, Ivan Gutman, Haruo Hosoya, Milan Randić and Nenad Trinajstić. Nowadays hundreds of researchers work in this area producing thousands articles annually.

A ***molecular graph*** is a simple graph such that its vertices correspond to the atoms and the edges to the bonds. Note that hydrogen atoms are often omitted. By IUPAC terminology, a topological index is a numerical value associated with chemical constitution purporting for correlation of chemical structure with

A.R. Ashrafi (✉)
Faculty of Science, Department of Mathematics and Computer Science, University of Kashan, Kashan 87317-51167, Iran
e-mail: ashrafi@kashanu.ac.ir

F. Cataldo et al. (eds.), *The Mathematics and Topology of Fullerenes*, Carbon Materials: Chemistry and Physics 4, DOI 10.1007/978-94-007-0221-9_2, various physical properties, chemical reactivity or biological activity (Kier and Hall 1976). There is not a one-to-one correspondence between chemical structures and topological indices, because several graphs may have the same topological index.

A ***graph*** is a collection of points and lines connecting a subset of them. The points and lines of a graph also called ***vertices*** and ***edges*** of the graph, respectively. If e is an edge of G, connecting the vertices u and v, then we write $e = uv$ and say "u and v are adjacent". Let G be a graph. The vertex and edge sets of G are denoted by $V(G)$ and $E(G)$, respectively. A ***path*** P in a graph G is a sequence $v_1, v_2, \ldots, v_r$ of vertices such that v_i and v_{i+1} are adjacent, $1 \leq i \leq r - 1$. A path graph is a graph consisting of a single path. A ***cycle*** graph C_n of order n is a graph with $V(G) = \{v_1, v_2, \ldots, v_n\}$ and $E(G) = \{v_1v_2, v_2v_3, \ldots, v_{n-1}v_n, v_nv_1\}$. An acyclic graph or tree is a graph without a subgraph isomorphic to cycle graphs. A ***connected graph*** is a graph such that there exists a path between all pairs of vertices. The ***distance*** $d(u, v) = d_G(u, v)$ between two vertices u and v is the length of a shortest (u,v)-path in G.

The ***Wiener index*** is the first distance-based topological index introduced by chemist Harold Wiener (Wiener 1947). It is widely used in QSPR and QSAR models, and it still represents an important source of inspiration for defining new topological indices. A tentative explanation of the relevance of the Wiener index in research of QSPR and QSAR is that it correlates with the van der Waals surface area of the molecule. The Wiener defined his index as the sum of all distances between any two carbon atoms in the molecule, in terms of carbon–carbon bonds. The Wiener index is principally defined for acyclic graphs (trees). It was in 1972 that Hosoya (Hosoya 1971) described its calculation using the distance matrix and proposed the name "Wiener index", see also (Hosoya 1988).

In recent years, several papers on methods for computing Wiener index of molecular graphs have been published. We encourage to interested readers to consult papers (Mohar and Pisanski 1988; Gutman et al. 1994; Gutman and Körtvélyesi 1995; Dobrynin et al. 2001, 2002) and references therein for background materials as well as basic computational techniques.

2.2 Wiener Index of Nanotubes

A ***nanostructure*** is an object of intermediate size between molecular and microscopic structures. It is a product derived through engineering at the molecular scale. The most important of these new materials are carbon nanotubes (Iijima 1991). Carbon nanotubes can be imagined as rolled sheets of graphite about different axes. There are three types of nanotubes: armchair, chiral and zigzag structures. Furthermore, nanotubes can be categorized as single-walled and multi-walled nanotubes. It is very difficult to produce the former type of nanotubes. Carbon nanotubes were discovered in 1991 by Iijima and Ichlhashi (1993) as multi walled structures and in 1993 as single walled carbon nanotubes (briefly denoted SWNT) independently by Iijima's group (Iijima and Ichlhashi 1993) and Bethune's group (Bethune et al. 1993) from IBM. SWNTs

The problem of computing Wiener index of nanotubes began by publishing two papers by M. V. Diudea and his co-workers (John and Diudea 2004; Diudea et al. 2004). In these papers, the authors gave closed formulas for the sum of all distances in "armchair" and "zig-zag" polyhex nanotubes, see Figs. 2.1 and 2.2. Diudea and his team continued their works towards the general form of nanostructures (Diudea and John 2001; Diudea and Kirby 2001; Diudea et al. 2003; Diudea 2002a, b).

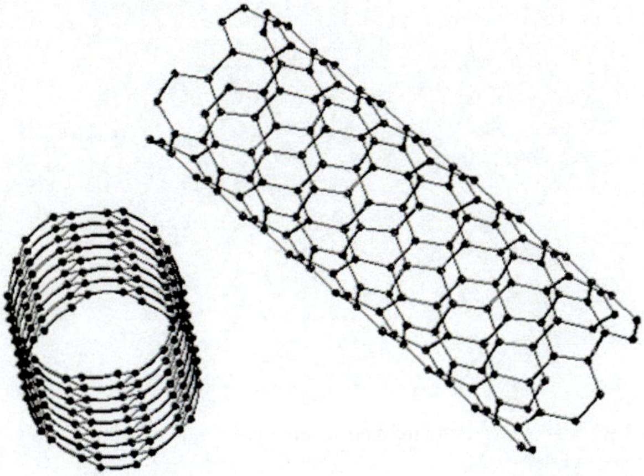

Fig. 2.1 An "armchair" polyhex nanotube

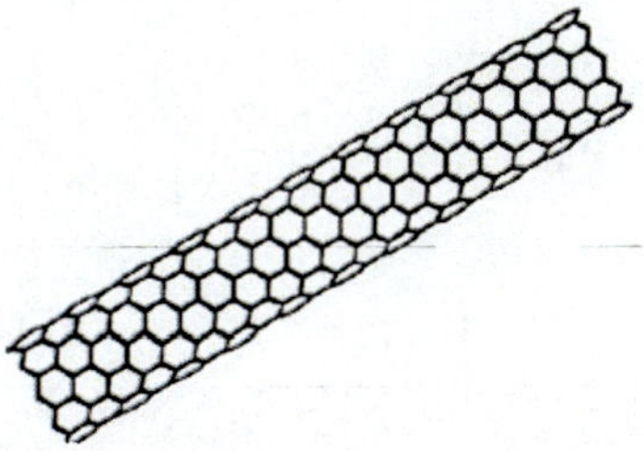

Fig. 2.2 The "zig-zag" polyhex nanotube

We begin by describing a method for calculation of the Wiener index of an armchair polyhex nanotube $T = TUVC_6[m, n]$, where n is twice the number of vertical crenels and m is the number of rows, see Fig. 2.3. This method can be applied to compute also the Wiener index of an achiral polyhex nanotorus. Let us consider an armchair lattice, as illustrated in Fig. 2.3. We first choose a base vertex b from the 2-dimensional lattice of T, Fig. 2.3, and assume that $x_{i,j}^{(1,1)}$ denotes the distance between (1,1) and (i,j). This defines a matrix

$$X_{m\times n}^{(1,1)} = [x_{i,j}^{(1,1)}]\,\text{where}\, x_{1,1}^{(1,1)} = 0,\ x_{1,2}^{(1,1)} = x_{2,1}^{(1,1)} = 1. \tag{2.1}$$

It is clear that by choosing different base vertices, we find different distance matrices for T. Suppose $s_k^{(p,q)}$ is the sum of *k*th row of $X_{m\times n}^{(p,q)}$, where (p,q) is the base vertex. Then $s_k^{(p,1)} = s_k^{(p,q)}$, $1 \le k \le m$, $1 \le p \le m$ and $1 \le q \le n$. On the other hand, by Eq. (2.1) and choosing a fixed column, we have:

$$s_k^{(i,j)} = \begin{cases} s_{i-k+1}^{(1,1)} & 1 \le k \le i \le m, \quad 1 \le j \le n \\ s_{k-i+1}^{(1,1)} & 1 \le i \le k \le m, \quad 1 \le j \le n \end{cases}. \tag{2.2}$$

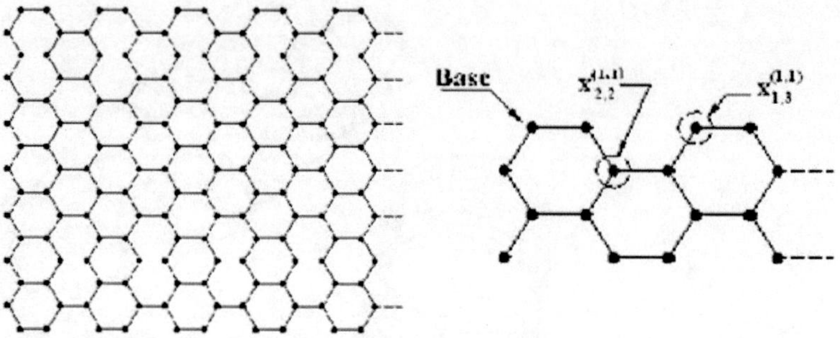

Fig. 2.3 The 2-dimensional fragments of an "armchair" polyhex nanotube and the scheme of its base vertex

We now define two new matrices $A_{(n/2+1)\times n}$ and $B_{(n/2+1)\times n}$ by $a_{1,1} = 0$, $a_{2,1} = 1$ and the following equations:

$$a_{1,j} = \begin{cases} a_{1,j-1} + 1 & j \le (n/2) + 1 \\ a_{1,j-1} - 1 & j > (n/2) + 1 \end{cases}; \qquad a_{2,j} = \begin{cases} a_{1,j} + 1 & j \le (n/2) + 1 \\ a_{1,j} - 1 & j > (n/2) + 1 \end{cases} \quad 2|j,$$

$$a_{2,j} = \begin{cases} a_{2,j-1} + 1 j \le (n/2) + 1 \\ a_{2,j-1} - 1 j > (n/2) + 1 \end{cases}; \qquad a_{1,j} = \begin{cases} a_{2,j} + 1 & j \le (n/2) + 1 \\ a_{2,j} - 1 & j > (n/2) + 1 \end{cases} \quad 2 \nmid j.$$

Other entries of this matrix is obtained from the first two rows by $a_{i,j} = a_{1,j}$, i is odd, and $a_{i,j} = a_{2,j}$, i is even. Define:

$$b_{n/2+1,j} = \begin{cases} n/2 + j - 1 & j \le n/2 + 1 \\ 3n/2 - j + 1 & j > n/2 + 1 \end{cases}.$$

Other entries of B is defined by $b_{i,j} = b_{i+1,j} - 1$, $i < n/2 + 1$. Therefore,

$$x_{i,j}^{(1,1)} = \begin{cases} c_{i,j} & i \le (n/2) + 1 \\ x_{i-1,j}^{(1,1)} + 1 & i > (n/2) + 1 \end{cases},$$

where $c_{i,j} = \mathrm{Max}\{a_{i,j}, b_{i,j}\}$. This computes the distance matrix $X^{(p,q)}_{m\times n}$ related to the vertex (p,q). Suppose $s_i^{(p,q)}$ is the sum of *i*th row of $X^{(p,q)}_{m\times n}$. Then by our calculations given above, we have:

$$s_i^{(1,q)} = \begin{cases} (n^2/2) + i^2 - 2i + (1/2)\left(1 - (-1)^{\left(\frac{n}{2}-1\right)}\right) & i \le n/2+1 \\ (n^2/4) + n\,(i-1) & i > n/2+1 \end{cases}, \quad 1 \le i \le m\,;\ 1 \le q \le n.$$

Suppose S_p is the sum of all entries of distance matrix $X^{(p,q)}_{m\times n}$. Then $S_1 = \sum_{i=1}^{m} s_i^{(1,q)}$ and $S_p = S_1 + \sum_{i=2}^{p} s_i^{(1,q)} - \sum_{i=m-p+2}^{m} s_i^{(1,q)}$. Thus

$$S_1 = \begin{cases} (m/6)\left(3n^2 + 2m^2 - 3m - 2\right) + \left[(-1)^{(n/2)}/4\right]\left[1 - (-1)^m\right] & m \le n/2+1 \\ (n/24)\left(n^2 + 12m^2 + 6mn + 3n - 12m - 4\right) + (1/4)\left[(-1)^{(n/2)} - 1\right] & m > n/2+1 \end{cases}.$$

If $m \le n/2+1$ then a direct calculation shows that

$$\begin{aligned} S_p &= (m/6)\left(3n^2 + 2m^2 + 3m - 2\right) - mp\,(m - p + 1) \\ &\quad -(1/4)\left[(-1)^{+(n/2p)}\right]\left[1 - (-1)^m\right]. \end{aligned}$$

Therefore, it is enough to consider case that $m > n/2+1$. To complete this case, we consider three sub cases as follows:

(I) $p \le n/2+1,\ p \le m - n/2 + 1$. In this case, we have:

$$\begin{aligned} S_p &= (n/24)\left(n^2 + 12m^2 + 6mn - 3n + 12m - 4\right) + (p^2/2)\,(n-1) \\ &\quad + (p/12)\left(3n^2 - 6n - 12mn - 4\right) + (p^3/3) + (1/4)\left[1 - (-1)^{(n/2+p)}\right]. \end{aligned}$$

(II) $m - n/2 + 1 < p \le n/2+1$. In this case, we have:

$$\begin{aligned} S_p &= (m/6)\left(3n^2 + 2m^2 + 3m - 2\right) \\ &\quad - mp\,(m+1) + mp^2 - (1/4)\left[(-1)^{n/2+p}\right]\left[1 - (-1)^m\right]. \end{aligned}$$

(III) $p > n/2+1$. In this case, $S_p = (n^3/12) - (n/3) + (mn/4)\,(n + 2m + 2) - np\,(m+1) + np^2$.

Suppose $W_{m\times n}$ denotes the Wiener index of the armchair polyhex nanotube T. We apply our calculations given above to compute the Wiener index of this nanotube. We have:

$$W_{m\times n} = \begin{cases} n\left(\sum\limits_{i=1}^{(m-1)/2} S_i + (1/2)S_{(m+1)/2}\right) & 2\mid m \\ n\sum\limits_{i=1}^{m/2} S_i & 2\nmid m \end{cases}$$

$$= \begin{cases} (m^2n/12)\left(3n^2+m^2-4\right) + (n/8)(-1)^{(n/2)}\left[1-(-1)^m\right] & m \le n/2+1 \\ (mn^2/24)\left(n^2+4m^2+3mn-8\right) - (n^3/192)\left(n^2-16\right) & \\ \qquad +(n/8)\left[(-1)^{(n/2)}-1\right] & m > n/2+1 \end{cases}$$

Therefore, we prove the following theorem:

Theorem 1 (Yousefi and Ashrafi 2008a) *The Wiener index of T is computed as follows*:

$$W_{m\times n} = \begin{cases} \frac{m^2n}{12}\left(3n^2+m^2-4\right) + \frac{n}{8}(-1)^{(n/2)}\left[1-(-1)^m\right] & : m \le \frac{n}{2}+1 \\ \frac{mn^2}{24}\left(n^2+4m^2+3mn-8\right) & \\ \qquad -\frac{n^3}{192}\left(n^2-16\right) + \frac{n}{8}\left[(-1)^{(n/2)}-1\right] & : m > \frac{n}{2}+1 \end{cases}$$

We now compute the distance matrix and Wiener index of the molecular graph of an zig-zag polyhex nanotube $S = \mathrm{TUHC_6}[m,n]$, *Fig.* 2.2. *Here m is the number of horizontal zig-zags and n is the number of columns. It is obvious that n is even and* $|V(T)| = mn$. *We first choose a base vertex b from the 2-dimensional lattice of T and assume that* x_{ij} *is the (i,j)th vertex of T, Fig.* 2.4.

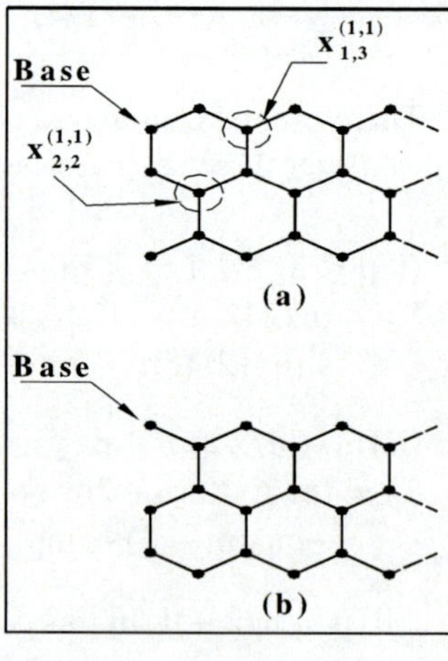

Fig. 2.4 Two basically different cases for the vertex b

Define $D^{(1,1)}_{m\times n} = [d^{(1,1)}_{i,j}]$, where $d^{(1,1)}_{i,j}$ is distance between (1,1) and (i,j), $i = 1, 2\ldots, m$ and $j = 1, 2, \ldots, n$. From Fig. 2.4, one can see that there are two separates cases for the (1,1)th vertex. For example in the case (a) of Fig. 2.4, $d^{(1,1)}_{1,1} = 0$,

$d_{1,2}^{(1,1)} = d_{2,1}^{(1,1)} = 1$ and in case (b), $d_{1,1}^{(1,1)} = 0$, $d_{1,2}^{(1,1)} = 1$, $d_{2,1}^{(1,1)} = 3$. In general, we assume that $D_{m\times n}^{(p,q)}$ is distance matrix of T related to the vertex (p,q) and $s_i^{(p,q)}$ is the sum of *i*th row of $D_{m\times n}^{(p,q)}$. Then there are two distance matrices related to the vertex (p,q) such that $s_i^{(p,2k-1)} = s_i^{(p,1)}$; $s_i^{(p,2k)} = s_i^{(p,2)}$; $1 \le k \le n/2$, $1 \le i \le m$, $1 \le p \le m$.

By Fig. 2.4, if b varies on a column of T then the sum of entries in the row containing base vertex is equal to the sum of entries in the first row of $D_{m\times n}^{(1,1)}$. On the other hand, one can compute the sum of entries in other rows by distance from the position of base vertex. Therefore,

$$s_k^{(i,j)} = \begin{cases} s_{i-k+1}^{(1,1)} & 1 \le k \le i \le m,\ 1 \le j \le n \\ s_{k-i+1}^{(1,2)} & 1 \le i \le k \le m,\ 1 \le j \le n \end{cases}$$

If $2 | (i+j)$

$$s_k^{(i,j)} = \begin{cases} s_{i-k+1}^{(1,2)} & 1 \le k \le i \le m,\ 1 \le j \le n \\ s_{k-i+1}^{(1,1)} & 1 \le i \le k \le m,\ 1 \le j \le n \end{cases}$$

If $2 \nmid (i+j)$

We now describe our algorithm to compute the distance matrix of a zig-zag polyhex nanotube. To do this, we define matrices $A_{m\times(1/2+1)}^{(a)} = [a_{ij}]$, $B_{m\times(n/2+1)} = [b_{ij}]$ and $A_{m\times(n/2+1)}^{(b)} = [c_{ij}]$ as follows:

$a_{1,1} = 0$	$a_{1,2} = 1$	$a_{i,j} = \begin{cases} a_{i,1} & 2 \nmid j \\ a_{i,2} & 2 \nmid j \end{cases}$	;	$a_{i,1} = a_{i-1,1} + 1$, $a_{i,2} = a_{i-1,2} + 1$,	$a_{i,2} = a_{i,1} + 1$; $2 \mid i$ $a_{i,1} = a_{i,2} + 1$; $2 \nmid i$
$c_{1,1} = 0$	$c_{1,2} = 1$	$c_{i,j} = \begin{cases} c_{i,1} & 2 \nmid j \\ c_{i,2} & 2 \mid j \end{cases}$	;	$c_{i,2} = c_{i-1,2} + 1$, $c_{i,1} = c_{i-1,1} + 1$,	$c_{i,1} = c_{i,2} + 1$; $2 \mid i$ $c_{i,2} = c_{i,1} + 1$; $2 \nmid i$
$b_{i,1} = i - 1$		; $b_{i,j} = b_{i,j-1} + 1$			

For computing the distance matrix of this nanotube we must compute matrices $D_{m\times n}^{(a)} = [d_{i,j}^{a}]$ and $D_{m\times n}^{(b)} = [d_{i,j}^{b}]$. But by our calculations, we can see that

$$d_{i,j}^{a} = \begin{cases} \mathrm{Max}\{a_{i,j}, b_{i,j}\} & 1 \le j \le n/2 \\ d_{i,n-j+2} & j > n/2+1 \end{cases} \text{ and } d_{i,j}^{b} = \begin{cases} \mathrm{Max}\{a_{i,j}, c_{i,j}\} & 1 \le j \le n/2 \\ d_{i,n-j+2} & j > n/2+1 \end{cases}.$$

This completes our calculations for the distance matrix of S. Suppose $s_i^{(p,q)}$ is the sum of ith row of $D_{m\times n}^{(p,q)}$. Then $s_i^{(p,2k-1)} = s_i^{(p,1)}$ and $s_i^{(p,2k)} = s_i^{(p,2)}$, where $1 \le k \le n/2$, $1 \le i \le m$ and $1 \le p \le m$. On the other hand,

$$s_i^{(1,2k-1)} = \begin{cases} \frac{n^2}{4} + (n+i-2)(i-1) & i \le \frac{n}{2}+1 \\ \frac{n}{2}(4i-5) & i \ge \frac{n}{2}+1 \end{cases}$$

$$,1 \le i \le m, \ 1 \le k \le \frac{n}{2}.$$

$$s_i^{(1,2k)} = \begin{cases} \frac{n^2}{4} + (n+i)(i-1) & i \le \frac{n}{2}+1 \\ \frac{n}{2}(4i-3) & i \ge \frac{n}{2}+1 \end{cases}$$

Suppose $S_p^{(a)}$ and $S_p^{(b)}$ are the sum of all entries of distance matrix $D_{m\times n}^{(p,q)}$ in two cases (a) and (b). Then

$$S_1^{(a)} = \begin{cases} (mn/4)(2m+n-2) + (m/3)(m-1)(m-2) & m \le n/2+1 \\ (mn/2)(2m-3) + (n/24)(n+2)(n+4) & m \le n/2+1 \end{cases},$$

$$S_1^{(b)} = \begin{cases} (mn/4)(2m+n-2) + (m/3)(m^2-1) & m \le n/2+1 \\ (mn/2)(2m-1) + (n/24)(n^2-4) & m \le n/2+1 \end{cases}.$$

If p is arbitrary then one can see that:

$$\begin{aligned} S_p^{(a)} &= S_1^{(a)} + \sum_{i=2}^{p} s_i^{(1,2)} - \sum_{i=m-p+2}^{m} s_i^{(1,1)} \\ S_p^{(b)} &= S_1^{(b)} + \sum_{i=2}^{p} s_i^{(1,1)} - \sum_{i=m-p+2}^{m} s_i^{(1,2)} \end{aligned}.$$

Thus it is enough to compute $S_p^{(a)}$ and $S_p^{(b)}$. When $m \le n/2$, one can see that:

$$\begin{aligned} S_p^{(a)} &= (mn/4)(2m+n+2) + (m/3)(m^2-1) - p(m^2+mn+n) + p^2(m+n) \\ S_p^{(b)} &= (mn/4)(2m+n+2) + (m/3)(m+1)(m+2) - p(m^2+mn+n+2m) \\ &\quad + p^2(m+n). \end{aligned}$$

To complete our argument, we must investigate the case of $m > n/2+1$. To do this, we consider three cases that $m > n/2+1$; $m \le n+1$, $m-n/2+1 < p \le n/2+1$ and $m > n+1$; $p > n/2+1$.

(I) $p \le \frac{n}{2}+1$ and $p \le m - \frac{n}{2}+1$. In this case we have:

$$\begin{aligned} S_p^{(a)} &= \frac{mn}{2}(2m+1) + \frac{n}{24}\left(n^2-4\right) + \frac{p}{12}\left(3n^2-24mn-12n-4\right) + \frac{3n}{2}p^2 + \frac{p^3}{3} \\ S_p^{(b)} &= \frac{mn}{2}(2m+3) + \frac{n}{24}(n-2)(n-4) + \frac{p}{12}\left(3n^2-24mn-24n+8\right) \\ &\quad + \frac{p^2}{2}(3n-2) + \frac{p^3}{3} \end{aligned}$$

(II) $m \le n+1$ and $m - \frac{n}{2} + 1 < p \le \frac{n}{2} + 1$. Therefore,

$$S_p^{(a)} = \frac{mn}{4}(2m+n+2) + \frac{m}{3}\left(m^2-1\right) - p\left(m^2+mn+n\right) + p^2(m+n)$$

$$S_p^{(b)} = \frac{mn}{4}(2m+n+2) + \frac{m}{3}(m+1)(m+2) - p\left(m^2+mn+n+2m\right) + p^2(m+n)$$

(III) $m > n+1$ and $p > \frac{n}{2} + 1$. In this case,

$$S_p^{(a)} = \frac{n}{12}\left(n^2-4\right) + \frac{n}{2}(m-2p)(2m+1) + 2np^2$$

$$S_p^{(b)} = \frac{n}{12}\left(n^2+8\right) + \frac{n}{2}(m-2p)(2m+3) + 2np^2$$

We assume again that $W_{m\times n} = W(TUHC_6[m,n])$. Then,

$$W_{m\times n} = \begin{cases} (n/2)\left[\sum_{i=1}^{(m-1)/2}\left(S_i^{(a)}+S_i^{(b)}\right) + (1/2)\left(S_{(m+1)/2}^{(a)} + S_{(m+1)/2}^{(b)}\right)\right] & 2 \nmid m \\ (n/2)\sum_{i=1}^{m/2}\left(S_i^{(a)}+S_i^{(b)}\right) & 2|m \end{cases}.$$

We now substitute the values of $S_p^{(a)}$ to compute the Wiener index of S, as follows:

Theorem 2 (Ashrafi and Yousefi 2007a) *The Wiener index of S is computed as follows:*

$$W_{m\times n} = \begin{cases} \frac{mn^2}{24}\left(4m^2+3mn-4\right) + \frac{m^2n}{12}\left(m^2-1\right) & m \le \frac{n}{2}+1 \\ \frac{mn^2}{24}\left(8m^2+n^2-6\right) - \frac{n^3}{192}\left(n^2-4\right) & m > \frac{n}{2}+1 \end{cases}.$$

We now present an algebraic method for computing Wiener index of molecular graphs. Let us recall some definitions and notations. An automorphism of a graph G is a permutation g of the vertex set V(G) with the property that, for any vertices u and v, g(u) and g(v) are adjacent if and only if u is adjacent to v. The set of all automorphisms of G, with the operation of the composition of permutations, is a permutation group on V(G), denoted by Aut*(G). Suppose G is a group and X is a set. G is said to act on X when there is a map* $\phi : G\times X \to X$ *such that all elements* $x \in X$ (i) $\varphi(e,x) = x$, *where e is the identity element of G, and,* (ii) $\varphi(g,\varphi(h,x)) = \varphi(gh,x)$ *for all g, h∈G. In this case, G is called a transformation group on X, X is called a G-set, and* φ *is called the group action. For simplicity we define* $gx = \varphi(g,x)$. *In a group action, a group permutes the elements of X. The identity does nothing, while a composition of actions corresponds to the action of the composition. For a given X, the set* $\{gx \mid g \in G\}$, *where the group action moves x, is called the group orbit* *of x. If G has exactly one orbit, then G is said to be vertex transitive. The following simple lemma is crucial for our algebraic method.*

Lemma 1 *Suppose G is a graph,* $A_1, A_2, \ldots, A_r$ *are the orbits of* Aut(*G*) *under its natural action on V(G) and* $x_i \in A_i$, $1 \le i \le r$. *Then* $W(G) = \sum_{j=1}^{r} \frac{|A_j|}{2} d(x_j)$, *where d(x) denotes the summation of topological distances between x and all vertices of G. In particular, if G is vertex transitive then* $W(G) = \frac{|V(G)|}{2} d(x)$, *for every vertex x.*

Proof It is easy to see that if vertices u and v are in the same orbit, then there is an automorphism φ such that $\varphi(u) = v$. So, by definition of an automorphism, for every vertex x,

$$\begin{aligned} d(u) &= \Sigma_{x \in V(G)} d(x,u) = \Sigma_{x \in V(G)} d(\varphi(x), \varphi(u)) \\ &= \Sigma_{x \in V(G)} d(\varphi(x), v) = \Sigma_{y \in V(G)} d(y, v) = d(v) \end{aligned}$$

Thus, $W(G) = W(G) = \sum_{j=1}^{r} \frac{|A_j|}{2} d(x_j)$. If G is vertex transitive then $r = 1$ and $|A_1| = |V(G)|$. Therefore, $W(G) = \frac{|V(G)|}{2} d(x)$, for each vertex x.

Apply our method on an toroidal fullerene (or achiral polyhex nanotorus) $R = R[p, q]$, Figs. 2.5 and 2.6. To compute the Wiener index of this nanotorus, we first prove its molecular graph is vertex transitive.

Lemma 2 *The molecular graph of a polyhex nanotorus is vertex transitive.*

Proof To prove this lemma, we first notice that p and q must be even. Consider the vertices u_{ij} and u_{rs} of the molecular graph of a polyhex nanotori $T = T[p, q]$, Fig. 2.6. Suppose both of i and r are odd or even and σ is a horizontal symmetry plane which maps u_{it} to u_{rt}, $1 \le t \le p$ and π is a vertical symmetry which maps u_{tj} to u_{ts}, $1 \le t \le q$. Then σ and π are automorphisms of T and we have $\pi\sigma(u_{ij}) = \pi(u_{rj}) = u_{rs}$. Thus u_{ij} and u_{rs} are in the same orbit under the action of Aut(G) on V(G). On the other hand, the map θ defined by $\theta(u_{ij}) = \theta(u_{(p+1-i)j})$ is a graph automorphism of T and so if "i is odd and r is even" or "i is even and r is odd" then again u_{ij} and u_{rs} will be in the same orbit of Aut(G), proving the lemma.

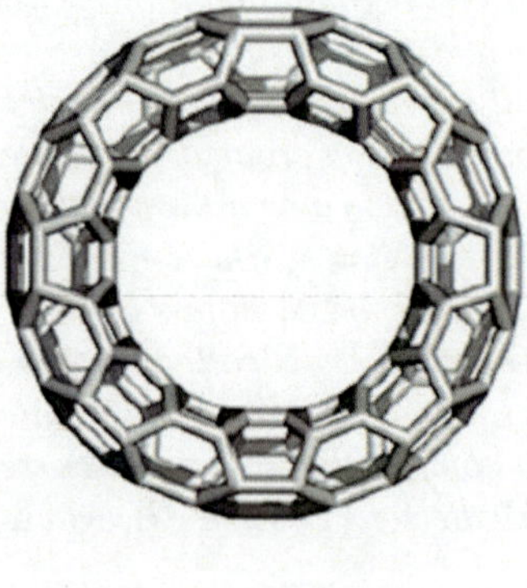

Fig. 2.5 A toroidal fullerene (or achiral polyhex nanotorus) R[p,q]

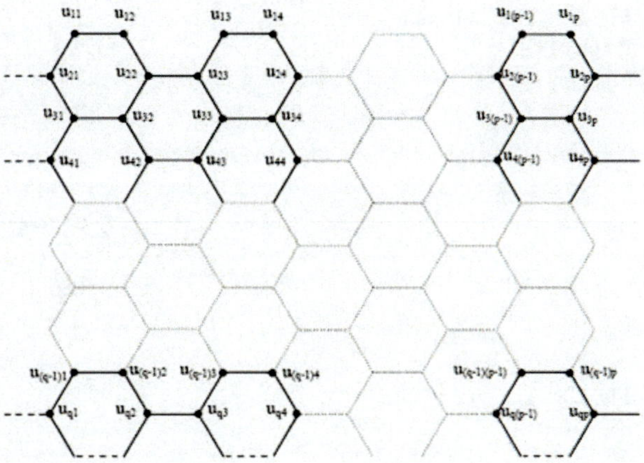

Fig. 2.6 A 2-dimensional lattice for an achiral polyhex nanotorus R[p,q]

We now apply Lemma 2, to compute the Wiener index of an achiral polyhex nanotorus. By Lemmas 1 and 2, in a polyhex nanotorus $T = T[p, q]$ we have $W(T) = (pq/2)d(x)$, for a fixed vertex x of T.

Theorem 3 (Yousefi et al. 2008c) *The Wiener index of an achiral polyhex nanotube $R = R[p, q]$ is computed as follows:*

$$W(R) = \begin{cases} \frac{pq^2}{24}(6p^2 + q^2 - 4) & q < p \\ \frac{p^2q}{24}(3q^2 + 3pq + p^2 - 4) & q \le p \end{cases}.$$

Using methods presented in this section, it is possible to compute the Wiener index of some other nanotubes as TUC_4C_8*(R) and* TUC_4C_8*(S) nanotubes and nanotori, see (Ashrafi and Yousefi 2007b, c; Yousefi and Ashrafi 2006, 2008b, Yousefi et al. 2008d; Iranmanesh and Ashrafi 2007) for more information on the problem.*

2.3 Wiener Index of Nanostar Dendrimers

The nanostar dendrimer is part of a new group of macromolecules that appear to be photon funnels just like artificial antennas. The topological study of these macromolecules is a new problem began by the present author (Ashrafi and Saati 2007; Ashrafi and Mirzargar 2008a, b, c; Karbasioun and Ashrafi 2009, Karbasioun et al. 2010). In this section two methods for computing Wiener index of nanostar dendrimers are presented.

Suppose G[n] denotes the molecular graph of nanostar dendrimer depicted in Figs. 2.7 and 2.8. We first calculate the distance matrix of the graph G[n] and then compute its Wiener index. At first, we introduce two concepts which are important in our calculations. Suppose G and H are graphs such that $V(H) \subseteq V(G)$ and $E(H) \subseteq$

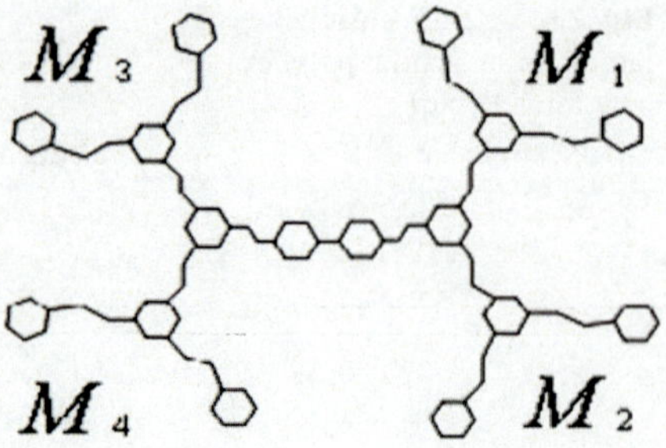

Fig. 2.7 The molecular graph of G[1]

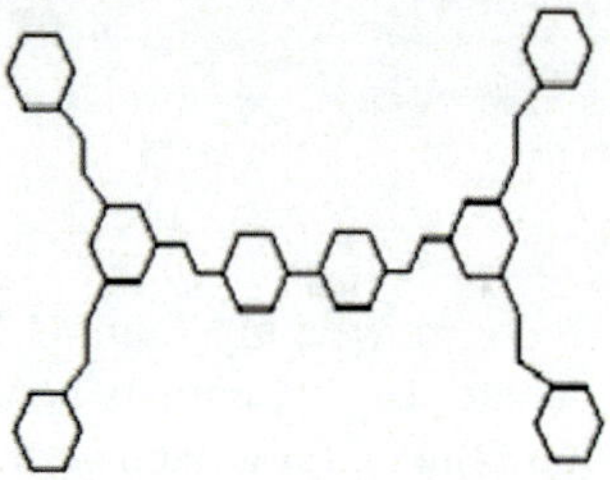

Fig. 2.8 The core of G[n]

E(G). Then we call H to be a subgraph of G. H is called isometric, if for each x, y $\in$ V(H), $d_H(x, y) = d_G(x, y)$.

In Fig. 2.9, four isometric subgraphs of G[n] are depicted. From this figure, it is clear that G[n] is constructed from the subgraphs isomorphic to B and the core, Fig. 2.8. To compute the Wiener index of G[n], we calculate matrices WA_1, WA_2, WA_3 and WB which are the Wiener matrices of the subgraphs A_1, A_2, A_3 and B, respectively. Suppose D_i and D_i' are 8×8 and 8×60 matrices in which each entry is equal to i and M is the distance matrix of the core.

To construct the Wiener matrix of G[n], it is enough to calculate the distance matrix between a subgraph isomorphic to B and core, distance matrix between two subgraphs isomorphic to B (see A_2 and A_3 in Fig. 2.9) and the distance matrix of the core. The distance matrix between a subgraph isomorphic to B and core is equal to the sum of the distance matrix of the subgraph A_1, WA_1, and the matrix D_i', where $i = l(P) - 1$ such that P is a minimum path connecting a vertex of core to a vertex of B and l(P) denotes the length of P. We now calculate the distance matrix between two subgraphs isomorphic to B. To do this, we assume that B_1 and B_2 are two subgraphs isomorphic to B and P is a minimum path connecting a vertex of B_1 to a vertex of B_2. Obviously, there are two separate cases that one of the end vertices of P is a vertex of a hexagon of G[n] or two end vertices of P are not belong to a hexagon. In the first case, the distance matrix $D(B_1, B_2)$ between B_1 and B_2 is equal to $WA_3 + D_i$ and for the second $D(B_1, B_2) = WA_2 + D_i$. From Fig. 2.7, one can partition the molecular graph of G[n] into a core together with six isomorphic subgraphs $M_1[n], \ldots, M_4[n]$. We name each of $M_1[n], \ldots, M_4[n]$, to be a branch of G and $M[n] = M_1[n] \cup \ldots \cup M_4[n]$.. Obviously, each of branches $M_i[n]$, $1 \le i \le 4$,

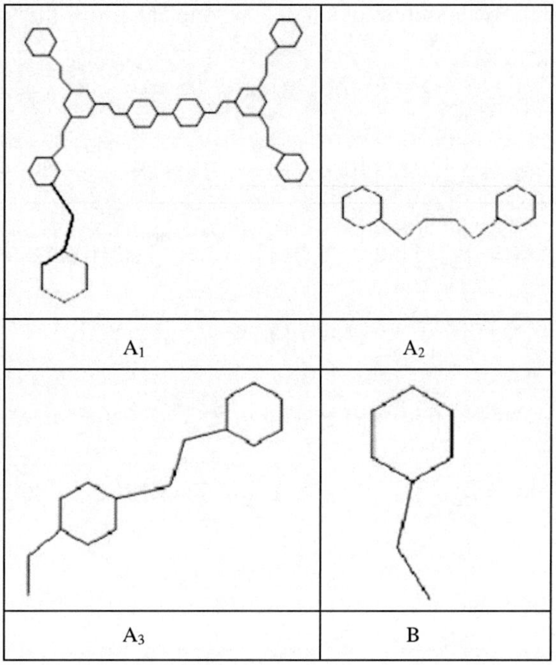

Fig. 2.9 Some subgraphs of G[n]

has exactly two isomorphic components $M_i^1[n]$ and $M_i^2[n]$. Moreover, the core and branches constitute a partition for G[n]. Every subgraph $M_i[n]$, $1 \leq i \leq 4$, has exactly $2^{n+1} - 2$ subgraphs isomorphic to B such that degree of vertices of their hexagons are 2. We name these subgraphs by Y^i_{1k}, $1 \leq i \leq 2^{n-1}$, $1 = 1,2$ and $k = 1,2,3,4$. We now define $s_1, \ldots, s_8$ as follows:

- s_1 is the summation of distances between vertices of Y^i_{11}, Y^j_{21} and Y^i_{12}, Y^j_{22}, as well as Y^i_{13}, Y^j_{23} and Y^i_{14}, Y^j_{24}, for each of i and j, $1 \leq i \neq j \leq 2^{n-1}$,
- s_2 is the summation of distances between vertices of Y^j_{23}, Y^i_{13} and Y^i_{11}, Y^j_{21} ; Y^j_{23}, Y^i_{13} and Y^i_{12}, Y^j_{22} ; Y^i_{14}, Y^j_{24} and Y^i_{11}, Y^j_{21} ; Y^i_{14}, Y^j_{24} and Y^i_{12}, Y^j_{22}, for each of i and j, $1 \leq i \neq j \leq 2^{n-1}$,
- s_3 is the summation of distances between vertices of Y^i_{1k} and Y^j_{2k}, for each of i, j and k, $1 \leq i \neq j \leq 2^{n-1}$ and $k = 1,2,3,4$,
- s_4 is the summation of distances between the vertices of $M_i^1[n]$ and $M_i^2[n-1]$,
- s_5 is the summation of distances between vertices of Y^i_{1k} and Y^j_{2k} in $M_k[n-1]$,
- s_6 is the summation of distances between vertices of Y^i_{13}, Y^j_{23} from $M_1[n-1]$ and Y^i_{14}, Y^j_{24} from $M_2[n-1]$,
- s_7 is the summation of distances between vertices of $M_2[n]$ and $M_1[1]$, as well as $M_3[n]$ and $M_4[1]$,
- s_8 is the summation of distances between other vertices of $M_i^1[n]$ and $M_i^2[1]$.

By definition of $s_1, \ldots, s_8$, one can prove the following equalities:

$$s_1 = \sum_{i=1}^{n} 2^{2i+1}.(10i+3) = -\frac{8}{9}.4^n + \frac{80}{3}.4^n.n + \frac{8}{9}$$

$$s_2 = \sum_{i=1}^{n} 2^{2i+2}.(10i+18) = \frac{704}{9}.4^n + \frac{160}{3}.4^n.n - \frac{704}{9}$$

$$s_3 = \sum_{j=1}^{n}\sum_{i=1}^{j} 4.2^{j-2+i}.(10i-7) = \frac{80}{3}.4^n + 68.2^n - \frac{124}{9} - \frac{488}{9}.4^n$$

$$s_4 = \sum_{l=3}^{n}\sum_{k=2}^{l-1}\sum_{i=1}^{l-k} 2^{l+i+1}.(5(l+i)-(5k+2))$$
$$= -\frac{968}{9}.4^n + \frac{80}{3}.4^n.n - 56.2^n + 156.2^n.n - 20.2^n.n^2 + \frac{1472}{9}$$

$$s_5 = \sum_{j=2}^{n}\sum_{i=1}^{j-1} 2^{i+j+1}.(5(i+j)-7) = 136.2^n - 40.2^n.n - \frac{488}{9}.4^n + \frac{80}{3}.4^n.n - \frac{736}{9}$$

$$s_6 = \sum_{j=2}^{n}\sum_{i=1}^{j-1} 2^{i+j+3}.(5.(i+j)+18)$$
$$= \frac{448}{9}.4^n + \frac{320}{3}.4^n.n - 256.2^n - 160.2^n.n + \frac{1856}{9}$$

$$s_7 = \sum_{j=3}^{n}\sum_{i=1}^{j-2} 3.2^{j+1}.(5i) = \sum_{j=2}^{n}\sum_{i=1}^{j-1} 2^{i+j+2}.(5(i+j)+3)$$
$$= -\frac{496}{9}.4^n + \frac{160}{3}.4^n.n + 112.2^n - 80.2^n.n - \frac{512}{9}$$

$$s_8 = \sum_{l=3}^{n}\sum_{k=2}^{l-1}\sum_{i=1}^{l-k} 3.2^{l+i}.(5(l+i)-(5k+2))$$
$$= \sum_{j=3}^{n}\sum_{i=1}^{j-2} 2^{j+2}.(5i) = 160.2^n - 100.2^n.n + 20.2^n.n^2 - 160$$

By a simple calculation with Maple, one can see that $s_1 + s_2 + \ldots + s_8 = 320.4^n.n + 164.2^n - 144.4^n - 224.2^n.n - 20$. Therefore we prove the following theorem,

Theorem 4 *The Wiener index of* $G = G[n]$ *is computed as follows*:

$$W(G) = -55424.2^n + 4480.2^n.n + 4096.4^n + 20480.4^n.n + 9048.2^{n+3} + 502.$$

We now present our final method for computing Wiener index of dendrimers. Let T[L] be a triangulane molecule containing $\lambda_L = 1 + 3 \times 1 + 3 \times 2 + 3 \times 2^2 + \cdots + 3 \times 2^{L-1}$ triangles. Then, $\lambda_L = 1 + 3(2^L - 1) = 3.2^L - 2$. We introduce an algorithm for constructing T[L] which is crucial in our calculations. Suppose

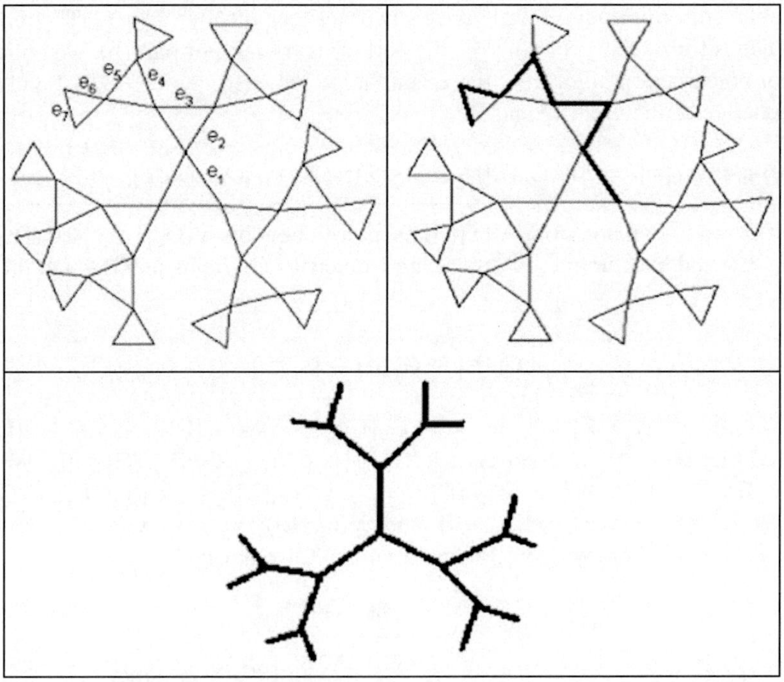

Fig. 2.10 Two shapes of the molecular graph of triangulane molecule T[4] and its associated dendrimer

K_3 denotes the graph of a triangle. In what follows, we assume that $K_3 \cong K_3^{mn}$, where m and n are positive integers. At first consider K_3^{11} and connect two new triangles K_3^{21} and K_3^{22} to two vertices of K_3^{11} and label the third vertex of K_3^{11} by X. Then we connect four triangles K_3^{31}, K_3^{32}, K_3^{33} and K_3^{34} to the vertices of degree 2 other than X of the graph constructed in last step. Continue this process to connect $K_3^{L1}, \ldots, K_3^{L2^{L-1}}$ to vertices of degree two in the graph constructed in the (L−1)th step of our algorithm. Thus, we obtain a graph called G^L. Suppose $G_u[L]$, $G_v[L]$ and $G_w[L]$ are three copies of G^L and X_u, X_v and X_w are vertices corresponding to X. We now join $G_u[L]$, $G_v[L]$, $G_w[L]$ by vertices X_u, X_v, X_w to a triangle K_3 and denote this new graph by T[L], $L < 0$, and $T[0] = K_3$. By this process, two graphs isomorphic to G^{L-m} will join to two vertices of K_3^{mn}, $0 < n \leq 2^{m-1}$ and $m < L$. We notice that the vertices correspond to X are denoted by X^{mn1}, X^{mn2} and their graphs by Gh^{mn1} and Gh^{mn2}.

Lemma 3 *Let T[L] denotes the molecular graph of a triangulane given above. Then,*

(a) Gh^{mn1}, Gh^{mn2} *are isometric subgraphs of T[L].*
(b) If $a \in G_u[L]$ *and* $b \in G_v[L]$ *then* $d_{T[L](a,b)} = d_{G_u[L]}(a, X_u) + d_{G_v[L]}(b, X_v) + 1$.

Proof Since a minimum path between two vertices of Gh^{mn1}(or Gh^{mn2}) contains only edges from Gh^{mn1}(or Gh^{mn2}), the part (a) is trivial. For part (b), we notice that by our construction algorithm, there is no edge between vertices $G_u[L]$ and $G_v[L]$ except edge connecting X_u and X_v.

Lemma 4 $W(T[L]) = 3W(G_u[L]){+}2^{L+1}-1)\times\left(2d_{G_u[L]}(X_u, G_u[L]) + (2^L + 1 - 1)\right)$.

Proof By our construction, V(T[L]) is partitioned by $V(G_u[L])$, $V(G_v[L])$ and $V(G_w[L])$ and by Lemma 3(a), these are isometric subgraphs of T[L]. On the other hand,

$$G_u[L] \cong G_v[L] \cong G_w[L]. \tag{2.3}$$

Therefore, $W(T[L]) = W(G_u[L]) + W(G_v[L]) + W(G_w[L]) + d_{T[L]}(G_u[L], (G_v[L]) + d_{T[L]}(G_u[L], (G_w[L]) + d_{T[L]}(G_w[L], (G_v[L])$. We now apply Eq. (2.3) to prove $W(T[L]) = 3W(G_u[L]) + 3d_{T[L]}(G_u[L], (G_v[L])$. Set $V(G_u[L]) = \{u_1, u_2, \ldots, u_m\}$ and $V(G_v[L]) = \{v_1, v_2, \ldots, v_m\}$, where $m = 2^{L+1} - 1$. Then by definition and Lemma 3, we have:

$$\begin{aligned} d_{T[L]}(G_u[L], G_v[L]) &= \textstyle\sum_{i=1}^{m} \sum_{j=1}^{m} d_{T[L]}(u_i, v_j) \\ &= \textstyle\sum_{i=1}^{m} \sum_{j=1}^{m} \big[d_{G_v[L]}(u_i, X_v) \\ &\quad + d_{G_u[L]}(v_i, X_u) + 1\big] \\ &= 2md_{Gu[L]}(X_u, G_u[L]) + m^2, \end{aligned}$$

which proves the lemma.

Lemma 5 $d_{G_u[L]}(X_u, G_u[L]) = (L-1)2^{L+1} + 2$.

Proof Induct on L. The case of $L = 1$ is trivial. Suppose $A = \left\{X^{L,1,1}, X^{L,1,2}, X^{L,2,1}, X^{L,2,2}, \ldots, X^{L,2^{L-1},1}, X^{L,2^{L-1},2}\right\}$ is the set of all vertices of degree 2 of the graph $G_u[L]$. Obviously, $|A| = 2^L$. By our construction, 2^L graphs $K_3^{L+1,1}, K_3^{L+1,2}, K_3^{L+1,3}, \ldots, K_3^{L+1,2^L}$ are connected to the vertices of A, respectively. Since the vertices of degree 2 in $G_u[L+1]$ other than X_u are connected only to one of $X^{L,i,j}$, distances of them from X_u is L+1. On the other hand, there are 2^{L+1} such vertices, and so $d_{G_u[L+1]}(X_u, G_u[L+1]) = d_{G_u[L]}(X_u, G_u[L]) = (L+1)2^{L+1} + 2 = L2^{L+2} + 2$.

Lemma 6 $W(G_u[L]) = (2L+5)2^{2L+1} + (4L+9)2^L + 1$.

Proof By Lemma 3, Gh^{111} and Gh^{112} are isometric subgraphs of T[L]. Hence these are isometric in $G_u[L]$. By our construction, vertices of Gh^{111} and Gh^{112} are disjoint and there is a unique edge of T[L] connecting $X^{111} \in V(Gh^{111})$ to $X^{112} \in V(Gh^{112})$. By a similar argument as Lemma 4 and formula given in Lemma 5, we have:

$$\begin{aligned} W(G_u[L]) &= 2\,W(G_u[L-1]) + d_{G_u[L]}(Gh^{111}, Gh^{112}) \\ &\quad + d_{G_u[L]}(X_u, G_u[L]) \\ &= 2\,W(G_u[L-1]) + (2^L - 2)[(L-2)2^L] \\ &\quad + (2^L - 1)^2 + (L-1)2^{L+1} + 2 \\ &= 2\,W(G_u[L-1]) + (n - 3/2)2^{2n+1} \\ &\quad + 2^{n+2} - 1. \end{aligned}$$

After solving this recurrence relation, $W(G_u[L]) = (2L-5)2^{2L+1} + (4L+9)2^L + 1$, proving the lemma.

We end this chapter by the following theorem:

Theorem 5 $W(T[L]) = (18L - 21)2^{2L+1} + 51 \times 2^L - 6.$

Proof The proof is follows from Lemmas 4-6.

References

Ashrafi AR, Mirzargar M (2008a) Int J Chem Mod 1:157
Ashrafi AR, Mirzargar M (2008b) Util Math 77:249
Ashrafi AR, Mirzargar M (2008c) Indian J Chem 47A:535
Ashrafi AR, Saati H (2007) J Comput Theor Nanosci 4:761
Ashrafi AR, Yousefi S (2007a) Nanoscale Res Lett 2:202
Ashrafi AR, Yousefi S (2007b) AIP Conf Proc 929:12
Ashrafi AR, Yousefi S (2007c) MATCH Commun Math Comput Chem 57:403
Balaban AT (1976) Chemical applications of graph theory. Academic, London
Bethune DS, Kiang CH, Devries MS, Gorman G, Savoy R, Vazquez J, Beyers R (1993) Nature 363:605
Bonchev D (1983) Information theoretic indices for characterization of chemical structures. Research Studies Press, Latchworth
Cyvin SJ, Gutman I (1988) Kekülé structures in benzenoid hydrocarbons. Lecture notes in chemistry, vol 46. Springer, Berlin
Diudea MV (2002a) Bull Chem Soc Jpn 75:487
Diudea MV (2002b) MATCH Commun Math Comput Chem 45:109
Diudea MV, John PE (2001) MATCH Commun Math Comput Chem 44:103
Diudea MV, Kirby EC (2001) Fuller Sci Technol 9:445
Diudea MV, Parv B, Kirby EC (2003) MATCH Commun Math Comput Chem 47:53
Diudea MV, Stefu M, Pârv B, John PE (2004) Croat Chem Acta 77:111
Dobrynin AA, Entringer R, Gutman I (2001) Acta Appl Math 66:211
Dobrynin AA, Gutman I, Klavzar S, Zigert P (2002) Acta Appl Math 72:247
Gutman I (2006) Mathematical methods in chemistry. Prijepolje Museum, Prijepolje.
Gutman I, Gaurilovic N, Nankovic D, Khadikar PV, Deshpande NV, Kale PP (1994) J Serb Chem Soc 59:519
Gutman I, Körtvélyesi T (1995) Z Naturforsch 50a:669
Gutman I, Polansky OE (1986) Mathematical concepts in organic chemistry. Springer, Berlin
Hosoya H (1971) Bull Chem Soc Jpn 44:2332
Hosoya H (1988) Discrete Appl Math 19:239
Iijima S (1991) Nature 354:56
Iijima S, Ichlhashi T (1993) Nature 363:603
Iranmanesh A, Ashrafi AR (2007) J Comput Theor Nanosci 4:514
John PE, Diudea MV (2004) Croat Chem Acta 77:127

Karbasioun A, Ashrafi AR (2009) Maced J Chem Chem Eng 28:49
Karbasioun A, Ashrafi AR, Diudea MV (2010) MATCH Commun Math Comput Chem 63:239
Kier, L. B., Hall, L. H. (1976). Molecular connectivity in chemistry and drug research. Research Studies Press, Latchworth
Mohar B, Pisanski T (1988) J Math Chem 2:267
Trinajstić N (1992) Chemical graph theory. CRC Press, Boca Raton, FL
Wiener H (1947) J Am Chem Soc 69:17
Yousefi S, Ashrafi AR (2006) MATCH Commun Math Comput Chem 56:169
Yousefi S, Ashrafi AR (2007) J Math Chem 42:1031
Yousefi S, Ashrafi AR (2008a) Studia Univ Babes-Bolyai CHEMIA, 53:111
Yousefi S, Ashrafi AR (2008b) Curr Nanosci 4:161
Yousefi S, Yousefi-Azari H, Ashrafi AR, Khalifeh MH (2008c) J Sci Univ Tehran 33:7
Yousefi S, Yousefi-Azari H, Khalifeh MH, Ashrafi AR (2008d) Int J Chem Mod 1:149

Chapter 3
C_{60} Structural Relatives – An Omega-Aided Topological Study

Aniela E. Vizitiu and Mircea V. Diudea

Abstract It was shown that the covering of C_{60} "Buckminsterfullerene" is basically *sumanenic*, with the empty π-electron faces being only pentagons. Four series of cages, tessellated by sumanenic patterns $S[r] = [r{:}(5,6)_{r/2}]$, were generated by sequences of map operations, and their topology described. Among these cages, which all show all_R[5] 2-factors, those belonging to the series designed on the dual pair Dodecahedron/ Icosahedron by iterating the P_5 operation and closing by *Le* operation, show a unique term Omega signature, thus being classified as the C_{60} series. C_{60} itself also shows the unique signature and all the members of its family show large HOMO-LUMO gap values, larger than of the cages belonging to the other three series herein built up. Coverings are given in terms of circulenes/flowers and the relation with the Omega and Ring polynomials is evidenced. Analytical formulas for the net parameters and for the used polynomials are given.

3.1 Introduction

Since the very beginning of the fullerene science (Endo et al. 1996; Fowler and Manolopolous 1994; Dresselhaus et al. 1996; Tanaka et al. 1999; Balaban 1997; Kroto et al. 1985; Kraetschmer et al. 1990) the aim of synthesis by "wet chemistry" of fullerenes with controlled tessellation has been a constant desiderate. Notably, in this respect, is the synthesis of dodecahedrane by Paquette et al. (1981).

The direct synthesis of fullerenes with desired covering, from appropriate precursors by pyrolysis, has been recently performed (Amsharov and Jansen 2008, 2009).

The idea of increasing aromaticity/stability of fullerenes tessellated by disjoint circulenes/flowers originates in the classical texts of Clar (1964, 1972), which postulated *disjoint benzenoid rings* (i.e., rings having six π-electrons localized in double-simple alternating bonds and separated from adjacent rings by formal single

M.V. Diudea (✉)
Faculty of Chemistry and Chemical Engineering, Babes-Bolyai University, 400028 Cluj, Romania
e-mail: diudea@chem.ubbcluj.ro

F. Cataldo et al. (eds.), *The Mathematics and Topology of Fullerenes*,
Carbon Materials: Chemistry and Physics 4, DOI 10.1007/978-94-007-0221-9_3,

bonds) as a criterion for the full aromatic conjugation (i.e., double-simple bond alternation).Molecular structures, showing such fully *resonant sextets* are expected to be extremely stable, according to the VB theory (Fowler and Pisanski 1994; Dias 1999; Cyvin and Gutman 1988).

A flower is symbolized as $[r{:}p_i]$Fw, with r, p_i being the folding of the polygonal core and its surrounding petals, respectively. Such flowers could appear either as intersect, joint or disjoint units.

A tiling is called Platonic if it consists of a single type of faces (see the Platonic solids). Archimedean is that tiling consisting of more than one type of faces. Platonic and Archimedean will refer here only to tessellation by flowers.

A set of disjoint faces, built up over all atoms of the molecule, is called a 2-*factor*. It is known (Clar 1964) that fullerenes showing a 2-factor structure also possess a Fries structure (Fries 1927), which is a Kekulé structure with the maximum possible (v/3) number of benzenoid faces. A Kekulé structure (Hosoya 1986; Veljan 2001; Zhang and Yan 2003; Shiu et al. 2002; John et al. 1995; Cyvin and Gutman 1988), or a perfect matching, is a set of pair-wise disjoint bonds, defined over all atoms of the molecule. The associate Fries structure will ensure the total resonance (i.e., conjugation) of the molecule.

Among several circulene/flower units so far reported (Sakurai et al. 2003, 2005; Yamamoto 1993) sumanene S[6]= $[6{:}(5,6)_3]$ is of higher interest; Fig. 3.1 illustrates the sumanenic patterns $S[r] = [r{:}(5,6)_{r/2}]$; $r = 6$, 8 and 10. Covering of the sphere by various polyhedral faces was achieved by means of operations on maps (Diudea et al. 2003; Diudea 2006, Vizitiu et al. 2006) as implemented in our original software CageVersatile (CVNET) (Stefu and Diudea 2005).

Note that any supra-face or Fw has its own intersected/superposed "counterpart", called here *co-Fw*. Of course, the attribute Fw and co-Fw are interchangeable. The covering by a sequence of operations is given in terms of flower patterns.

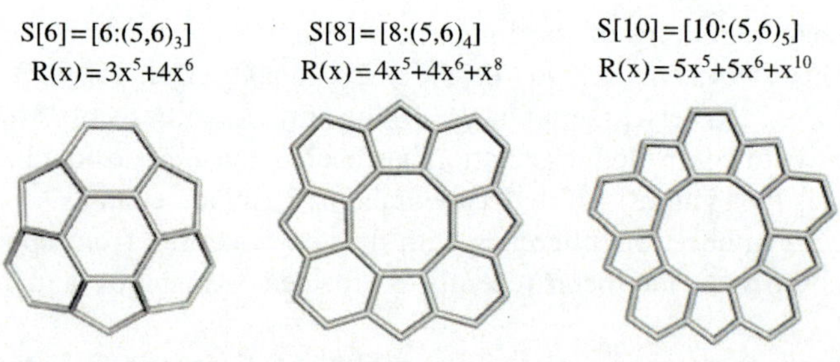

Fig. 3.1 Sumanenic circulene/flower patterns: ring polynomials are given in the *top* of figures

The symbols used for naming cages in the hereafter text will include the actual number of atoms, the starting Platonic cage and the map operation sequence (by its m-factor, multiplying the Platonic pattern) used in their construction. When obtained by the Stone-Wales (Stefu and Diudea 2005) SW edge-rotation, the suffix RO is added.

Recall the five Platonic polyhedra and their symbols herein used: *T* (tetrahedron), *O* (octahedron), *C* (cube), *D* (dodecahedron, and *I* (icosahedron). The dual pairs are: *T*&*T* (self-dual); *O*&*C* and *D*&*I*.

It was shown (Stone and Wales 1986) that the covering of "Buckminsterfullerene" C_{60} is basically *sumanenic*, with the empty π-electron faces being only pentagonal. The immediate consequence is the corresponding Fries structure will include all the hexagonal faces of the molecule. The higher hypothetical analogues with sumanenic tessellation must also show only pentagonal empty π-electron faces. Recall that the structure of sumanene molecule was explicitly related to the "Buckminsterfullerene" tessellation (Sakurai et al. 2003, 2005).

The structure of the paper is as follows. The second section gives definitions and examples of the main map operations used within this work. The third section gives a panel of the most used sequences of map operations leading to sumanenic covering/tessellation. The fourth section introduces in the theory of counting polynomials while the fifth section presents in detail the topology of the C_{60} family and its structural relatives. Conclusions and references will close the paper.

3.2 Operations on Maps

A map *M* is a combinatorial representation of a (closed) surface (Pisanski and Randić 2000; Fowler and Manolopolous 1994). Operations on maps are topological-geometrical transformations allowing to transform or to relate a given polygonal structure. Several operations on maps are known and used for various purposes.

Dualization Du of a map starts by locating a point in the center of each face (Fig. 3.2a) (Diudea 2003; Pisanski and Randić 2000; Diudea 2004). Next, two such points are joined if their corresponding faces share a common edge (Fig. 3.2b).

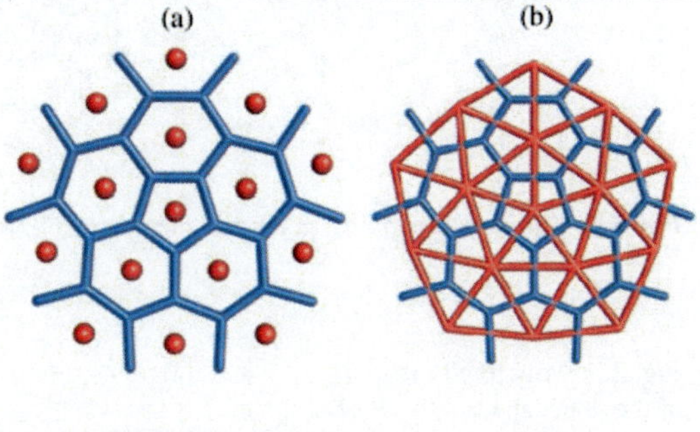

Fig. 3.2 Dualization of a fullerene patch

It is the (Poincaré) *dual Du*(*M*). The vertices of *Du*(*M*) represent the faces of *M* and vice-versa (Pisanski and Randić 2000). Thus the following relations exist: $Du(M)$; $v = f_0$; $e = e_0$; $f = v_0$.

Dual of the dual returns the original map: $Du(Du(M)) = M$. Tetrahedron is self dual while the other Platonic polyhedra form pairs: *Du*(Cube) = Octahedron;

Du(Dodecahedron) = Icosahedron (Fig. 3.3). It is also known the Petrie dual (Pisanski and Randić 2000).

Note that all the operation parameters herein presented refer to regular maps (e.g., the Platonic solids); a subscript zero indicates a parent map parameter.

Fig. 3.3 The duals of the five platonic polyhedra

Polygonal P_k mapping ($k = 3,\ 4,\ 5$) of a face is achieved as follows: (Diudea 2004; Diudea and Nagy 2007) add a new vertex in the center of the face. Put k–3 points on the boundary edges (Fig. 3.4). Connect the central point with one vertex (the end points included) on each edge. In this way the parent face is covered by triangles ($k = 3$), quadrilaterals ($k = 4$) and pentagons ($k = 5$). The P_3 operation is also called *stellation* or (centered) *triangulation*. The resulting map shows the relations: $P_k(M)$, $v = v_0 + (s-3)e_0 + f_0$; $e = se_0$; $f = s_0 f_0$, so that the Euler's relation holds (see below). Figure 3.5 gives examples of the P_k operations realization.

(a) (b) (c)

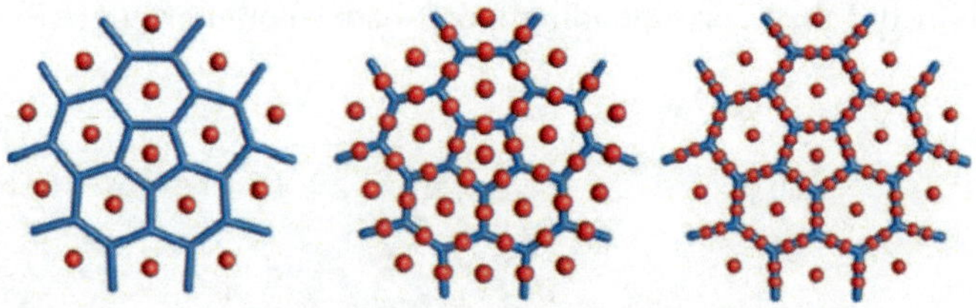

Fig. 3.4 Polygonal mapping of a fullerene patch; P_3 (**a**); P_4 (**b**) and P_5 (**c**)

Fig. 3.5 Polygonal mapping of the dodecahedron by $P_3(D)$ (**a**); $P_4(D)$ (**b**) and $P_5(D)$ (**c**)

(a) (b) (c)

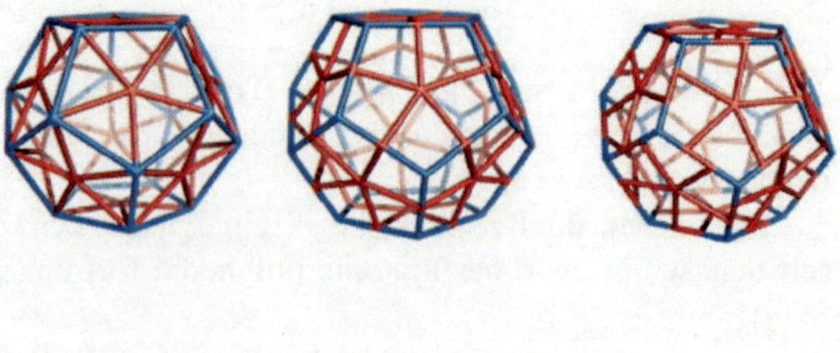

Medial Med of a map is achieved (Diudea 2003; Pisanski and Randić 2000; Diudea 2004) by putting a new vertex in the middle of each original edge. Join two vertices if the original edges span an angle (and are consecutive within a rotation path around their common vertex in *M*), Fig. 3.6. Medial is a 4-valent graph and $Med(M) = Med(Du(M))$. The transformed parameters are: $Med(M)$; $v = e_0$; $e = 2e_0; f = f_0 + v_0$.

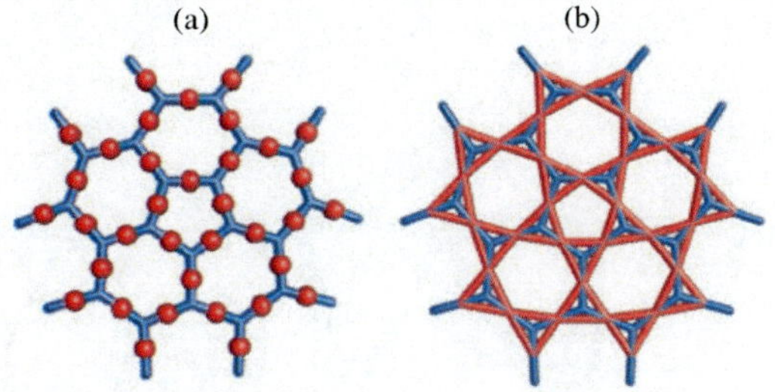

Fig. 3.6 Medial of a fullerene patch

Medial operation rotates parent *s*-gonal faces by π/s. Points in the medial represent original edges, thus this property can be used in topological analysis of edges in the parent polyhedron. Similarly, the points in dual give information on the topology of parent faces. Figure 3.7 illustrates the *medial* operation performed on the five Platonic polyhedra.

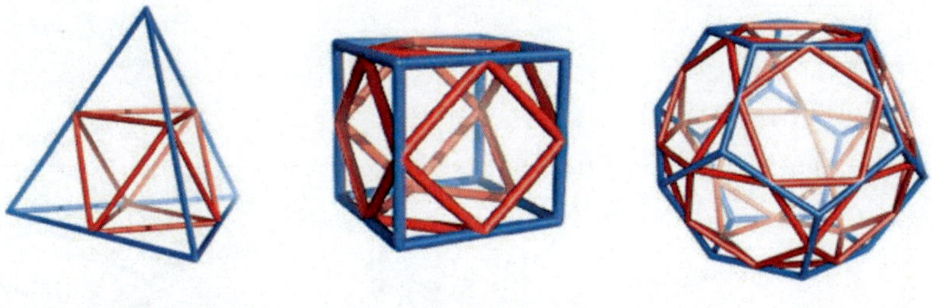

Fig. 3.7 The medials of the five platonic polyhedra

Truncation Tr is achieved (Pisanski and Randić 2000; Vizitiu et al. 2009) by cutting off the neighborhood of each vertex by a plane close to the vertex, such that it intersects each edge incident to that vertex (Fig. 3.8).

Truncation is similar to the medial, the transformed parameters being: $Tr(M)$; $v = 2e_0 = d_0v_0$; $e = 3e_0$; $f = f_0 + v_0$. This was the main operation used by Archimedes in building up the well-known 13 (Archimedean) solids (Vaissiere et al. 2001). Figure 3.9 illustrates the realization of this operation on the icosahedron.

Snub Sn is a composite operation that can be written as: (Stone and Wales 1986; Diudea 2004) $Sn(M) = Du(P_5(M))$. The dual of a snub is the $P_5(M)$ transform: $Du(Sn(M) = P_5(M)$. Similar to the medial operation, $Sn(M) = Sn(Du(M))$. In case

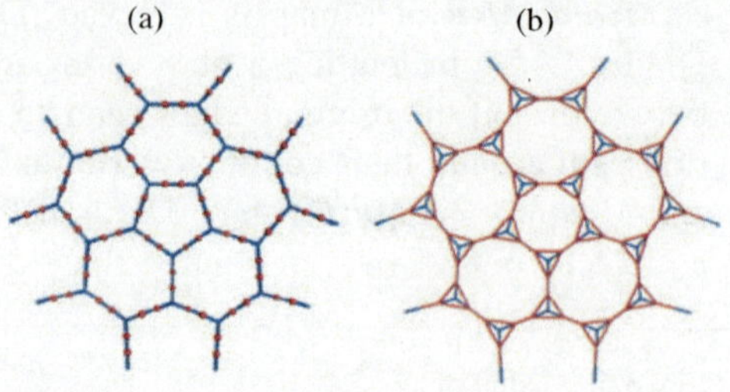

Fig. 3.8 Truncation of a patch of a fullerene

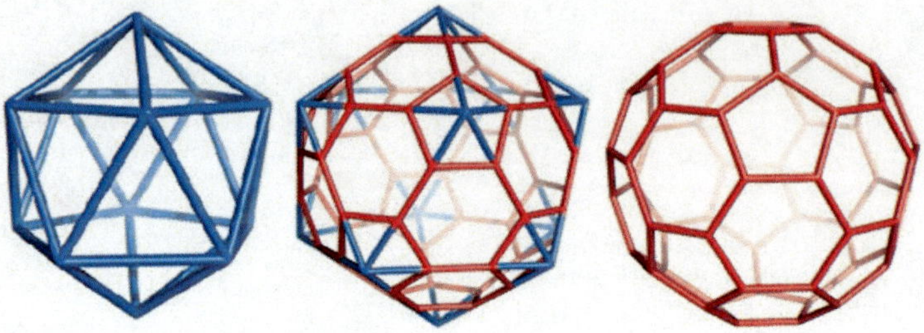

Fig. 3.9 Truncation of the icosahedron

of $M = T$, the snub $Sn(M) = I$. The snub $Sn(M)$ is always a pentavalent graph. The transformed parameters are: $Sn(M)$; $v = s_0 f_0 = d_0 v_0$; $e = 5e_0$; $f = v_0 + 2e_0 + f_0$. Figure 3.10 presents the realization of this operation on the Dodecahedron.

Fig. 3.10 Snub of the dodecahedron

Leapfrog *Le* is a composite operation, (Fowler 1986; Fowler and Steer 1987; Fowler and Rogers 1998a, b; Eberhard 1891) which can be written as: $Le(M) = Du(P_3(M)) = Tr(Du(M))$. A sequence of stellation (i.e., P_3)-dualization rotates the parent s-gonal faces by π/s. Leapfrog operation is illustrated, for a fullerene patch, in Fig. 3.11.

Quadrupling *Q* (i.e., Chamfering (Goldberg 1937)) is another composite operation, achieved by the sequence (Diudea 2004): $Q(M) = RE(Tr_{P_3}(P_3(M)))$, where *RE* means the (old) edge deletion (the blue lines, in Fig. 3.12) of the truncation Tr_{P3} of each central vertex introduced by P_3 capping. *Q* insulates the parent faces always by hexagons.

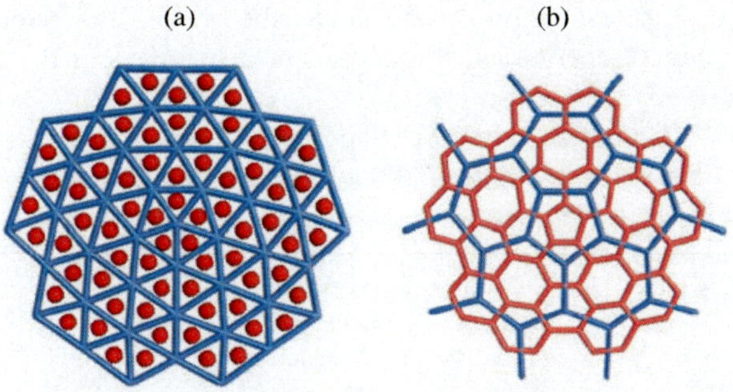

Fig. 3.11 Stellation (**a**) and dualization (**b**) of a patch of a fullerene

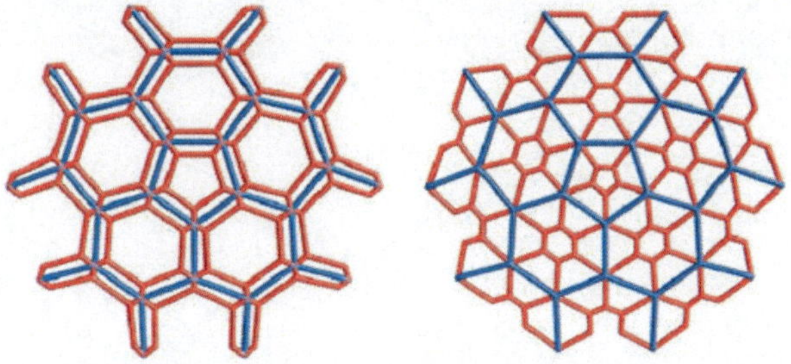

Fig. 3.12 Chamfering of a patch of a fullerene

Capra *Ca* – the goat, is the Romanian corresponding of the *leapfrog* English children game. It is a composite operation (Stone and Wales 1986; Goldberg 1937; Diudea 2003), necessarily coming from the Goldberg's (1937) multiplying factor $m = (a^2 + ab + b^2)$; $a \geq b$; $a + b > 0$, predicted as follows: *Le*: (1,1); $m = 3$; *Q*: (2,0); $m = 4$; *Ca*: (2,1); $m = 7$.

The transformation can be written as: $Ca(M) = Tr_{P_5}(P_5(M))$ with Tr_{P5} being the truncation of each central vertex introduced by P_5 mapping. *Ca* insulates any face of *M* by its own hexagons, which are not shared with any old face (in contrast to *Le* or *Q*). It is an intrinsic chiral operation (Fig. 3.13). It rotates the parent edges by $\pi/(3s/2)$.

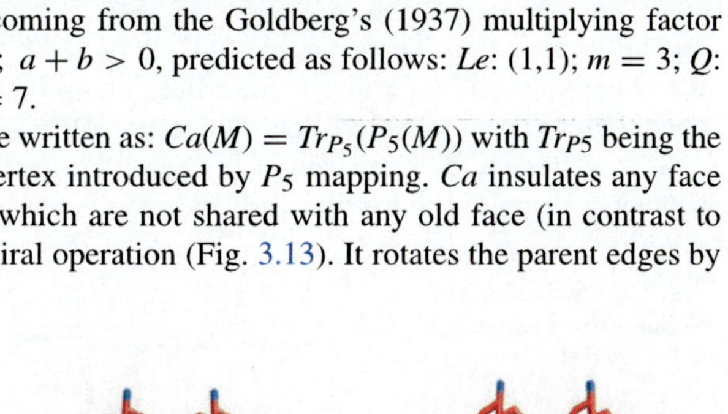

Fig. 3.13 Chiral lattices performed by the *Ca* operation

Its realization on the Dodecahedron is illustrated in Fig. 3.14, along with *Le* and *Q* operations; the cages are also given in the Schlegel projection (Schlegel 1893).

Ca is also called S_1 (Septupling) operation, differing from its S_2 twin operation (see below) (Diudea 2005a).

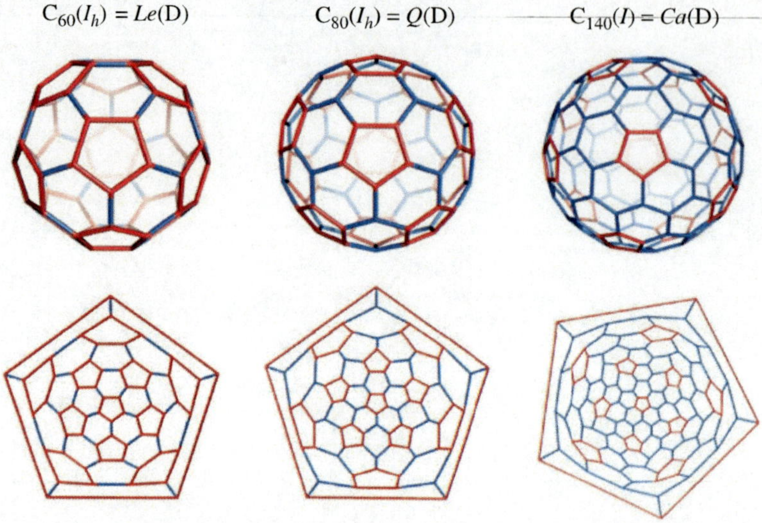

Fig. 3.14 Molecular realization of the three main composite map operations, given both as 3D objects and schlegel projections (the *bottom row*)

The S_2 operation (Diudea et al. 2007, 2004, 2005) is a simpler one (Fig. 3.15); it can be achieved by putting four vertices on each edge of the parent map *M* (E_4 operation) and next join these new vertices in order (−1, +3): $S_2 = J_{(-1,+3)}(E_4(M))$.

It insulates the double sized parent faces by pentagons and parent vertices by pentagon d_0-multiples; the transformed object is non-chiral.

Fig. 3.15 Septupling S_2 operation on a square face, up to the open structure

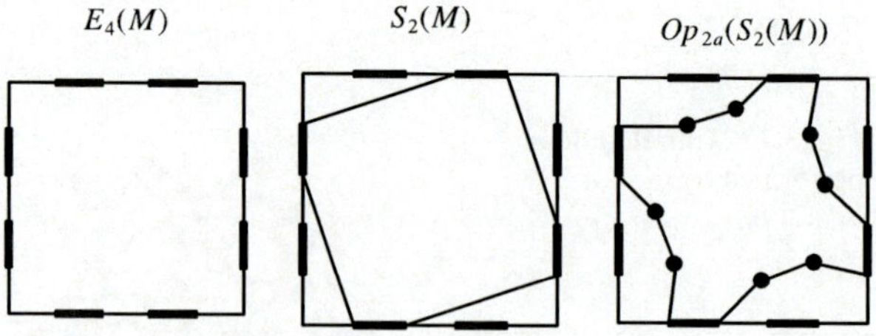

Chirality in S_2 can be obtained by the opening operation Op_{2a}, achieved by putting two points on alternative edges of the double sized parent face boundary (Fig. 3.15).

In case of a closed cage, the transformed lattice parameters are identical to those of S_1: $S_1(M)$ & $S_2(M)$; $v = v_0(2d_0 + 1)$; $e = 7e_0$; $f = f_0(s_0 + 1)$, differences appearing in case of open objects. Observe that both the septupling operations keep the parent vertices.

1. The iterative application of S_2 reveals the fractal fashion of the covering (Fig. 3.16). The fractal characteristic can be seen even in the algebraic form of lattice parameters (Diudea and Nagy 2007).

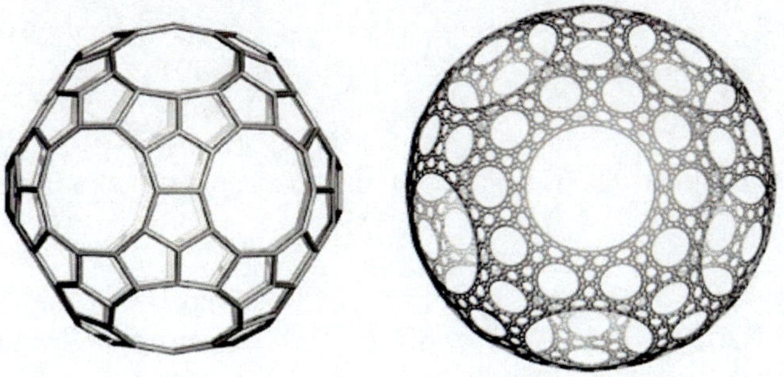

Fig. 3.16 Iterative S_2 operation on dodecahedron: observe the fractal covering in case of 3-times repetition (**b**)

The only fullerene constructible by S_2 is C_{28}, when applied on the Tetrahedron.

3.3 Coverings by Sequences of Map Operations

Sumanenic flowers S[r] can be generated by several sequences of map operations, as follows (Diudea 2005b).

3.3.1 Sequence $Tr_5(Ca_f(Q(M)))$

In the above sequence, Q is the quadrupling map operation, Ca_f represents *Capra* operation performed so that the original faces of M remain untransformed and Tr_s is the truncation of selected vertices. The sequence leads to a Platonic disjoint corazulenic [r:$(7(5d))_r$]Fw, chiral (by virtue of Ca/S_1) covering; the co-Fw is a sumanenic pattern S[r] $=$ [r:$(5,6)_{r/2}$]Fw which forms an Archimedean covering with R[p] rings (Diudea and Nagy 2007; Schlegel 1893).

The above sequence shows a multiplication factor $m = 8d$, where d is the vertex degree in the parent Platonic. As an example of molecular realization, the covering of 96T-8d is a disjoint Platonic of [3:$(7(5d))_3$]Fw (Fig. 3.17a) while the co-Fw forms a disjoint Archimedean covering: S[6]&R[3] (Fig. 3.17b).

Another example is the stable corazulenic cage: $C_{192} = 192O\text{-}8d$. It shows a Platonic, disjoint corazulenic $[4{:}(7(5d))_4]$Fw covering of which co-Fw forms a disjoint Archimedean S[6]&R[4] covering. One of the most important of its valence structure shows the maximum possible 32 Kekulé benzenoid rings, thus being identified as the Fries valence structure (Fries 1927).

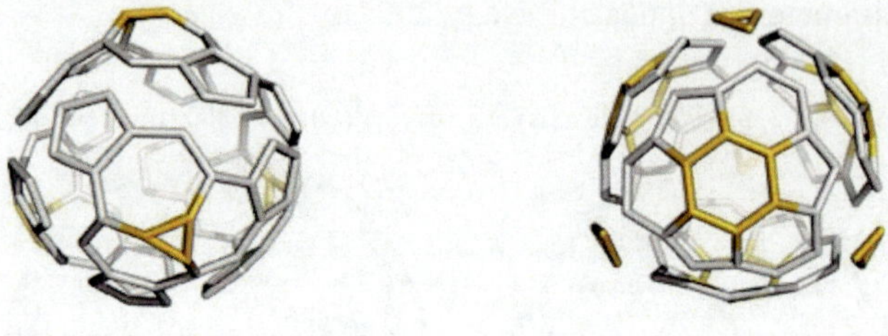

Fig. 3.17 Corazulenic disjoint pattern and its co-Fw; cage designed by the sequence $Tr_s(Ca_f(Q(M)))$

Notably is the alternation, with respect to the vertices of the cube, of triphenylenic and tripentylenic types of S[6]Fws, as shown by the length of bonds in the optimized molecular structure (Fig. 3.18). The tripentylene $[6{:}(0,5)_3]$Fw can be viewed as an analogue of the triphenylene $[6{:}(0,6)_3]$Fw. The numerical Kekulé count (John et al. 2007a; Diudea 2006; Diudea et al. 2007, 2009; Euler 1758; Randić 2004; Nagy et al. 2009) of π-electrons are given in the top of figures.

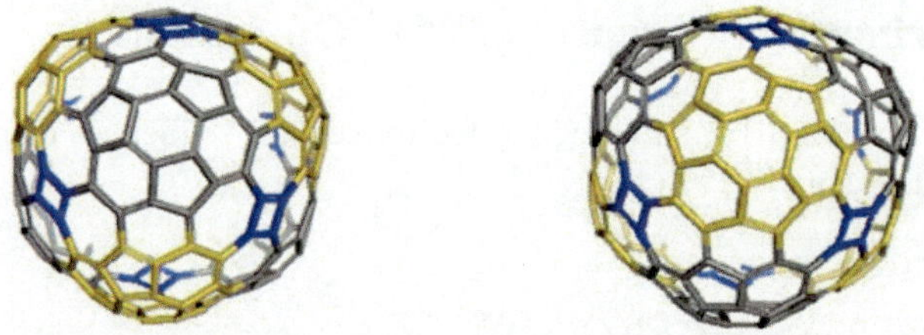

Fig. 3.18 The "Sumanenic-Kekulé" valence structure of C_{192} with two distinct alternating S[6]Fw and their π-electron population; cage designed by the sequence $Tr_s(Ca_f(Q(M)))$

3.3.2 Sequence $Tr_s(Ca_{3,2c}(M))$

In the above sequence, $Ca_{3,2c}$ is the generalized *Capra* operation $Ca_{3,2}$ with the faces of original map cut-off. It provides joint corazulenic flowers $[r{:}(7(5c))_r]$Fw, which can also read $[r{:}(7(5d))_r]$Fw; the co-Fw is of corazenic type $[r{:}(5,7)_{r/2}]$Fw. The two corazulenic patterns of this covering both transforms, by

SW, into sumanenic joint patterns. The multiplication factor is $m = 5d$; when applied on the medial *Med*, $m = 10d$. The sequence is exemplified for the cage 120*O*-5d (Figs. 3.19 and 3.20) (Diudea and Nagy 2007; Schlegel 1893).

(a) 120*O*-5d; $6[4{:}(7(5c/d))_4]$Fw (b) 120*O*-5d; $8[6{:}(5,7)_3]$co-Fw

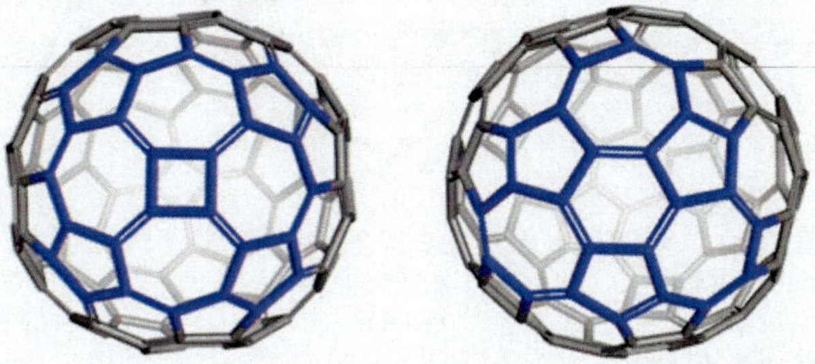

Fig. 3.19 The two corazulenic flowers tessellating the 120*O*-5d cage, designed by $Tr_s(Ca_{3,2c}(M))$

(a)120*O*–5dRO; 6S[8]&8R[6] (b) 120*O*–5dRO; 6R[8]&8S[6] co–Fw

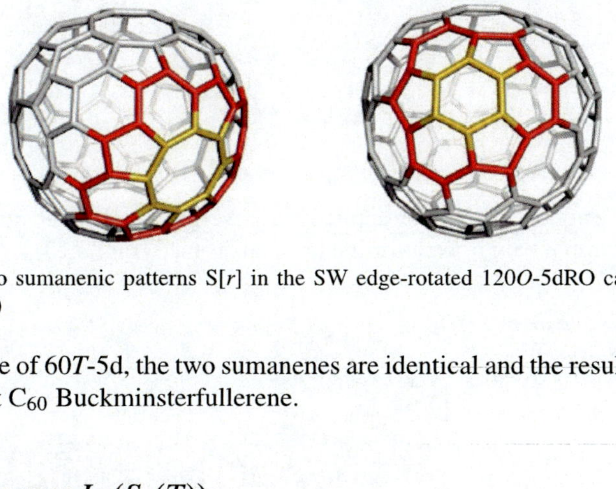

Fig. 3.20 Two sumanenic patterns S[*r*] in the SW edge-rotated 120*O*-5dRO cage, designed by $Tr_s(Ca_{3,2c}(M))$

In the case of 60*T*-5d, the two sumanenes are identical and the result is 60T-5dRO which is just C_{60} Buckminsterfullerene.

3.3.3 Sequence $Le(S_2(T))$

This sequence is non-commutative (because of the S_2 operation) and provides a disjoint sumanenic pattern S[*r*], in a Platonic covering (Diudea and Nagy 2007; Schlegel 1893).

The co-Fw forms an Archimedean joint of coronenic $[r{:}6_r]$Fw, and pentylenic $[p{:}(0,5))_{p/2}]$Fw, patterns. The multiplication factor is $m = 7d$ or, in case of the sequence involving the medial operation, $m = 14d$). Figures 3.21 and 3.22 illustrate the above covering.

These cages are relatively stable structures, with the most one being 168*O*-3d, an almost spherical cage (Fig. 3.22b).

(a) $84T$–7d; 4S[6]
Platonic

(b) $84T$–7d; $[(6{:}6_6)\&(6{:}(0,5)_3]$co–Fw
Archimedean

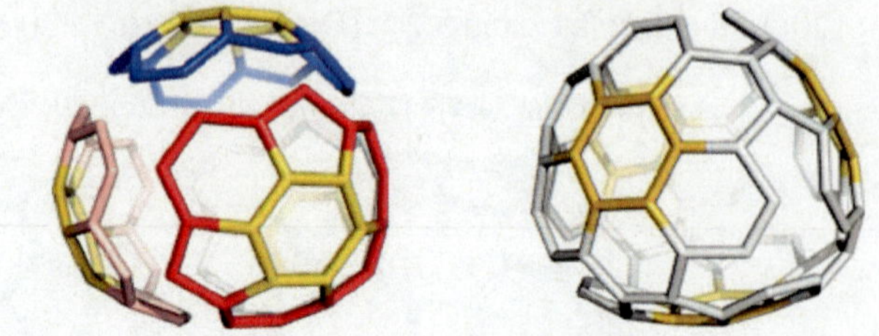

Fig. 3.21 Platonic disjoint sumanenic S[6] covering (**a**) and archimedean joint coronenic and pentylenic co-Fw

(a) $168C$–7d; 8S[6]
Platonic

(b) $168O$–7d; 6S[8]
Platonic

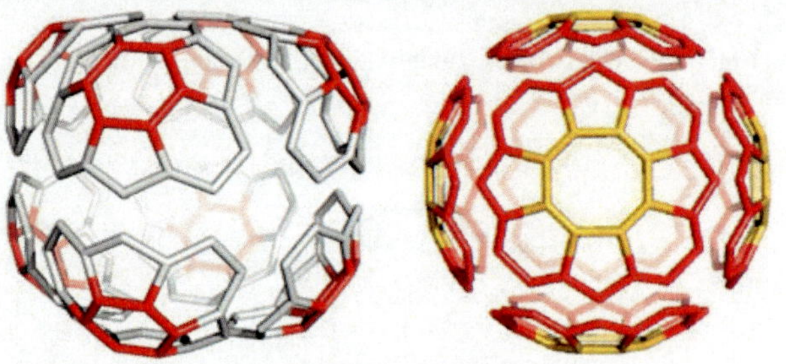

Fig. 3.22 Platonic disjoint sumanenic covering S[r] on the transforms of cube (**a**) and octahedron (**b**) by the sequence $Le(S_2(T))$

The above sequences can be applied on coverings embedded in any type of surface, an example being given for the toroidal surface (Fig. 3.23).

Fig. 3.23 Sumanenic disjoint S[8] covering by $Le(S_2(M)$; 64S[8]; $v = 1792$; $M =$ T(4,4) [4,16]

3.3.4 Sequence $Le(P_5(M))/Le(P_5(Med(M)))$

These sequences show a multiplication factor of $m = 5d$ (or $m = 10d$, in case the medial operation Med(M) is included) (Diudea 2008). Note $Le(P_5(Med(M))) = RO(TR_s(Ca_{3,2c}(M)))$, so that the SW operation is already included. Also note $Le(P_5(M)) = Tr(Sn(M)$, Sn being the Snub operation. The Med(M) operation induces a four-valent atom and further the derived cages will contain structural features originating in this type of atom (see below).

Because the sequence is commutative, a dual-pair will provide one and the same transform by $Le(P_5(M))/Le(P_5(Me(M)))$. In the above series, the P_5-operation can be iterated k-times: $Le((P_5(\mathrm{M}))^k)$thus generating series of cages of similar tessellation.

The most important, in the sumanenic covering, is the corresponding 2-*factor, consisting of only pentagons* (Diudea and Nagy 2007, 2008). It means that the empty faces (in the most important Kekulé valence structure) are only pentagons while the hexagonal ones will all participate to the Fries (most benzenoid) structure. This aspect, we believe, is one of the structural characteristics of the C_{60} family (see below). This is supported by the large HOMO-LUMO gap of all the cages belonging to this family. Figures 3.24 and 3.25 illustrate the sumanenic tessellation of C_{60} and 300*D*/*I*-5d, which is the next member of the C_{60} family.

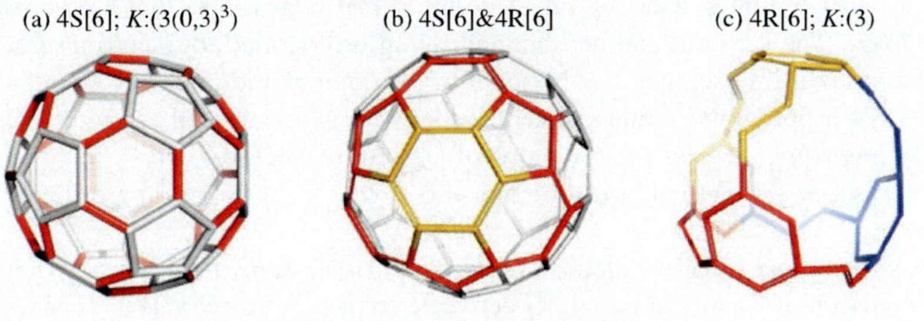

Fig. 3.24 C_{60}; Sumanenic S[6] patterns in a tetrahedral embedding

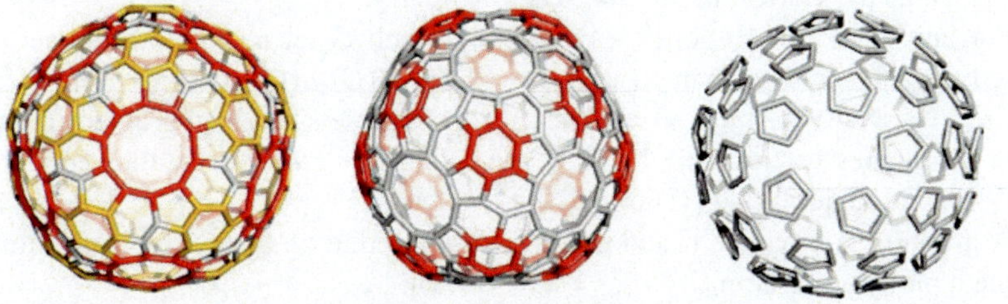

Fig. 3.25 Tessellation of 300*D*/*I*-5d cage, designed by $Le(P_5(M))$; S-core in *color*/*black*

3.4 Counting Polynomials

A counting polynomial (Diudea et al. 2007) is a representation of a graph $G(V, E)$, with the exponent k showing the extent of partitions $p(G)$, $\cup\, p(G) = P(G)$ of a graph property $P(G)$ while the coefficient $p(k)$ are related to the occurrence of the partition of extent k.

$$P(x) = \sum_k p(k) \cdot x^k \tag{3.1}$$

Let G be a connected graph, with the vertex set $V(G)$ and edge set $E(G)$. Two edges $e = (u, v)$ and $f = (x, y)$ of G are called *codistant* (briefly: $e\ co\ f$) if the notation can be selected such that (John et al. 2007a, b)

$$d(v, x) = d(v, y) + 1 = d(u, x) + 1 = d(u, y)\,, \tag{3.2}$$

where d is the usual shortest-path distance function. The above relation *co* is reflexive ($e\ co\ e$) and symmetric ($e\ co\ f$) for any edge e of G but in general is not transitive.

A graph is called a *co-graph* if the relation *co* is also transitive and thus an equivalence relation.

Let $C(e) := \{f \in E(G);\ f\ co\ e\}$ be the set of edges in G that are codistant to $e \in E(G)$. The set $C(e)$ can be obtained by an orthogonal edge-cutting procedure: take a straight line segment, orthogonal to the edge e, and intersect it and all other edges (of a polygonal plane graph) parallel to e. The set of these intersections is called an *orthogonal cut* (*oc* for short) of G, with respect to e.

If G is a *co-graph* then its orthogonal cuts $C_1, C_2, ..., C_k$ form a partition of $E(G)$: $E(G) = C_1 \cup C_2 \cup ... \cup C_k, \quad C_i \cap C_j = \varnothing,\ i \neq j$.

A subgraph $H \subseteq G$ is called *isometric*, if $d_H(u, v) = d_G(u, v)$, for any $(u, v) \in H$; it is *convex* if any shortest path in G between vertices of H belongs to H. The n-cube Q_n is the graph whose vertices are all binary strings of length n, two strings being adjacent if they differ in exactly one position (Harary 1969). The distance function in the n-cube is the Hamming distance: the distance between two vertices of Q_n is equal to the number of positions in which they differ. A hypercube can also be expressed as the Cartesian product: $Q_n = \square_{i=1}^{n} K_2$.

For any edge $e = (u, v)$ of a connected graph G let n_{uv} denote the set of vertices lying closer to u than to v: $n_{uv} = \{w \in V(G) | d(w, u) < d(w, v)\}$. It follows that $n_{uv} = \{w \in V(G) | d(w, v) = d(w, u) + 1\}$. The sets (and subgraphs) induced by these semicubes vertices, n_{uv} and n_{vu}, are called *semicubes* of G; the semicubes are *opposite* and disjoint ones (Diudea 2008).

A graph G is bipartite if and only if, for any edge of G, the opposite semicubes define a partition of G: $n_{uv} + n_{vu} = v = |V(G)|$.

The relation *co* is related to $\sim$ Djoković and Combin (1973), and Θ (Winkler 1984) relations (Klavžar 2008): in aconnected bipartite graph, $co\ = \sim = \Theta$. For two edges $e = (u, v)$ and $f = (x, y)$ of G the theta relation is defined as: $e\Theta f$ if $d(u, x) + d(v, y) \neq d(u, y) + d(v, x)$. A connected graph G is a *co-graph* if and only if it is a *partial cube*, and all its are convex; relation co/Θ is then transitive (Diudea et al. 2009).

Two edges e and f of a plane graph G are in relation *opposite, e op f,* if they are opposite edges of an inner face of G. Then $e\ co\ f$ holds by the assumption that faces are isometric. The relation *co* is defined in the whole graph while *op* is defined only in faces/rings (see below), thus being included in relation *co*. Note that John et al. (Diudea et al. 2008; John et al. 2007a, b) implicitly used the "*op*" relation in defining the Cluj-Ilmenau index *CI*. Also note that in the previous papers, *ops* were denoted *qoc* (quasi orthogonal cuts) to say the transitivity relation is not, in general, obeyed.

Relation *op* will partition the edges set of *G* into *opposite* edge *strips ops*, as follows. (i) Any two subsequent edges of an *ops* are in *op* relation; (ii) Any three subsequent edges of such a strip belong to adjacent faces; (iii) In a plane graph, the inner dual of an *ops* is a path, an open or a closed one (however, in 3D networks, the ring/face interchanging will provide *ops* which are no more paths); (iv) The *ops* is taken as maximum possible, irrespective of the starting edge. The choice about the maximum size of face/ring, and the face/ring mode counting, will decide the length of the strip.

The Ω-polynomial (Diudea 2006) is defined on the ground of opposite edge strips *ops* $S_1, S_2, ..., S_k$ in the graph. Denoting by *m*, the number of *ops* of cardinality/length $s=|S|$, then we can write

$$\Omega(x) = \sum_{s} m \cdot x^s \tag{3.3}$$

The set of rings in a molecular structure can be represented by a counting polynomial which we call here the *Ring polynomial*, *R*(*x*): in this case, *p*(*k*) of relation (1) represents the number of *k*-fold rings (Diudea 2008).

The first derivative (in $x = 1$) can be taken as a graph invariant or a topological index:

$$\Omega'(1) = \sum_{s} m \cdot s = e = |E(G)| \tag{3.4}$$

$$R'(1) = 2e \tag{3.5}$$

In a set of isomeric cages, the number of edges is constant, so that in view of discriminating such cages, the second derivative is recommended. In this respect, an index, called Cluj-Ilmenau (John et al. 2007a)*CI*(*G*), was defined on Ω(*x*):

$$CI(G) = \{[\Omega'(1)]^2 - [\Omega'(1) + \Omega''(1)]\} \tag{3.6}$$

1. In tree graphs, the Omega polynomial simply counts the non-opposite edges, being included in the term of exponent $c = 1$. The coefficient of the term of exponent $c = 1$ has found applications as a topological index, called n_p, the number of *pentagon fusions*, appearing in small fullerenes as a destabilizing factor. This index accounts for more than 90% of the variance in heat of formation HF of fullerenes C_{40} and C_{50} (Diudea and Nagy 2007; Diudea et al. 2007, 2009).

3.5 Topology of the $Le((P_5(\mathrm{M}))^k)$ Designed Cages

There are four series of structures showing an all_R[5] 2-factor and all are designable by the $Le((P_5(\mathrm{M}))^k)/Le((P_5(Med(\mathrm{M})))^k)$sequences (see Table 3.1). The criterion which enabled discrimination of these series was the Omega polynomial (Diudea 2008).

The first term of the series $Le((P_5(\mathrm{M}))^k)$; M = D/I, is just $60T$-5d = C_{60} Buckminsterfullerene. It is covered by four joint sumanenic S[6] flowers, in a tetrahedral Archimedean disposition 4S[6]&4R[6] (Fig. 3.24). It is counted at iteration $k=0$ (Table 3.1) because the series is referred to the pair Dodecahedron/Icosahedron while Tetrahedron T precedes D by just one P_5 iteration (i.e., $P_5(\mathrm{T}=\mathrm{D})$). The dual pair D/I gives identical objects under this operation sequence and this is also true for the pair Cube/Octahedron C/O.

Since P_5 is a pro-chiral operation (Stone and Wales 1986), the objects designed at $k > 0$ all show chiral pairs, which is not the case for $k = 0$. This operation introduces vertices of degree $d = 5$ which provide $2 \times d = 10$ fold polygons by the consecutive *Le* operation.

The Omega polynomial shows a unique term for the series $Le((P_5(\mathrm{M}))^k)$; M = D/I, this fact being singular among the four series herein discussed. The polynomial coefficient $p(x^3;\Omega)$ is related to the number of hexagons in the sumanenic covering, as provided by the coefficients of Ring polynomial, $p(x^6;R(x))$: $p(x^3;\Omega(x)) = 2p(x^6;R(x)) - 10$.

Analytical formulas for the counting polynomials are given in Table 3.1, which also includes the number of atoms/vertices v, formula of counting, function of the k iteration. The number of edges e comes immediately from the first derivative of Omega polynomial in $x=1$ (relation 4). Next, the lattice parameters can be checked for consistency according to the theorem of Euler (Diudea et al. 2009):

$$v - e + f = 2(1 - g) \tag{3.7}$$

where v, e, f, and g being the number of vertices, edges, faces, and genus, respectively. The *genus* is the number of handles attached to the (or holes performed in a) sphere to make it homeomorphic to the actual surface; $g = 0$ for the graphs embedded on the sphere, and 1 for those embedded in the torus or tube. The faces f are identical to the rings in case of convex polyhedra. An *embedding* is the representation of a graph on a surface such that no crossing lines appear (John et al. 2007a). Examples for some lower terms of the series of the discussed structures are listed.

The cage tessellation is given in terms of flower covering: Tess(k) is the tessellation of the actual iteration of $(P_5(\mathrm{M}))^k$ and is deducible from the $(k-1)$th ring polynomial. Patterns represent joint flowers and are completed by some rings, in an Archimedean covering. For any Tess(k) two complementary coverings can be figured out; they are derivable from each other by simply changing R by S (i.e., "ring" by "sumanene") together with their counting. The number of involved faces always fits that given by the ring polynomial. Note that the symmetry of Fw-covering in C_{60} is tetrahedral while in the higher terms is icosahedral, the relation being discussed above. Tessellations for some lower terms of the series, for which the ring polynomial is also given, are listed in Table 3.1 (Diudea 2005).

The π-electron local distribution of C_{60}, in terms of numerical Kekulé valence structure (Euler 1758), is K: $(3(0,3)^3)$&(3). This electronic distribution corresponds to the most important geometric Kekulé valence structure, as evaluated from the optimized inter-atomic distances by our Nano Studio software (Nagy et al. 2009). For the higher terms, the counting is K: $(3(0,3)^3)$&(5).

Table 3.1 Topology of C_{60} structural relatives

	Cage type M	Ring polynomial/ Tess(k)	Omega polynomial	v	k	Deg(M)/ GAP(β)	m/ LUMO
1	$Le((P_5(\mathrm{M}))^k)$	$p(x^5) = 12 \cdot 5^k$	$6 \cdot 5^{k+1} \cdot x^3$	$v_k = 12 \cdot 5^{k+1}$	–	3(D)	$d \cdot 5^k$
	Dodecahedron D Icosahedron I	$p(x^6) = 5(1 + 3 \cdot 5^k)$ $p(x^{10}) = 3(5^k - 1)$				5(I)	
	Examples	$12x^5 + 20x^6$ $60x^5 + 80x^6 + 12x^{10}$ $300x^5 + 380x^6 + 72x^{10}$ $1500x^5 + 1880x^6 + 372x^{10}$	$30x^3$ $150x^3$ $750x^3$ $3750x^3$	60 300 1500 7500	0 1 2 3	0.757 0.529 0.474 –	–
	Tess(0)	(a) 4S[6]&4R[6] (b) 4R[6]&4S[6]		–	–	–	–
	Tess(1)	(a) 12S[10]&20R[6] (b) 12R[10]&20S[6]		–	–	–	–
	Tess(2)	(a) 60S[10]&80R[6]&12R[10] (b) 60R[10]&80S[6]&12S[10]					
	Tess(3)	(a) 300S[10]&380R[6]&72R[10] (b) 300R[10]&380S[6]&72S[10]					
2	$Le((P_5(\mathrm{M}))^k)$	$p(x^5) = 24 \cdot 5^{k-1}$	$12x^2 + 12(3 + 5(5^{k-1} - 1))x^3 + 12x^4$	$v_k = 24 \cdot 5^k$	–	3(C)	$d \cdot 5^k$
	Cube C Octahedron O	$p(x^6) = 2(1 + 3 \cdot 5^k)$ $p(x^8) = 6$				4(O)	

Table 3.1 (continued)

	Cage type M	Ring polynomial/ Tess(k)	Omega polynomial	v	k	Deg(M)/ GAP(β)	m/ LUMO
		$p(x^{10}) = 6(5^{k-1} - 1)$					
	Examples	$24x^5 + 32x^6 + 6x^8$	$12x^2 + 36x^3 + 12x^4$	120	1	0.364	NBO
		$120x^5 + 152x^6 + 6x^8 + 24x^{10}$	$12x^2 + 276x^3 + 12x^4$	600	2	0.350	
		$600x^5 + 752x^6 + 6x^8 + 144x^{10}$	$12x^2 + 1476x^3 + 12x^4$	3000	3	0.376	
	Tess(1)	(a) 8S[6]&6R[8] (b) 8R[6]&6S[8]					
	Tess(2)	(a) 24S[10]&32R[6]&6R[8] (b) 24R[10]&32S[6]&6S[8]					
	Tess(3)	(a) 120S[10]&152R[6]&6R[8]&24R[10] (b) 120R[10]&152S[6]&6S[8]&24S[10]					
3	$Le((P_5(Med(\mathrm{M})))^k)$ Dodecahedron D Icosahedron I	$p(x^5) = 24 \cdot 5^k$ $p(x^6) = 5(6 \cdot 5^k - 2)$ $p(x^8) = 30$ $p(x^{10}) = 6(5^k - 3)$	$60x^2 + 60(3 + 5(5^{k-1} - 1))x^3 + 60x^4$	$v_k = 24 \cdot 5^{k+1}$	–	3(D) 5(I)	$5d \cdot 5^k$
	Examples	$120x^5 + 140x^6 + 30x^8 + 12x^{10}$	$60x^2 + 180x^3 + 60x^4$	600	1	0.302	NBO
		$600x^5 + 740x^6 + 30x^8 + 132x^{10}$	$60x^2 + 1380x^3 + 60x^4$	3000	2	0.345	
	Tess(1)	(a) 12S[10]&20S[6]&30R[8] (b) 12R[10]&20R[6]&30S[8]					
	Tess(2)	(a)120S[10]&140R[6]&30R[8]&12R[10] (b)120R[10]&140S[6]&30S[8]&12S[10]					

Table 3.1 (continued)

	Cage type M	Ring polynomial/ Tess(k)	Omega polynomial	v	k	Deg(M)/ GAP(β)	m/ LUMO
4	$Le((P_5(Med(\mathrm{M})))^k)$	$p(x^5) = 48 \cdot 5^{k-1}$	$36x^2 + 48(1 + 5(5^{k-1} - 1)/2)x^3 + 36x^4$	$v_k = 48 \cdot 5^k$	–	3(C)	$2d \cdot 5^k$
	Cube C	$p(x^6) = 4(3 \cdot 5^k - 1)$				4(O)	
	Octahedron O	$p(x^8) = 18$					
		$p(x^{10}) = 12(5^{k-1} - 1)$					
	Examples	$48x^5 + 56x^6 + 18x^8$	$36x^2 + 48x^3 + 36x^4$	240	1	0.248	NBO
		$240x^5 + 296x^6 + 18x^8 + 48x^{10}$	$36x^2 + 528x^3 + 36x^4$	1200	2	0.336	
		$1200x^5 + 1496x^6 + 18x^8 + 288x^{10}$	$36x^2 + 2928x^3 + 36x^4$	6000	3	0.375	
	Tess(1)	(a) 12S[8]&8R[6]&6R[8]					
		(b) 12R[8]&8S[6]&6S[8]					
	Tess(2)	(a) 48S[10]&56R[6]&18R[8]					
		(b) 48R[10]&56S[6]&18S[8]					
	Tess(3)	(a) 240S[10]&296R[6]&18R[8]&48R[10]					
		(b) 240R[10]&296S[6]&18S[8]&48S[10]					

At the level of the simple Hűckel theory, the members of C_{60} family (i.e., the series designed by $Le((P_5(\mathrm{M}))^k)$; M = D/I) are closed π-electron shells, showing well defined HOMO-LUMO gaps (in β units). The gap is the highest for C_{60} and decreases slowly to the higher terms of family. At comparable number of atoms, the cages of the other three series show a lower gap (Table 3.1, the next last column).

The other three related series, all showing an all_R[5] 2-factor, resents a more complicated Omega signature (see Table 3.1). The series #2 is derived from the dual pair C/O by the general sequence $Le((P_5(\mathrm{M}))^k)$. Because of d(O)=4, by *Le* six octagons will appear in the resulting covering, in addition to the decagons induced by the P_5 operation. Thus, the Omega polynomial consists of three terms, the corresponding analytical formula, at various k values being given in Table 3.1. In fact, the building of series #1 and #2 follow the same rule: $Le((P_5(\mathrm{M}))^k)$ and the difference in Omega signature is due to octagons. This also caused the drop in the HOMO-LUMO gap. At least the terms at $k = 1$, in both series, show all equivalent sumanenic flowers of a given folding.

Apparition of octagons can be induced by the medial *Med* operation; in case of M = D/I, by applying the sequence $Le((P_5(Med(\mathrm{M})))^k)$ results in the series #3, which shows a translation of the #2 Omega signature by a factor of 5, also evident in the number of vertices and multiplication factor m (Table 3.1, last column).

Finally, the sequence $Le((P_5(Med(\mathrm{M})))^k)$ applied to the pair C/O results in the #4 series, with an even more complicated Omega signature (see Table 3.1).

The first terms of the last three series show the LUMO orbital as being a non bonding orbital NBO (Table 3.1, last column).

The counting polynomials were performed by our original Nano Studio software (Nagy et al. 2009).

3.6 Conclusions

Four series of cages, tessellated by sumanenic flowers/circulenes were generated by sequences of map operations and their topology was analyzed in terms of Omega and Ring counting polynomials. Among these cages, all showing all_R[5] 2-factors, those belonging to the series designed on the dual pair Dodecahedron/Icosahedron by iterating the P_5 operation and finally closed by *Le* operation, show a unique term Omega signature, thus being classified as the C_{60} family. C_{60} also shows the unique signature and all the members of its family show large HOMO-LUMO gap values, larger than those of the cages belonging to the other three series herein discussed. The unique Omega signature and the large gap value we consider as pertinent criteria for the members of C_{60} family. However, the series #1 and #2 are built up by the same sequence $Le((P_5(\mathrm{M}))^k)$ of map operations, thus no essential structural difference exists between the two series, except the Omega signature.

Coverings were given in terms of sumanenic flowers and the relation among the Ring and Omega polynomials was evidenced. Analytical formulas for the net parameters and polynomials were presented.

Cages consisting of a number of Carbon atoms larger than 100 have been detected experimentally. We believe the newly proposed fullerene structures (even of non-classical tessellation), can be of interest in the future experiments.

References

Amsharov K Yu, Jansen M (2008) J Org Chem 73:2931
Amsharov K Yu, Jansen M (2009) Chem Commun 2691
Balaban AT (1997) From chemical topology to three-dimensional geometry. Plenum Press, New York, NY
Clar E (1964) Polycyclic hydrocarbons. Academic, London
Clar E (1972) The aromatic sextet. Wiley, New York, NY
Cyvin SJ, Gutman I (1988) Kekulé structures in benzenoid hydrocarbons. Springer, Berlin
Dias JR (1999) J Chem Inf Comput Sci 39:144
Diudea MV (2003) Studia Univ"Babes-Bolyai" 48(2):3
Diudea MV (2004) Forma (Tokyo) 19:131
Diudea MV (2005a) J Chem Inf Model 45:1002
Diudea MV (2005b) Phys Chem Chem Phys 7:3626
Diudea MV, (2006) Carpath J Math, 22:43
Diudea MV (2008) Int J Chem Model 1(3/4):335
Diudea MV, Cigher S, John PE, (2008) MATCH Commun Math Comput Chem 60:237
Diudea MV, Cigher S, Vizitiu AE, Florescu MS, John PE (2009) J Math Chem 45:316
Diudea MV, John PE, Graovac A, Primorac M, Pisanski T (2003) Croat Chem Acta 76:153
Diudea MV, Klavžar S (2010) Acta Chim Sloven 57:565
Diudea MV, Nagy CsL (2007) Periodic nanostructures. Springer, Berlin
Diudea MV, Ştefu M, John PE, Graovac A (2006) Croat Chem Acta 79:355
Diudea MV, Vizitiu AE, Janežič D (2007) J Chem Inf Model 47:864
Djoković DŽ, Combin J (1973) Theory Ser B 14:263
Dresselhaus MS, Dresselhaus G, Eklund PC (1996) Science of fullerenes and carbon nanotubes. Academic, San Diego, CA
Eberhard V (1891) Zur morphologie der polyeder. Teubner Leipzig, S.180ff
Endo M, Iijima S, Dresselhaus MS (1996) Carbon nanotubes. Pergamon, Oxford
Euler L (1758) Novi comment Acad Sci I Petropolit 4:109
Fowler PW (1986) Phys Lett 131:444
Fowler PW, Manolopolous DE (1994) An atlas of fullerenes. Oxford University Press, London
Fowler PW, Pisanski T (1994) J Chem Soc Faraday Trans 90:2865
Fowler PW, Rogers KM (1998a) J Chem Soc Faraday Trans 94:1019
Fowler PW, Rogers KM (1998b) J Chem Soc Faraday Trans 94:2509
Fowler PW, Steer JI (1987) J Chem Soc Chem Commun 1403
Fries K (1927) J Liebigs Ann Chem 454:121
Goldberg M (1937) Tohoku Math J 43:104
Harary F (1969) Graph theory. Addison-Wesley, Reading, MA
Hosoya H (1986) Comput Math Appl 12B:271
John PE, Khadikar PV, Singh J (2007a) J Math Chem 42:37
John PE, Sachs H, Zheng M (1995) J Chem Inf Comput Sci 35:1019
John PE, Vizitiu AE, Cigher S, Diudea MV (2007b) MATCH Commun Math Comput Chem 57:479
Klavžar S (2008) MATCH Commun Math Comput Chem 59:217
Kraetschmer W, Lamb LD, Fostiropoulos K, Huffman DR (1990) Nature 347:354
Kroto H, Heath JR, O'Brian SC, Curl RF, Smalley RE (1985) Nature 318:162

Nagy CsL, Diudea MV (2009) Nano studio software. "Babes-Bolyai" University, Cluj
Paquette LA, Balogh DW, Usha R, Koutz D (1981) Science 211:575
Pisanski T, Randić M (2000) In geometry at work. M. A. A. Notes 53:174
Randić M (2004) J Chem Inf Comput Sci 44:365
Sakurai H, Daiko T, Hirao T (2003) Science 301:1878
Sakurai H, Daiko T, Sakane H, Amaya T, Hirao T (2005) J Am Chem Soc 127:11580
Schlegel V (1893) Verhandlungen der kaiserlichen leopoldinisch-carolinischen deutshen akademie der naturforscher 44:337
Shiu WC, Lam PCB, Zhang F, Zhang H (2002) J Math Chem 31:405
Stefu M, Diudea MV (2005) cageversatile 1.5, Babes-Bolyai University, Cluj
Stone AJ, Wales DJ (1986) Chem Phys Lett 128:501
Tanaka K, Yamabe T, Fukui K (1999) The science and technology of carbon nanotubes Elsevier, Amsterdam
Vaissiere B, De L, Fowler PW, Deza M (2001) J Chem Inf Comput Sci 41:376
Veljan D (2001) Combinatorics and discrete mathematics. Algorithm, Zagreb
Vizitiu AE, Diudea MV, Nikolić S, Janežić D (2006) J Chem Inf Model 46:2574
Vizitiu AE, Nagy CsL, Stefu M, Katona G, Diudea MV, Parv, B, Vukičević D (2009) J Math Chem 45:513
Winkler PM (1984) Discrete Appl Math 8:209
Yamamoto K (1993) Pure Appl Chem 65:157
Zhang F, Yan W (2003) MATCH Commun Math Comput Chem 48:117

Chapter 4
Local Combinatorial Characterization of Fullerenes

Tamás Réti, István László, and Ante Graovac

Abstract We present a general method which enables a possible classification of fullerenes by means of local topological invariants. In this study fullerenes are considered as bifaced simple (trivalent) polyhedra. The method proposed is based on the combinatorial analysis of the first neighbor environments (coronas) of vertices and/or edges of bifaced simple polyhedra. For this purpose, we used the so-called line-corona detectors (LC detectors) which are simple connected acyclic graphs (trees) having only 1- and 3-valent vertices. It is demonstrated that by performing certain matching operations with appropriately defined LC detectors, a finite set of local, algebraically independent topological invariants can be obtained by which various fullerene structures can be partitioned into disjoint classes of equivalence. We found also linear interdependencies between similar parameters previously defined in the scientific literature. Discriminating performance of computed topological descriptors have been tested on the set of C_{40} fullerene isomers.

4.1 Introduction

Methods for topological characterization of fullerene isomers have made steady progress over the past decade. This can be explained by the fact that combinatorial properties of fullerenes play a key role in classifying their structures and in predicting their various physical and chemical properties.

This study was motivated primarily by the concept outlined in two papers (Balaban et al. 1995; Alcami et al. 2007). Both of them are focused on the combinatorial characterization and the classification of fullerenes using local topological parameters (graph invariants).

T. Réti (✉)
Szechenyi István University, Egyetem tér 1, H-9026 Győr, Hungary
e-mail: reti.tamas@bgk.bmf.hu

F. Cataldo et al. (eds.), *The Mathematics and Topology of Fullerenes*,
Carbon Materials: Chemistry and Physics 4, DOI 10.1007/978-94-007-0221-9_4,

Balaban et al. (1995) reported extensive computations of a number of topological parameters, and as a result, a classification of 1812 C_{60} isomers into disjoint subclasses has been performed on the basis of 4 different vertex types. Alcami et al. classified the traditional fullerene isomers by partitioning their edges, taking into considerations 9 possible distinct arrangements of 5- and 6-gons adjacent to end-vertices of a given edge (Alcami et al. 2007).

Starting with the extension of the concept detailed in Balaban et al. (1995) and Alcami et al. (2007), the aim of our investigations was to develop a general method which enables a more efficient classification of fullerenes by means of local topological invariants. It will be demonstrated that analyzing the first neighbor environments of vertices and/or edges, it is possible to generate a finite set of algebraically independent, local topological descriptors satisfying the requirements formulated.

4.2 Basic Notions and Definitions

Although we are primarily interested in bifaced simple polyhedra (trivalent polyhedra having only two types of faces), some of our results hold for 3-valent 2-connected finite graphs embedded on a surface (an orientable compact two-dimensional manifold Ψ) such that every face has at least 3 sides (sphere, torus, double torus).

For easier formulation of our results we need to introduce some definitions and recall some known equations (Grünbaum 1967; Brehm and Schulte 2004; Fowler and Manolopoulos 1995; Deza et al. 2000). We start by recalling the Euler's equation given as $V - M + F = 2 - 2G$ where G is the genus of the orientable compact two-dimensional manifold, V, M and F stand for the number of vertices, edges and faces of the embedded graph, respectively (Brehm and Schulte 2004). The number of faces is $F = \sum F_n$ where F_n is the number of n-gonal (n-sided) faces, for $n \geq 3$. There exist polyhedra whose vertices do not all have the same valency. Consequently, we may define an average valency [r] as follows:

$$[r] = \frac{1}{V}\sum_{r} rV_r \tag{4.1}$$

where V_r is the number of vertices having valency r, and $V = \sum V_r$. The number M of edges is related to the number F_n of n-sided faces, the number V of vertices, and the average valency [r]

$$2M = \sum_{n} nF_n = \sum_{r} rV_r = [r]V \tag{4.2}$$

From the Euler formula and Eq. (4.2) it follows that

$$2\sum_{r\geq 3}(3 - r)V_r = 12(1 - G) + \sum_{n\geq 3}(n - 6)F_n \tag{4.3}$$

For $G = 0$, by Steinitz's Theorem, Ψ is combinatorially isomorphic to a 3-dimensional convex polytope, if and only if the graph of Ψ is 3-connected. This implies that for simple polyhedra (if $G = 0$ and $V_r = 0$ for $r > 3$ hold) from Eq. (4.3) one obtains

$$3F_3 + 2F_4 + F_5 = 12 + \sum_{n \geq 7} (n - 6)F_n \geq 12 \quad (4.4)$$

From the previous considerations it follows that any polyhedron contains at least a triangle, or a quadrilateral or a pentagon, i.e. there is no polyhedron whose faces are all hexagons, or polygons with six or more sides. Moreover, for a simple polyhedron containing triangles, or quadrilaterals, or pentagons or hexagons, equality $3F_3 + 2F_4 + F_5 = 12$ holds, independently of the number of hexagonal faces. Consequently, the set of simple polyhedra composed of triangles, or quadrilaterals, or pentagons is finite. It is easy to see that there exist only six bifaced simple polyhedra including triangles, or quadrilaterals or pentagons (Grünbaum 1967).

In a simple (trivalent) polyhedron a q-gonal face is called isolated, if it is surrounded only by non-q-gonal faces. Let IS(q) denote the number of q-gonal isolated faces. It is obvious that IS(q) is a topological invariant for which $0 \leq IS(q) \leq F_q$ holds. A simple polyhedron is called q-isolated, if its all q-gonal faces are adjacent only to non-q-gonal faces. In this particular case $IS(q) = F_q$. A simple polyhedron is called completely isolated, if equality $\sum IS(q) = \sum F_q = F$ is fulfilled. There exist several completely isolated trivalent polyhedra. A simple example is the great rhombicosidodecahedron (truncated icosidodecahedron), which is 4-, 6- and 10-isolated. The great rhombicosidodecahedron has 62 faces: 30 squares, 20 hexagons and 12 decagons, consequently $IS(4) + IS(6) + IS(10) = 62$ holds.

In what follows, we do not distinguish between the polyhedron and its corresponding finite 3-connected planar graph.

4.3 Fullerenes, Fulleroids and Bifaced Polyhedra – Classification

Polyhedra are generally used as geometric models for carbon molecules called fullerenes consisting only of pentagons and hexagons (Fowler and Manolopoulos 1995). In chemistry, according to the classical definition, a fullerene is an all-carbon molecule where vertices represent the atoms of carbon, and edges between vertices realize the bonds between pairs of carbon atoms. Fullerenes C_k with vertex number k exist for all even $k \geq 20$ except $k = 22$, where the number of pentagons is 12 ($F_5 = 12$) and the number of hexagons is $F_6 = (V/2) - 10$.

Call bifaced (two-faced) any polyhedron, whose faces are F_α α-gons and F_β β-gons only, where $3 \leq \alpha < \beta$, $F_\alpha > 0$ and $F_\beta > 0$. An important consequence of Eq. (4.4) is that if a bifaced polyhedron is simple, this implies that α can be only 3 or 4, or 5. In the following the set of bifaced simple polyhedra (BS polyhedra) consisting of α- and β-gons is denoted by S(α, β) where $3 \leq \alpha \leq 5$ and $\alpha < \beta$.

A bifaced simple polyhedron is called α-isolated, (β-isolated) if each α-gons (β-gons) is isolated, that is each of them is surrounded only by β-gons (α-gons), respectively. For example, the Buckminster fullerene represents the smallest 5-isolated S(5,6) polyhedron with 60 vertices. Since there are no edge-neighbor triangular faces in BS polyhedra, this implies that all S(3, β) polyhedra are 3-isolated.

It is also obvious that there are no completely isolated S(α, β) polyhedra composed only of α-isolated and β-isolated faces. From this it follows that for BS polyhedra the inequality $0 \leq \mathrm{IS}(\alpha) + \mathrm{IS}(\beta) < \mathrm{F}$ holds.

In Fig. 4.1, Schlegel diagrams of six different BS polyhedra are shown, they are denoted by $H_v(\alpha, \beta)$. As can be seen, polyhedra $H_6(3,4)$, $H_{12}(3,5)$ and $H_{12}(3,6)$ are 3-isolated, $H_{24A}(4,6)$ is 4-isolated, while $H_{10}(4,5)$ is 5-isolated. Polyhedra $H_{24A}(4,6)$ and $H_{24B}(4,6)$ are so-called structural isomers. Since there are only six BS polyhedra including triangles, or quadrilaterals, or pentagons, in the following we assume that $3 \leq \alpha \leq 5$ and $\beta \geq 6$ (Fowler and Manolopoulos 1995).

From geometric point of view, there exist several generalizations of conventional fullerenes (Deza et al. 2000; Laszlo and Rassat 2001; Fowler et al. 1996, 2001; Fowler and Heine 2001; Albertazzi et al. 1999; Dress and Brinkmann 1996; Delgado-Friedrichs and Deza 2000; Jendrol' and Trenker 2001; Deza et al. 1998; Gan et al. 2009). Deza et al. (2000) defined a fullerene in a wider sense as a finite, trivalent map on a closed, unbounded surface with a non-negative Euler-characteristic, where faces are pentagons and hexagons. According to this concept, the only surfaces admitting finite fullerene maps are as follows: the sphere, the torus, the Klein bottle and the real projective plane (Deza et al. 2000).

Using semi-empirical models, Fowler et al. (1996, 2001; Fowler and Heine 2001) calculated the relative energies of hypothetical C_{40} cages (generalized polyhedral

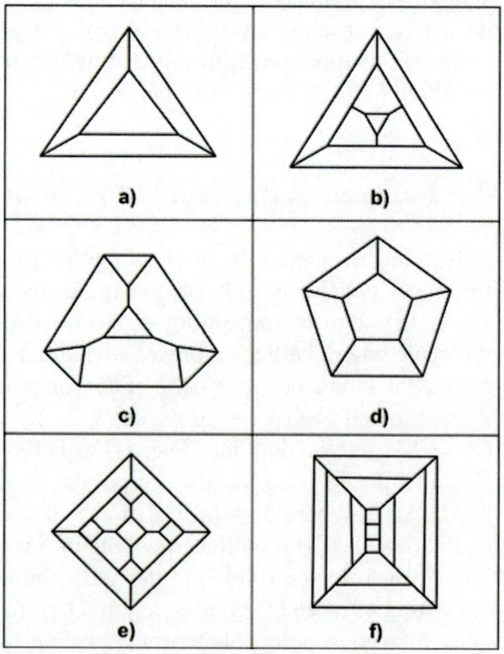

Fig. 4.1 Schlegel diagrams of six bifaced simple polyhedra (**a**) $H_6(3,4)$, (**b**)$H_{12}(3,5)$, (**c**) $H_{12}(3,6)$, (**d**) $H_{10}(4,5)$, (**e**) $H_{24A}(4,6)$ and (**f**) $H_{24B}(4,6)$

fullerenes with 40 vertices) that can be assembled from square, pentagonal, hexagonal and heptagonal faces. It has been pointed out that it is possible to reduce the number of pentagons by introducing some square faces or heptagonal faces into a fullerene. Calculations performed have suggested that whilst introduction of square faces is highly unfavorable, carbon cages including a heptagonal face fall within the traditional fullerene energy range (Albertazzi et al. 1999).

Dress and Brinkmann (1996) introduced the notion of fulleroid defined as follows: a fulleroid is a tiling of the sphere such that all its vertices have valency 3, while all its faces are pentagons ($\alpha = 5$) and β-gons, where $\beta > 5$. It follows that fulleroids are BS polyhedra of type S(5, β).

During the last two decades, the combinatorial structure and symmetry of fullerenes and of fulleroids has been studied deeply (Dress and Brinkmann 1996; Delgado-Friedrichs and Deza 2000; Jendrol' and Trenkler 2001; Deza et al. 1998; Babić et al. 1993; Graver 2005; Jendrol' and Kardoš 2007; Kardoš 2007; Fowler 2003, 2002; Austin et al. 1995; Fajtlowitz and Larson 2003; Zhang and Zhang 2001; Torrens 2002; Došlić 2002a, b; 2007; 2008; Hu 2003; Reti and Bitay 2007). Babić et al. studied fullerenes with up to 70 vertices and divided them according to the symmetry group (Babić et al. 1993). Graver published a catalogue of all fullerenes with ten or more symmetries and verified that there are infinitely many fullerenes with icosahedral or tetrahedral symmetry (Graver 2005). Jendrol' and Kardoš found a necessary and sufficient condition for existence of O_h(5, β) fulleroids, where O_h stands for the full symmetry group of regular octahedron (Jendrol' and Kardoš 2007). Recently, Kardoš presented a necessary and sufficient condition for existence of fulleroids characterized by tetrahedral symmetry types (Kardoš 2007). Detailed investigations have been performed to find graph theoretical descriptors by which the relative stability of fullerene isomers can be more efficiently predicted (Balaban et al. 1995; Alcami et al. 2007; Albertazzi et al. 1999; Austin et al. 1995; Fajtlowitz and Larson 2003; Zhang and Zhang 2001; Fowler 2002; Torrens 2002; Došlić 2002a, b, 2007, 2008; Hu 2003; Reti and Bitay 2007).

Similarly, there are also many interesting results concerning bifaced simple polyhedra. A systematic investigation of BS polyhedra was inspired by Grünbaum's book "Convex Polytopes" (Grünbaum 1967). Since then, the subject has grown into an active field of research (Goldberg 1937; Grünbaum and Zachs 1974; Deza and Grishukhin 2001; Brinkmann and Deza 2000; Deza and Dutour 2005; Deza and Grishukin 2002; Dutour-Sikiric et al. 2008; Deza et al. 2009; Deza 2000).

It is known that the sets of BS polyhedra of types S(3,6), S(4,6), S(5,6), S(4,β) and S(5,β) are infinite (Grünbaum 1967; Deza et al. 1998; Dutour-Sikiric et al. 2008). Moreover it has been verified that there are infinitely many BS polyhedra of types S(5,7) and S(5,8) (Delgado-Friedrichs and Deza 2000).

In this study fullerenes are considered as a subset of BS polyhedra. In order to investigate their local combinatorial structure, we introduce the notion of the line-corona detector (LC detector), which serves as an efficient tool to analyze the correspondences between the first neighbors of vertices, edges and faces, respectively.

4.4 Line-Corona Detectors

The objective of using LC detectors is to analyze the intrinsic, local topological properties of BS polyhedra and to obtain new topological invariants designated primarily to classification purposes.

It will be demonstrated that by means of LC detectors it is possible to generate a set of topological invariants (graph invariants) by which BS polyhedra of type S(α,β) can be classified into a finite number of equivalence classes.

An LC detector is a simple connected acyclic graph (tree) having only 1- and 3-valent vertices. In order to obtain topological invariants some matching operations should be carried out by an appropriately defined LC detector.

As a result of matching operations performed on trivalent polyhedral graphs, the edges of a selected LC detector will be incident (adjacent) exactly to P faces (denoted by D_1, D_2, $\ldots D_P$) of the trivalent polyhedron investigated. For trivalent bifaced polyhedra, faces $D_j (j = 1, 2, ..P)$ can be α- and β- gons, only. By definition, the set of actual faces D_1, D_2, .. $D_j \ldots D_P$ (where D_j are α-gons or β-gons) is called a line corona generated by the selected LC detector. It is easy to see that the possible number of different line coronas can not be larger than 2^P.

Various LC detectors represented by acyclic graphs of small size can be easily constructed. Let L(ν,P) denote a LC-detector characterized by ν vertices and a P-component vector of faces [D_1, D_2, $\ldots D_P$]. In Fig. 4.2, four different LC detectors (namely L(4,3), L(6,4), L(10,6) and L(14,8)) can be seen. The simplest type L(4,3) shown in Fig. 4.2a includes 4 vertices and 3 edges. The corresponding 4 line coronas denoted by $C_{\alpha,\alpha,\alpha}$, $C_{\alpha,\alpha,\beta}$, $C_{\alpha,\beta,\beta}$ and $C_{\beta,\beta,\beta}$ are depicted in Fig. 4.3. For C_{60} fullerene isomers (case of $\alpha = 5$ and $\beta = 6$) the properties of line coronas $C_{5,5,5}$, $C_{5,5,6}$, $C_{5,6,6}$ and $C_{6,6,6}$ were investigated by Balaban et al. (1995). It was verified that the 1812 structural isomers of C_{60} fullerenes could be partitioned into

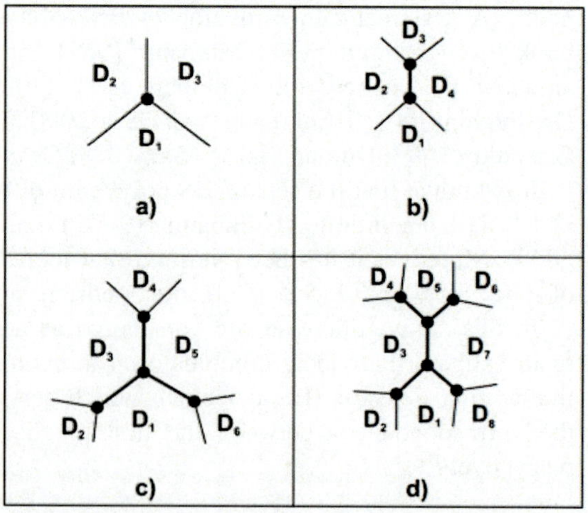

Fig. 4.2 LC-detectors of types L(4,3) (**a**), L(6,4) (**b**), L(10,6) (**c**) and L(14,8) (**d**)

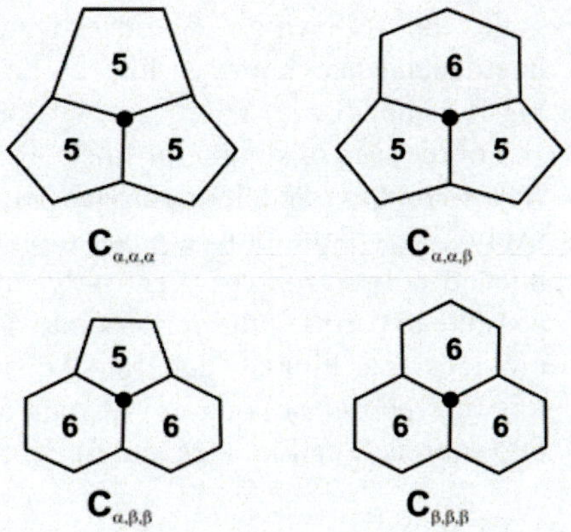

Fig. 4.3 Four possible types of vertex coronas for case of $\alpha = 5$ and $\beta = 6$

42 equivalence classes on the basis of two vertex coronas $C_{5,5,5}$ and $C_{5,5,6}$ (Balaban et al. 1995).

Figure 4.2b shows a LC detector of type L(6,4) with 6 vertices and 5 edges. It is easy to see that in this case, due to the symmetry of face arrangements, 9 different line coronas can be obtained. Figure 4.4 indicates the 9 different line coronas denoted by $E_1, E_2 \ldots E_9$. Starting with the concept outlined in Ref. (Cioslowski et al. 2002) Alcami et al. developed a model by which the enthalpy of formation (QE) of traditional fullerenes $C_k(k \leq 72)$ can be estimated as a multi-linear function of topological invariants characterizing the 9 line corona configurations (Alcami et al. 2007).

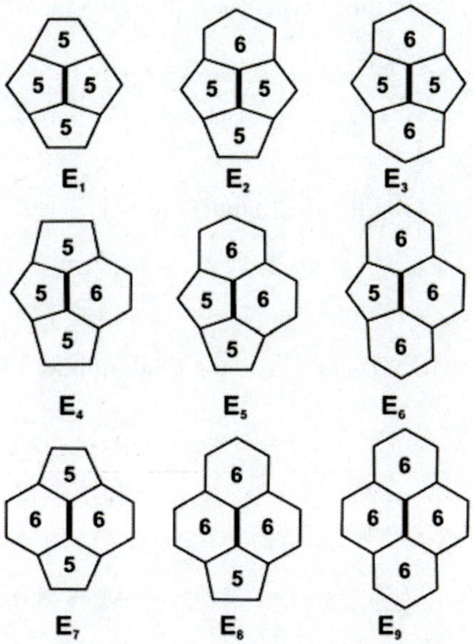

Fig. 4.4 Nine types of edge-coronas for case of $\alpha = 5$ and $\beta = 6$

In the following, our investigations will be focused on the analysis of interdependence between line coronas of types L(4,3) and L(6,4) depicted in Fig. 4.2a and b.

For the sake of simplicity, line coronas $C_{\alpha,\alpha,\alpha}$, $C_{\alpha,\alpha,\beta}$, $C_{\alpha,\beta,\beta}$ and $C_{\beta,\beta,\beta}$ are called *vertex-coronas*, and line coronas E_1, E_2 ... E_9 are called *edge-coronas*, respectively. These definitions are based on the following observations: For any trivalent bifaced polyhedron (trivalent polyhedral graph) the LC detector L(4,3) makes it possible to partition the vertices into 4 disjoint classes, while the LC detector L(6,4) provides a partition of edges into 9 disjoint subsets. It should be noted that LC detectors of type L(10,6) and L(14,8) are also applicable to classify vertices and edges into subsets (See Fig. 4.2c and d).

4.5 Some Combinatorial Properties of Polyhedra

In what follows some propositions are presented. As we have mentioned, they concern primarily the intrinsic relationships between vertex- and edge-coronas of BS polyhedra (trivalent bifaced polyhedral graphs).

Proposition 1 *For any polyhedral graph*

$$\sum_{n} n^{j+1} F_n = \sum_{n} \sum_{k \le n} e(n,k) W(n,k,j) \tag{4.5}$$

where e(n,k) stands for the number of common edges between n- and k-sided neighbor faces, j is an arbitrary integer, W(n,k,j) are positive edge weights, defined as $W(n,k,j) = n^j + k^j$.

Proof Since e(n, k) = e(k, n) we have

$$M = \sum_{n} e(n,n) + \frac{1}{2} \sum_{\substack{n,k \\ n \ne k}} e(n,k) = \sum_{n} \sum_{k \le n} e(n,k) \tag{4.6}$$

On the other hand,

$$\sum_{n} n^{j+1} F_n = \sum_{n} n^j A_n \tag{4.7}$$

where $A_n = nF_n$ is the total number of edges of n-sided cells, for which

$$\sum_{n} A_n = 2M. \tag{4.8}$$

Since

$$A_n = 2e(n,n) + \sum_{k \ne n} e(n,k). \tag{4.9}$$

this implies that

$$\sum_{n} n^{j+1} F_n = \sum_{n} n^{j} \left\{ 2e(n,n) + \sum_{k \neq n} e(n,k) \right\} = \sum_{n} \sum_{k \leq n} e(n,k) \left\{ n^{j} + k^{j} \right\} \quad (4.10)$$

Corollary 1.1 *Consider a BS polyhedron including* α*- and* β*-sided faces, and denote by* $m_k = m(E_k)$ *the number of different coronas* $E_k (k = 1, 2, \ldots 9)$ *for which* $\sum m_k = M$ *holds (See Fig. 4.4) From Eq. (4.10), if* $j = 1$*, one obtains*

$$\alpha^2 F_\alpha + \beta^2 F_\beta = 2\alpha e(\alpha, \alpha) + (\alpha + \beta) e(\alpha, \beta) + 2\beta e(\beta, \beta) \quad (4.11)$$

Moreover, from Fig. 4.4 it is easy to see

$$e(\alpha, \alpha) = m_1 + m_2 + m_3 \quad (4.12)$$

$$e(\alpha, \beta) = m_4 + m_5 + m_6 \quad (4.13)$$

$$e(\beta, \beta) = m_7 + m_8 + m_9 \quad (4.14)$$

and

$$M = \sum_{k=1}^{9} m_k = e(\alpha, \alpha) + e(\alpha, \beta) + e(\beta, \beta) \quad (4.15)$$

Remark 1.1 *It is obvious that a BS polyhedron is* α*-isolated if and only if* $e(\alpha, \alpha) = m_1 + m_2 + m_3 = 0$ *holds. (If* $e(\alpha, \alpha) = 0$ *this implies that* $m_4 = m_5 = 0$*.) Similarly, a BS polyhedron is* β*-isolated if and only if* $e(\beta, \beta) = m_7 + m_8 + m_9 = 0$ *is fulfilled.*

Remark 1.2 *Parameter* $e(\alpha,\alpha)$ *is considered as a generalization of the pentagon adjacency index* N_P *which is generally used for stability prediction of traditional fullerene isomers consisting of pentagons and hexagons. The pentagon adjacency index* $N_P = e(5, 5)$ *is equal to the total number of fused pentagon pairs (Fowler and Manolopoulos 1995; Albertazzi et al. 1999; Campbell et al. 1996).*

Remark 1.3 *Based on the concept outlined in Campbell et al. (1996), it is easy to show that the topological invariant* $e(\alpha,\alpha)$ *can be decomposed in the following form:*

$$e(\alpha, \alpha) = \frac{1}{2} \sum_{j=0}^{\alpha} j p(j) \quad (4.16)$$

where $p(j)$ *stands for the number of those* α*-faces that have exactly* j *edge-neighbor* α*-faces, for* $j = 0, 1, 2, \ldots \alpha$*. From the definition it follows that* $\sum p(j) = F_\alpha$*. A similar linear decomposition can be performed for parameter* $e(\beta,\beta)$*.*

Proposition 2 *For any trivalent polyhedral graph*

$$\sum_{n} n^{j+1} F_n = \sum_{n} \sum_{k \leq n} g(n,k) W(n,k,j) \quad (4.17)$$

where g(n,k) is the total number of edges whose end-vertices incident to n-sided and k-sided faces, j is an arbitrary integer, $W(n,k,j) = n^j + k^j$ *are positive edge weights.*

Proof The proof is an exact analogue of the proof of Proposition 1.

Remark 2.1 *Formulas (4.5) and (4.17) are of similar form. The main difference between them is that identity represented by Eq. (4.5) holds for any polyhedra, but Eq. (4.17) is valid only for simple polyhedra.*

Remark 2.2 *It is easy to see that Eqs. (4.5) and (4.17) are the generalizations of the well known identity* $\sum \mathrm{nF_n} = 2\mathrm{M}$.

Corollary 2.1 *Consider a BS polyhedron including* α*- and* β*-sided faces. If* $j = -1$*, then from Eq. (4.17) it follows that*

$$\mathrm{F} = \frac{\mathrm{M}+6}{3} = \frac{2}{\alpha}\mathrm{g}(\alpha,\alpha) + \left(\frac{1}{\alpha} + \frac{1}{\beta}\right)\mathrm{g}(\alpha,\beta) + \frac{2}{\beta}\mathrm{g}(\beta,\beta) \tag{4.18}$$

where

$$\mathrm{g}(\alpha,\alpha) = \mathrm{m_1} + \mathrm{m_4} + \mathrm{m_7} \tag{4.19}$$

$$\mathrm{g}(\alpha,\beta) = \mathrm{m_2} + \mathrm{m_5} + \mathrm{m_8} \tag{4.20}$$

$$\mathrm{g}(\beta,\beta) = \mathrm{m_3} + \mathrm{m_6} + \mathrm{m_9} \tag{4.21}$$

and

$$\mathrm{M} = \sum_{\mathrm{k}=1}^{9} \mathrm{m_k} = \mathrm{g}(\alpha,\alpha) + g(\alpha,\beta) + \mathrm{g}(\beta,\beta) \tag{4.22}$$

Proposition 3 (Jucovič 1974) *For simple (trivalent) polyhedra the following inequality is valid*:

$$\begin{aligned} 120 \le\ & 20\mathrm{e}(3,3) + 25\mathrm{e}(3,4) + 16\mathrm{e}(3,5) + 10\mathrm{e}(3,6) + (20/3)\mathrm{e}(3,7) \\ & + 5\mathrm{e}(3,8) + (5/2)\mathrm{e}(3,9) + 2\mathrm{e}(3,10) + 20\mathrm{e}(4,4) + 11\mathrm{e}(4,5) + 5\mathrm{e}(4,6) \\ & + 5\mathrm{e}(4,7) + 5\mathrm{e}(4,8) + 3\mathrm{e}(4,9) + 8\mathrm{e}(5,5) + 2\mathrm{e}(5,6) + 2\mathrm{e}(5,7) + 2\mathrm{e}(5,8) \end{aligned} \tag{4.23}$$

Proof Inequality (4.23) represents a reformulated form of Jucovič result which is based on the extension of Kotzig's theorem to simplicial polyhedra. Consequently, using the duality concept, formula (4.23) is a simple corollary of Jucovič inequality (Jucovič 1974).

Remark 3.1 *The only simple polyhedron for which e(3,3) is a positive integer, is the tetrahedron, where* $M = e(3,3) = 6$.

Corollary 3.1 *Consider BS polyhedra of types S(3,6), S(3,7), S(3,8), S(3,9), S(3,10), S(4,6), S(4,7), S(4,8), S(4,9), S(5,6), S(5,7), and S(5,8). From Eq. (4.23) it follows that for these BS polyhedra*

$$\begin{aligned}&120 \leq 10\mathrm{e}(3,6),\ 120 \leq (20/3)\mathrm{e}(3,7),\ 120 \leq 5\mathrm{e}(3,8),\ 120 \leq (5/2)\mathrm{e}(3,9),\\&120 \leq 2\mathrm{e}(3,10),\ 120 \leq 20\mathrm{e}(4,4) + 5\mathrm{e}(4,6),\ 120 \leq 20\mathrm{e}(4,4) + 5\mathrm{e}(4,7),\\&120 \leq 20\mathrm{e}(4,4) + 5\mathrm{e}(4,8),\ 120 \leq 20\mathrm{e}(4,4) + 3\mathrm{e}(4,9),\ 120 \leq 8\mathrm{e}(5,5) + 2\mathrm{e}(5,6),\\&120 \leq 8\mathrm{e}(5,5) + 2\mathrm{e}(5,7),\ 120 \leq 8\mathrm{e}(5,5) + 2\mathrm{e}(5,8).\end{aligned} \tag{4.24}$$

hold.

4.6 Characterization of Combinatorial Structure of BS Polyhedra

In this section we restrict our considerations to BS polyhedra.

Proposition 4 (Jendrol' 1977; Jendrol' 1999; Jendrol' and Skupien 2001)

(a) *There exists no BS polyhedron of type S* (3, β) *with* $\beta > 10$,
(b) *There exists no* 4*-isolated BS polyhedron of type S*(4, β) *with* $\beta > 7$,
(c) *There exists no* 5*-isolated BS polyhedron of type S*(5, β) *with* $\beta > 6$.

Proof This statement follows immediately from a strengthening of Kotzig's theorem. In a trivalent polyhedron a common edge incident to a-sided and b-sided faces (where $b \geq a \geq 3$) is called the (a,b)-edge. It has been proved (Jendrol' 1977, 1999; Jendrol' and Skupien 2001) that every trivalent polyhedron contains an (a,b)-edge where $a = 3$, $3 \leq b \leq 10$, or $a = 4$, $4 \leq b \leq 7$, or $a = 5$, $5 \leq b \leq 6$.

Remark 4.1 *There exists a* 3*-isolated BS polyhedron of type S*(3, 10)*, a known example is the truncated dodecahedron composed of* 20 *triangles and* 12 *decagons. Similarly, there exist* 4*-isolated BS polyhedra of type S*(4, 7) *(Deza 2000).*

Remark 4.2 *There are infinitely many* 3*-isolated BS polyhedra of types S*(3, 6) *with four triangular faces, consequently, for these S*(3, 6) *polyhedra IS*(3)=4 *holds (Grünbaum 1967; Deza et al. 1998). It is conjectured that the set of BS polyhedra of type S*(3, β) *is finite if* $\beta \neq 6$.

Remark 4.3 *There are infinitely many BS polyhedra of types S*(4, β)*. A classical example is the family of* β*-isolated* β*-gonal prisms with* β *quadrangular faces and two* β*-gonal faces. There exists another infinite set of S*(4, β) *polyhedra composed of four* β*-gonal faces and* $F_4 = 2\beta - 6$ *quadrangular faces, for which IS*(4)=*IS*(β)=0 *holds (Dutour-Sikiric et al. 2008).*

Remark 4.4 *The number of* 5*-isolated S*(5,6) *polyhedra (fullerenes) is infinite. These are the so-called IPR fullerenes, for which IS*(5)=12 *is fulfilled (Fowler and*

Manolopoulos 1995). It is easy to prove that the set of 6-isolated S(5,6) polyhedra (fullerenes) is finite. It is also known that there exists an infinite family of β-isolated $S(5, \beta)$ polyhedra with 2β pentagonal faces and two β-gonal faces.

Proposition 5 *Denote by $V_{\alpha\alpha\alpha}$, $V_{\alpha\alpha\beta}$, $V_{\alpha\beta\beta}$ and $V_{\beta\beta\beta}$ the number of vertices belonging to vertex coronas $C_{\alpha,\alpha,\alpha}$, $C_{\alpha,\alpha,\beta}$, $C_{\alpha,\beta,\beta}$ and $C_{\beta,\beta,\beta}$ of a BS polyhedron (See Fig. 4.3). Then we have for $V_{\alpha\alpha\alpha}$, $V_{\alpha\alpha\beta}$, $V_{\alpha\beta\beta}$ and $V_{\beta\beta\beta}$*

$$V_{\alpha\alpha\alpha} = (2m_1 + m_2)/3 \tag{4.25}$$

$$V_{\alpha\alpha\beta} = m_2 + 2m_3 \tag{4.26}$$

$$V_{\alpha\beta\beta} = 2m_7 + m_8 \tag{4.27}$$

$$V_{\beta\beta\beta} = (m_8 + 2m_9)/3 \tag{4.28}$$

Proof The edge number M of any BS polyhedron having α- and β-sided faces can be partitioned into 4 disjoint sets, and calculated as a sum of four terms as follows:

$$M = m(\alpha, \alpha, \alpha) + m(\alpha, \alpha, \beta) + m(\alpha, \beta, \beta) + m(\beta, \beta, \beta) \tag{4.29}$$

where

$$m(\alpha, \alpha, \alpha) = 3V_{\alpha\alpha\alpha}/2 \tag{4.30}$$

$$m(\alpha, \alpha, \beta) = 3V_{\alpha\alpha\beta}/2 \tag{4.31}$$

$$m(\alpha, \beta, \beta) = 3V_{\alpha\beta\beta}/2 \tag{4.32}$$

$$m(\beta, \beta, \beta) = 3V_{\beta\beta\beta}/2 \tag{4.33}$$

by definition. From Fig. 4.4 it is clear

$$3V_{\alpha\alpha\alpha}/2 = m_1 + m_2/2 \tag{4.34}$$

$$3V_{\alpha\alpha\beta}/2 = 3m_2/2 + 3m_3 \tag{4.35}$$

$$3V_{\alpha\beta\beta}/2 = 3m_7 + 3m_8/2 \tag{4.36}$$

$$3V_{\beta\beta\beta}/2 = m_8/2 + m_9 \tag{4.37}$$

From Eqs. (4.30), (4.31), (4.32), (4.33), (4.34), (4.35), (4.36), and (4.37), we obtain formulas (4.25), (4.26), (4.27), and (4.28).

Corollary 5.1

$$e(\alpha, \alpha) = (3V_{\alpha\alpha\alpha} + V_{\alpha\alpha\beta})/2 = m_1 + m_2 + m_3 \tag{4.38}$$

$$e(\beta,\beta) = (3V_{\beta\beta\beta} + V_{\alpha\beta\beta})/2 = m_7 + m_8 + m_9 \tag{4.39}$$

$$e(\alpha,\beta) = M - e(\alpha,\alpha) - e(\beta,\beta) = V_{\alpha\alpha\beta} + V_{\alpha\beta\beta} \tag{4.40}$$

Corollary 5.2

$$e(\beta,\beta) = M + e(\alpha,\alpha) - (3V_{\alpha\alpha\alpha} + 2V_{\alpha\alpha\beta} + V_{\alpha\beta\beta}) \tag{4.41}$$

Proof

$$\begin{aligned} e(\beta,\beta) &= M - e(\alpha,\alpha) - e(\alpha,\beta) = M + e(\alpha,\alpha) - 2e(\alpha,\alpha) - V_{\alpha\alpha\beta} - V_{\alpha\beta\beta} \\ &= M + e(\alpha,\alpha) - (3V_{\alpha\alpha\alpha} + V_{\alpha\alpha\beta}) - V_{\alpha\alpha\beta} - V_{\alpha\beta\beta} \end{aligned} \tag{4.42}$$

The following three lemmas also concern BS polyhedra.

Lemma 1

$$\alpha^2 F_\alpha + \beta^2(F - F_\alpha) = 3\alpha V_{\alpha\alpha\alpha} + (2\alpha+\beta)V_{\alpha\alpha\beta} + (\alpha+2\beta)V_{\alpha\beta\beta} + 3\beta V_{\beta\beta\beta} \tag{4.43}$$

Proof By substituting parameters $e(\alpha,\alpha)$, $e(\alpha,\beta)$ and $e(\beta,\beta)$ represented by Eqs. (4.38), (4.39), and (4.40) into Eq. (4.11) we have Eq. (4.43)

Lemma 2

$$\begin{aligned} &(\beta^2 - 6\alpha)V_{\alpha\alpha\alpha} + (\beta^2 - 4\alpha - 2\beta)V_{\alpha\alpha\beta} + (\beta^2 - 2\alpha - 4\beta)V_{\alpha\beta\beta} \\ &+(\beta^2 - 6\beta)V_{\beta\beta\beta} = 2F_\alpha(\beta^2 - \alpha^2) - 4\beta^2 \end{aligned} \tag{4.44}$$

Proof Consider Eq. (4.43) and identity given by

$$2F - 4 = V = V_{\alpha\alpha\alpha} + V_{\alpha\alpha\beta} + V_{\alpha\beta\beta} + V_{\beta\beta\beta} \tag{4.45}$$

By eliminating F from Eqs. (4.43) and (4.45), this leads to identity Eq. (4.44).

Lemma 3

$$F_\alpha = \frac{4\beta + (\beta-6)V}{2(\beta-\alpha)} = \frac{(\beta-6)}{3(\beta-\alpha)}M + \frac{2\beta}{\beta-\alpha} \tag{4.46}$$

Proof Taking into consideration that $F = F_\alpha + F_\beta = 2 + V/2$, $3V = 2M = \alpha F_\alpha + \alpha F_\beta$ and $F - M + V = 2$, we have

$$3V = \alpha F_\alpha + \beta(F - F_\alpha) = \alpha F_\alpha + \beta(2 - F_\alpha + V/2) = (\alpha-\beta)F_\alpha + 2\beta + \beta V/2 \tag{4.47}$$

From this, Eq. (4.46) yields. A consequence of Eq. (4.46) is that there are only 3 exceptional cases when F_α is independent of V. The 3 particular cases are as follows: i) $\alpha = 3$, $\beta = 6$ with $F_3 = 4$, ii) $\alpha = 4$, $\beta = 6$ with $F_4 = 6$, iii) $\alpha = 5$, $\beta = 6$ with

$F_5 = 12$. This implies that, BS polyhedra of type S(3,6), S(4,6), and S(5,6) contain 4 triangles, 6 quadrilaterals, and 12 pentagons, respectively.

Proposition 6

$$3V_{\alpha\alpha\alpha} + 2V_{\alpha\alpha\beta} + V_{\alpha\beta\beta} = \alpha F_\alpha \quad (4.48)$$

Proof Eq. (4.44) can be rewritten in the following form:

$$6(\beta-\alpha)V_{\alpha\alpha\alpha}+4(\beta-\alpha)V_{\alpha\alpha\beta}+2(\beta-\alpha)V_{\alpha\beta\beta}+V(\beta^2-6\beta) = 2(\beta-\alpha)(\beta+\alpha)F_\alpha-4\beta^2 \quad (4.49)$$

From this one obtains

$$3V_{\alpha\alpha\alpha} + 2V_{\alpha\alpha\beta} + V_{\alpha\beta\beta} = \frac{2(\beta-\alpha)(\beta+\alpha)F_\alpha - 4\beta^2 - V\beta(\beta-6)}{2(\beta-\alpha)} \quad (4.50)$$

By using formula (4.46), we have

$$3V_{\alpha\alpha\alpha} + 2V_{\alpha\alpha\beta} + V_{\alpha\beta\beta} = (\beta+\alpha)F_\alpha - \frac{4\beta^2 + V\beta(\beta-6)}{2(\beta-\alpha)} = \alpha F_\alpha \quad (4.51)$$

Remark 6.1 *If* $3 \le \alpha \le 5$ *and* $\beta \ge 6$*, the minimum value of* αF_α *is equal to* 12. *For example, this is fulfilled for polyhedron* $H_{12}(3,6)$ *shown in Fig.* 4.1c.

Corollary 6.1 *Using Eqs.* (4.25), (4.26), (4.27), *and* (4.28) *we obtain*

$$\alpha F_\alpha = 3V_{\alpha\alpha\alpha} + 2V_{\alpha\alpha\beta} + V_{\alpha\beta\beta} = 2m_1 + 3m_2 + 4m_3 + 2m_7 + m_8 \quad (4.52)$$

This implies that

$$m_8 = \alpha F_\alpha - 2m_1 - 3m_2 - 4m_3 - 2m_7 \quad (4.53)$$

Corollary 6.2 *Because* $2M = 3V = \alpha F_\alpha + \alpha F_\beta$*, this implies that*

$$\beta F_\beta = 3V - \alpha F_\alpha = V_{\alpha\alpha\beta} + 2V_{\alpha\beta\beta} + 3V_{\beta\beta\beta} = m_2 + 2m_3 + 4m_7 + 3m_8 + 2m_9 \quad (4.54)$$

Corollary 6.3 *Since* $M = (\alpha F_\alpha + \alpha F_\beta)/2$*, from Eqs.* (4.52) *and* (4.54) *one obtains*

$$M = m_1 + 2m_2 + 3m_3 + 3m_7 + 2m_8 + m_9 \quad (4.55)$$

Corollary 6.4 *Using Eqs.* (4.53) *and* (4.55) *we have*

$$\begin{aligned} m_9 &= M - m_1 - 2m_2 - 3m_3 - 3m_7 - 2m_8 \\ &= M - 2\alpha F_\alpha + 3m_1 + 4m_2 + 5m_3 + m_7 \end{aligned} \quad (4.56)$$

Proposition 7 *For any BS polyhedron*

$$e(\beta,\beta) = M + e(\alpha,\alpha) - \alpha F_\alpha = M - \alpha F_\alpha + m_1 + m_2 + m_3 \tag{4.57}$$

Proof This results follows directly from Eqs. (4.12), (4.41), and (4.51).

Corollary 7.1 *From Eqs.* (4.12), (4.13), (4.15), *and* (4.57) *we have*

$$\begin{aligned} m_4 + m_5 + m_6 &= e(\alpha,\beta) = M - e(\alpha,\alpha) - e(\beta,\beta) \\ &= \alpha F_\alpha - 2e(\alpha,\alpha) = \alpha F_\alpha - 2m_1 - 2m_2 - 2m_3 \end{aligned} \tag{4.58}$$

This implies that

$$m_6 = \alpha F_\alpha - 2m_1 - 2m_2 - 2m_3 - m_4 - m_5 \tag{4.59}$$

Proposition 8 *For any BS polyhedron*

$$g(\beta,\beta) = M + g(\alpha,\alpha) - \alpha F_\alpha = M - \alpha F_\alpha + m_1 + m_4 + m_7 \tag{4.60}$$

Proof Starting with Eq. (4.18), one obtains on the one hand,

$$\beta M = \frac{6\beta g(\alpha,\alpha)}{\alpha} + 3\beta\left(\frac{1}{\alpha} + \frac{1}{\beta}\right)g(\alpha,\beta) + 6g(\beta,\beta) - 6\beta. \tag{4.61}$$

On the other hand, from Eq. (4.22)

$$6M = 6g(\alpha,\alpha) + 6g(\alpha,\beta) + 6g(\beta,\beta) \tag{4.62}$$

Now, by eliminating $g(\beta,\beta)$ from Eqs. (4.61) and (4.62) we have

$$g(\alpha,\beta) = \frac{\alpha}{3(\alpha-\beta)}\left\{(6-\beta)M - 6\beta - 6g(\alpha,\alpha)(1-\beta/\alpha)\right\}. \tag{4.63}$$

Since $M = (3/2)V$, now, using the relation between V and F_α represented by formula (4.46), Eq. (4.63) can be transformed into the following form:

$$\begin{aligned} m_2 + m_5 + m_8 = g(\alpha,\beta) &= \frac{\alpha(6-\beta)}{3(\beta-\alpha)}M + \frac{2\alpha\beta}{\beta-\alpha} - 2g(\alpha,\alpha) \\ &= \alpha F_\alpha - 2g(\alpha,\alpha) = \alpha F_\alpha - 2m_1 - 2m_4 - 2m_7 \end{aligned} \tag{4.64}$$

From this

$$m_8 = \alpha F_\alpha - 2m_1 - m_2 - 2m_4 - 2m_7 - m_5 \tag{4.65}$$

This implies that

$$\begin{aligned} m_3 + m_6 + m_9 = g(\beta,\beta) &= M - g(\alpha,\alpha) - g(\alpha,\beta) \\ &= M + g(\alpha,\alpha) - \alpha F_\alpha = M - \alpha F_\alpha + m_1 + m_4 + m_7 \end{aligned} \tag{4.66}$$

Proposition 9 *For any BS polyhedron*

$$m_5 = 2m_2 + 4m_3 - 2m_4 \tag{4.67}$$

Proof From Eqs. (4.53), and (4.65) one obtains

$$m_8 = \alpha F_\alpha - 2m_1 - 3m_2 - 4m_3 - 2m_7 \tag{4.68}$$

$$m_8 = \alpha F_\alpha - 2m_1 - m_2 - 2m_4 - 2m_7 - m_5 \tag{4.69}$$

From the formulas above it follows the identity (4.67).

Corollary 9.1 *From Eqs.* (4.59) *and* (4.67) *we have*

$$\begin{aligned} m_6 &= \alpha F_\alpha - 2m_1 - 2m_2 - 2m_3 - m_4 - m_5 \\ &= \alpha F_\alpha - 2m_1 - 4m_2 - 6m_3 + m_4 \end{aligned} \tag{4.70}$$

Now, we introduce two definitions concerning the notions of independent and maximal finite sets of edge parameters.

Definition 1 A subset S of edge parameters with $K \leq 9$ elements is called an independent subset if any m_j in S can not be expressed as a function of parameters m_k included in S, where $k = 1, 2, \ldots K$ and $k \neq j$. Consequently, this means that all elements of S are algebraically independent.

Definition 2 An independent subset S_T of edge parameters with K elements is called maximal if each edge parameter which is not included in S_T can be generated as a function of independent parameters in S_T. This means that for any $m_j \notin S_T$, $S_T \bigcup m_j$ is not an independent set.

Proposition 10 *Consider a set of BS polyhedra of type $S(\alpha, \beta)$. If α, β, M are fixed, then parameters (m_1, m_2, m_3, m_4 and m_7) represent an independent and maximal subset. This implies that knowing parameters α, β, M, m_1, m_2, m_3, m_4, m_7, from these data edge parameters m_5, m_6, m_8, and m_9 can be computed.*

Proof

(i) As we have shown, from edge parameters (m_1, m_2, m_3, m_4 and m_7) the remaining four parameters can be generated. Parameter m_5 can be calculated using the general formula (4.67) which is valid for any BS polyhedra, independently of α, β and M. Moreover, from Eqs. (4.53) and (4.70) it follows that

$$\begin{aligned} m_8 &= \alpha F_\alpha - 2m_1 - 3m_2 - 4m_3 - 2m_7 \\ &= \frac{\alpha(\beta - 6)}{3(\beta - \alpha)} M + \frac{2\alpha\beta}{\beta - \alpha} - 2m_1 - 3m_2 - 4m_3 - 2m_7 \end{aligned} \tag{4.71}$$

$$\begin{aligned} m_6 &= \alpha F_\alpha - 2m_1 - 4m_2 - 6m_3 + m_4 \\ &= \frac{\alpha(\beta-6)}{3(\beta-\alpha)}M + \frac{2\alpha\beta}{\beta-\alpha} - 2m_1 - 4m_2 - 6m_3 + m_4 \\ &= m_4 + 2m_7 + m_8 - m_2 - 2m_3 \end{aligned} \tag{4.72}$$

Finally, according to Eq. (4.56)

$$\begin{aligned} m_9 = \quad & M - \sum_{k=1}^{8} m_k = M - 2\alpha F_\alpha + 3m_1 + 4m_2 + 5m_3 + m_7 \\ &= M - \frac{2\alpha(\beta-6)}{3(\beta-\alpha)}M - \frac{4\alpha\beta}{\beta-\alpha} + 3m_1 + 4m_2 + 5m_3 + m_7 \end{aligned} \tag{4.73}$$

(ii) Now, we verify that the five edge parameters (m_1, m_2, m_3, m_4, m_7) form an algebraically independent and maximal set for BS polyhedra of type $S(\alpha, \beta)$ having a fixed edge number $M = 3V/2$.

To prove this, consider the three C_{40} fullerene isomers C40: 10, C40: 12 and C40: 13 whose Schlegel diagrams are illustrated in Fig. 4.5. Their edge-parameters are summarized in Table 4.1. As can be stated, in all cases, there are only two isomers having 4 identical edge parameters from set (m_1, m_2, m_3, m_4, m_7), and there exists a fifth one which is different. This means that the set of five independent edge parameters is maximal in the sense that their numbers ($K = 5$) can not be decreased (or increased).

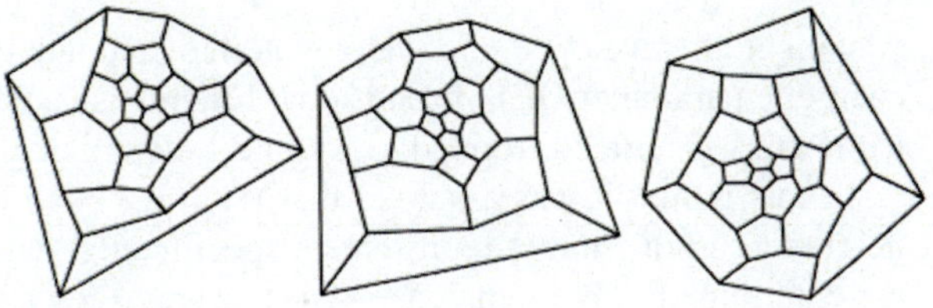

Fig. 4.5 Schlegel diagrams of three C_{40} isomers

Table 4.1 Topological parameters of five C_{40} fullerene isomers

Isomer	Topological parameters									
	m_1	m_2	m_3	m_4	m_5	m_6	m_7	m_8	m_9	Np
C40:10	1	7	5	8	18	8	5	7	1	13
C40:12	1	7	5	7	20	7	5	7	1	13
C40:13	1	7	5	8	18	8	4	9	0	13
C40:35	0	0	11	10	24	4	5	6	0	11
C40:36	0	0	11	10	24	4	5	6	0	11

Corollary 10.1 *For BS polyhedra of types S(α,6) where α =* 3, 4 and 5 we have

$$m_6 = \frac{12\alpha}{6-\alpha} - 2m_1 - 4m_2 - 6m_3 + m_4 \tag{4.74}$$

$$m_8 = \frac{12\alpha}{6-\alpha} - 2m_1 - 3m_2 - 4m_3 - 2m_7 \tag{4.75}$$

$$m_9 = M - \frac{24\alpha}{6-\alpha} + 3m_1 + 4m_2 + 5m_3 + m_7 \tag{4.76}$$

Corollary 10.2 *As a particular case, for traditional fullerene isomers* (*where* $\alpha = 5$ *and* $\beta = 6$) *we have*

$$m_5 = 2m_2 + 4m_3 - 2m_4 \tag{4.77}$$

$$m_6 = 60 - 2m_1 - 4m_2 - 6m_3 + m_4 \tag{4.78}$$

$$m_8 = 60 - 2m_1 - 3m_2 - 4m_3 - 2m_7 \tag{4.79}$$

$$m_9 = M - 120 + 3m_1 + 4m_2 + 5m_3 + m_7 \tag{4.80}$$

4.7 Application

4.7.1 Energetic Characterization of Fullerenes

Alcami et al. developed a model devoted to estimate the enthalpy of formation (the energetic parameter QE) of traditional fullerenes $C_k (k \leq 72)$ on the basis of 9 edge parameters generated from edge coronas (Alcami et al. 2007).

In this model it was assumed that (i) every edge (i.e. every bond between two neighbor carbon atoms) represents a specific edge-energy value, (ii) edge energies are determined only by the edge-types, more exactly, by the local configurations of pentagons and hexagons occurring in edge coronas, (iii) QE can be estimated as a weighted linear function of edge-parameters ($m_1, m_2, \ldots, m_9$), where the positive weights are identical to the specific edge-energy values $\varepsilon_j (1 \leq j \leq 9)$ belonging to the 9 distinct edge-coronas (See Fig. 4.4).

From the model outlined it follows that QE can be calculated as

$$QE = \varepsilon_1 m_1 + \varepsilon_2 m_2 + \ldots + \varepsilon_9 m_9 \tag{4.81}$$

Specific edge–energy values ε_j given in Alcami et al. (2007) are as follows: $\varepsilon_1 = 19.8$, $\varepsilon_2 = 17.6$, $\varepsilon_3 = 10.3$, $\varepsilon_4 = 15.7$, $\varepsilon_5 = 12.4$, $\varepsilon_6 = 7.8$, $\varepsilon_7 = 6.2$, $\varepsilon_8 = 4.7$ and $\varepsilon_9 = 1.7$.

By using formulas (4.77), (4.78), (4.79), and (4.80) linear interdependencies can be found between these nine parameters, and QE can be calculated by the following simplified equation:

$$QE = \mu_{01} + \mu_{02}M + \mu_1 m_1 + \mu_2 m_2 + \mu_3 m_3 + \mu_4 m_4 + \mu_5 m_7 \tag{4.82}$$

where

$$\mu_{01} = 60\varepsilon_6 + 60\varepsilon_8 - 120\varepsilon_9 = 546 \tag{4.83}$$

$$\mu_{02} = \varepsilon_9 = 1.7 \tag{4.84}$$

$$\mu_1 = \varepsilon_1 - 2\varepsilon_6 - 2\varepsilon_8 + 3\varepsilon_9 = -0.1 \tag{4.85}$$

$$\mu_2 = \varepsilon_2 + 2\varepsilon_5 - 4\varepsilon_6 - 3\varepsilon_8 + 4\varepsilon_9 = 3.9 \tag{4.86}$$

$$\mu_3 = \varepsilon_3 + 4\varepsilon_5 - 6\varepsilon_6 - 4\varepsilon_8 + 5\varepsilon_5 = 2.8 \tag{4.87}$$

$$\mu_4 = \varepsilon_4 - 2\varepsilon_5 + \varepsilon_6 = -1.3 \tag{4.88}$$

$$\mu_5 = \varepsilon_7 - 2\varepsilon_8 + \varepsilon_9 = -1.5 \tag{4.89}$$

We can conclude that for C_k fullerene isomers the energetic parameter QE can be directly calculated by a multi-linear function of 6 variables instead of 9 ones. If the number of edges is fixed by M (for example, in the case of C_{60} isomers, where $M = 90$) the number of independent variables will be equal to 5.

Assuming that values of QE are known for a given set of fullerene isomers, the coefficients of Eq. (4.82) can be easily estimated by means of a multi-linear regression analysis. Consequently, Eq. (4.82) can be used to predict the stability of fullerene isomers on the basis of edge parameters m_k.

It should be noted that there are some limitations concerning the discriminating power of edge parameters. In practice, this means that there exist fullerene isomers for which the set of edge parameters is identical, although their combinatorial structures are different. For example, among the forty C_{40} isomers, fullerenes isomer pairs C40:35 and C40:36 have identical edge parameters. See Table 4.1. A similar problem arises for the combinatorial characterization of so-called IPR fullerenes where all pentagons are isolated (Fowler and Manolopoulos 1995).

4.7.2 A Case Study Concerning the Classification of C_{40} Isomers

We investigated and compared the discriminating performances of four different types of topological descriptors (invariants) designated to classify fullerene isomers into a finite number of equivalence classes. For this comparative test the set of C_{40} isomers has been selected. It is interesting to note that a special property of the forty C_{40} isomers is that equalities $Np = e(5,5) = e(6,6)$ and $g(5,5) = g(6,6)$ hold for them.

To rank the selectivity of various topological descriptors quantitatively, we introduced a discriminating index I_D. By definition, $I_D = 100^*N_C/N_t$ where N_C and N_t are the number of equivalence classes and the total number of fullerene isomers, respectively. (In this case, $N_t = 40$). Using Density Functional Tight-Binding

Table 4.2 Topological parameters of forty C_{40} isomers

	Topological parameters									
Isomer	m_1	m_2	m_3	m_4	m_7	Np	V_{555}	V_{556}	Ω	Energy, (eV)
C40:38	0	0	10	8	10	10	0	20	2.727	–342,031
C40:39	0	0	10	10	10	10	0	20	2.727	–341,631
C40:31	0	6	5	5	11	11	2	16	2.500	–341,438
C40:29	0	6	5	6	11	11	2	16	2.500	–341,345
C40:26	0	3	8	8	8	11	1	19	2.250	–341,094
C40:24	0	3	8	9	8	11	1	19	2.250	–341,022
C40:37	0	0	11	10	6	11	0	22	2.083	–340,636
C40:40	0	12	0	0	12	12	4	12	2.308	–340,580
C40:14	1	7	4	5	9	12	3	15	2.077	–340,476
C40:36	0	0	11	10	5	11	0	22	2.000	–340,431
C40:30	0	9	3	6	9	12	3	15	2.077	–340,304
C40:25	0	6	6	8	7	12	2	18	1.923	–340,277
C40:22	0	6	6	9	6	12	2	18	1.846	–340,230
C40:35	0	0	11	10	5	11	0	22	2.000	–340,196
C40:21	0	6	6	10	7	12	2	18	1.923	–340,151
C40:27	0	6	6	8	6	12	2	18	1.846	–340,126
C40:15	1	4	7	6	6	12	2	18	1.846	–339,943
C40:17	1	10	2	3	7	13	4	14	1.714	–339,884
C40:34	0	3	9	10	4	12	1	21	1.692	–339,827
C40:28	0	6	6	9	7	12	2	18	1.923	–339,777
C40:16	2	8	3	2	7	13	4	14	1.714	–339,645
C40:20	0	3	9	12	3	12	1	21	1.615	–339,627
C40:9	2	8	3	4	8	13	4	14	1.786	–339,614
C40:10	1	7	5	8	5	13	3	17	1.571	–339,558
C40:12	1	7	5	7	5	13	3	17	1.571	–339,370
C40:13	1	7	5	8	4	13	3	17	1.500	–339,347
C40:19	1	10	2	4	7	13	4	14	1.714	–339,292
C40:23	0	6	7	12	3	13	2	20	1.429	–338,690
C40:6	2	8	4	7	3	14	4	16	1.267	–338,624
C40:18	1	10	3	6	4	14	4	16	1.333	–338,341
C40:5	3	9	2	5	7	14	5	13	1.533	–338,332
C40:32	0	12	2	8	2	14	4	16	1.200	–338,270
C40:8	4	10	1	2	4	15	6	12	1.188	–338,113
C40:33	0	12	2	8	4	14	4	16	1.333	–337,922
C40:4	3	9	3	6	3	15	5	15	1.125	–337,348
C40:7	2	11	2	6	3	15	5	15	1.125	–337,330
C40:11	2	8	5	8	1	15	4	18	1.000	–336,642
C40:2	4	10	2	6	2	16	6	14	0.941	–336,489
C40:3	6	12	0	4	0	18	8	12	0.632	–335,193
C40:1	10	10	0	0	0	20	10	10	0.476	–333,806

(DFTB) method (Porezag et al. 1995) we calculated the total energy values Q_C characterizing the relative stability of isomers. These energies and the corresponding topological parameters are summarized in Table 4.2. From the comparative investigation the following conclusions can be drawn:

(a) By using the traditional pentagon adjacency index Np, ($Np = m_1 + m_2 + m_3 = (3V_{555} + V_{556})/2$) it is possible to classify the 40 isomers into 9 subclasses, consequently the discriminating performance is $I_D = 100*9/40 = 22.5\%$.
(b) Performing the classification by two independent vertex numbers (V_{555}, V_{556}) where $V_{555} = (2m_1+m_2)/3$ and $V_{556} = m_2+2m_3$, we have $I_D = 100*19/40 = 47.5\%$.
(c) The third topological descriptor Ω was defined as

$$\Omega = \frac{31 + m_7}{1 + Np} - 1 = \frac{31 + m_7}{1 + m_1 + m_2 + m_3} - 1 \qquad (4.90)$$

Using parameter Ω we obtain $I_D = 100*24/40 = 60.0\%$. As shown in Table 4.2, Ω correlates highly with the calculated total energy values Q_C characterizing the relative stability of isomers. Additionally, it is worth noting that for the topological parameter Ω the inequality $0 \le \Omega \le 60$ holds. Since $0 \le Np \le 30$ and $0 \le m_7 \le 30$, this implies that $\Omega = 0$ for fullerene C_{20} (dodecahedron) and $\Omega = 60$ for the Buckminster fullerene, only.
(d) Finally, we have the best result when the vector $\nu = [m_1, m_2, m_3, m_4, m_7]$ of five independent edge-parameters was selected for classification purposes. In this case, $I_D = 100*39/40 = 97.5\%$ yields.

4.8 Final Remarks, Conclusions

We have developed general relations between local combinatorial parameters of fullerenes represented by bifaced polyhedral graphs and we obtained the following properties.

(i) Edge-parameters m_k can be easily calculated using a simple computer program designated to the structural analysis of Schlegel diagrams of S(5,6) polyhedra.
(ii) The 5 component vector $\nu = [m_1, m_2, m_3, m_4, m_7]$ of edge-parameters offers a better discrimination than the single-valued graph invariants (Np, Ω) when partitioning the fullerene isomers.
(iii) The structural similarity (or dissimilarity) of fullerene isomers can be quantitatively measured by introducing a distance function $d(\nu_A, \nu_B)$ where ν_A and ν_B stand for the corresponding edge-parameter vectors of isomers H_A and H_B, respectively. (It follows that fullerene isomers H_A and H_B belong to the same class of equivalence if and only if $d(\nu_A, \nu_B) = 0$.)

(iv) Because the 5 edge parameters included in vector ν are algebraically independent and have a high discriminating performance, it is expected that a function of 5 variables can be constructed by which the relative stability of fullerene isomers can be more efficiently predicted.

Acknowledgements I. L. thanks for the support of the Hungarian state grant OTKA K73776.

References

Albertazzi E, Domene C, Fowler PW, Heine T, Seifert G, Van Alsenoy C, Zerbetto F (1999) Phys Chem Chem Phys 1:2913
Alcami M, Sanchez G, Diaz-Tendero S, Wang Y, Martin F (2007) J Nanosci Nanotechnol 7:1329
Austin SJ, Fowler PW, Manolopoulos DE, Zerbetto F (1995) Phys Lett 235:146
Babić D, Klein DJ, Sah CH (1993) Chem Phys Lett 211:235
Balaban AT, Liu X, Klein DJ, Babić D, Schmalz TG, Seitz WA, Randić M (1995) J Chem Comput Sci 35:396
Brehm U, Schulte E (2004) In: Goodman JE, O' Rourke J (eds) Handbook of discrete and computational geometry, 2nd edn. CRC Press, Boca Raton, FL, p 431
Brinkmann G, Deza M (2000) J Chem Inf Comput Sci 40:530
Campbell EEB, Fowler PW, Mitchell D, Zerbetto F (1996) Chem Phys Lett 250:544
Cioslowski J, Rao N, Moncrieff D (2002) J Am Chem Soc 122:8265
Delgado-Friedrichs O, Deza M (2000) DIMACS Series Discrete Math Theor Comp Sci 51:97
Deza M (2000) Face-regular polyhedra and tilings with two combinatorial types of faces. In: Proceedings of a conference honoring Professor Dijen K. Ray-Chaudhuri on the occasion of his 65th birthday. The Ohio State University, Columbus, OH, p 49
Deza A, Deza M, Grishukhin V (1998) Discrete Math 192:41
Deza M, Dutour M (2005) Combin Probab Comput 14:31
Deza M, Dutour-Sikiric M, Fowler PW (2009) MATCH Commun Math Comput Chem 61:589
Deza M, Fowler PW, Rassat A, Rogers KM (2000) J Chem Phys 40:550
Deza M, Grishukhin V (2001) J Stat Plann Inference 95:175
Deza M, Grishukhin V (2002) Bull Inst Combin Appl 34:99
Došlić T (2002a) J Math Chem 31:187
Došlić T (2002b) Croat Chem Acta 75:869
Došlić T (2007) J Math Chem 42:183
Došlić T (2008) J Math Chem 43:647
Dress WM, Brinkmann G (1996) MATCH Commun Math Comput Chem 33:87
Dutour-Sikiric M, Deza M, Shtogrin M (2008) Discrete Appl Math 156:1518
Fajtlowitz S, Larson CE (2003) Chem Phys Lett 377:485
Fowler PW (2002) Croat Chem Acta 75:401
Fowler PW (2003) MATCH Commun Math Comput Chem 48:87
Fowler PW, Heine T (2001) J Chem Soc Perkin Trans 2:487
Fowler PW, Heine T, Manolopoulos D, Mitchell D, Orlandi G, Schmidt R, Seifert G, Zerbetto F (2001) J Phys Chem 100:6984
Fowler PW, Manolopoulos DE (1995) An atlas of fullerenes. Calendron Press, Oxford
Fowler PW, Heine T, Mitchell D, Orlandi G, Schmidt R, Seifert G, Zerbetto F (1996) J Chem Soc Faraday Trans 62:2203
Gan LH, Liu J, Hui Q, Shao SQ, Liu ZH (2009) Chem Phys Lett 472:224
Goldberg M (1937) Tohoku Math J 43:104
Graver JE, (2005) DIMACS Series Discrete Math Theor Comp Sci 69:167
Grünbaum B (1967) Convex polytopes. Wiley, Sydney

Grünbaum B, Zachs J (1974) Discrete Math 10:93
Hu QN, Liang ZZ, Fang KT (2003) J Data Sci 1:361
Jendrol' S (1977) Geom Dedicata 68:91
Jendrol' S (1999) Discrete Math 196:177
Jendrol' S, Kardoš F (2007) Discrete Appl Math 155:2181
Jendrol' S, Skupien Z (2001) Discrete Math 236:167
Jendrol' S, Trenkler M (2001) J Math Chem 29:235
Jucovič E (1974) Geom Dedicata 3:233
Kardoš F (2007) J Math Chem 41:101
Laszlo I, Rassat A (2001) Int J Quant Chem 84:136
Porezag D, Frauenheim T, Köhler T, Seifert G, Kashner R (1995) Phys Rev B 51:12947
Reti T, Bitay E (2007) Mater Sci Forum 537–538:439
Torrens F (2002) Internet Electron J Mol Des 1:351
Zhang H, Zhang F (2001) J Math Chem 30:343

Chapter 5
Computation of Some Topological Indices of C_{60} and C_{80} Fullerenes by GAP Program

Ali Iranmanesh

Abstract In this chapter, we give a GAP program for computing the Schultz index, Modified Schultz index, Wiener index, hyper Wiener index, Balaban index and Zagreb indices for any graph and by this algorithm we compute the Schultz polynomial and Schultz index of C_{60} and C_{80} fullerene by GAP program. Also we compute the Wiener index, hyper Wiener index and Wiener polynomial of C_{80} fullerene by this program and finally we compute Balaban and Zagreb indices for IPR C_{80} fullerene isomers by GAP program.

5.1 Introduction

One of the main distinctive characteristics of modern chemistry is the use of theoretical tools for the molecular modeling of physicochemical processes, chemical reactions, medicinal and toxicological events, etc., in which chemicals are involved. The success of the molecular modeling is judged by the insights that it offers on the nature of the processes studied, which permit better comprehension and a rational modification of them. These properties, measured experimentally, are almost invariably expressed in quantitative terms, think for instance of boiling point, refraction index, transition state energy, percentage of inhibition of some enzymatic activity, lethal dose, and so forth. The paradigm for the modeling of such properties is the relationship that exists between them and the molecular structure of chemical. This fact presupposes for the first challenge in the molecular modeling: the properties are expressed as numbers while the molecular structure is not. The way to solve this problem is by using molecular descriptors that are numbers representing information about different molecular features, to describe quantitatively the properties under study. These models are known as quantitative structure-property (QSPR) and quantitative structure-activity relationships (QSAR), depending on the physicochemical or biological nature of the properties studied, respectively.

A. Iranmanesh (✉)
Department of Mathematics, Tarbiat Modares University, P.O. Box: 14115-137, Tehran, Iran
e-mail: iranmanesh@modares.ac.ir

F. Cataldo et al. (eds.), *The Mathematics and Topology of Fullerenes*,
Carbon Materials: Chemistry and Physics 4, DOI 10.1007/978-94-007-0221-9_5,

Topological indices are numerical descriptors derived from the associate graphs of chemical compounds. Some indices based on the distances in graph are widely used in establishing relationships between the structure of molecules and their physico-chemical properties. Usage of topological indices in chemistry began in 1947 when the chemist Harold Wiener introduced Wiener index to demonstrate correlations between physicochemical properties of organic compounds and the index of their molecular graphs (Wiener 1947). Wiener originally defined his index (W) on trees and studied its use for correlations of physicochemical properties of alkanes, alcohols, amines and analogous compounds (Khadikar and Karmarkar 2002). Starting from the middle of the 1970s, the Wiener index gained much popularity and, since then, new results related to it are constantly being reported. For a review, historical details and further bibliography on the chemical applications of the Wiener index see (Gutman 1994; Gutman and Potgieter 1997; Nikolić et al. 1995).

Let G be a connected graph. The vertex-set and edge-set of *G* denoted by *V(G)* and *E(G)* respectively. The distance between the vertices u and v, $d(u, v)$, in a graph is the number of edges in a shortest path connecting them. Two graph vertices are adjacent if they are joined by a graph edge. The degree of a vertex $i \in V(G)$ is the number of vertices joining to i and denoted by δ_i.

The Wiener index of G is

$$W(G) = \frac{1}{2} \sum_{\{u,v\} \subseteq V(G)} d(u,v) \tag{5.1}$$

The Wiener polynomial of G is

$$W(G,x) = \frac{1}{2} \sum_{\{u,v\} \subseteq V(G)} x^{d(u,v)} \tag{5.2}$$

The Hyper Wiener index of G is

$$WW(G) = \frac{1}{2} \sum_{\{u,v\} \subseteq V(G)} d(u,v) + d(u,v)^2 \tag{5.3}$$

Wiener polynomial was first introduced by Hosoya (1988). Some authors call these Polynomials Hosoya's polynomials as an honor of Haruo Hosoya. Many papers have been devoted to compute the Wiener polynomial for different types of graphs. More information can be found in Gutman (1999), Lepovic and Gutman (1998), and Sagan et al. (1996).

Observe that the degree of the Wiener polynomial is equal to the diameter of G. Also, notice that

$$W(G) = W'(G,1), \qquad WW(G) = W'(G,1) + W''(G,1) \tag{5.4}$$

The Balaban index of a molecular graph G is defined by Balaban (1982, 1983). The Balaban index of a graph G is denoted by J(G) and defined as $J(G) =$

$\frac{m}{\mu+1} \sum_{ij \in E(G)} \frac{1}{\sqrt{d(i)d(j)}}$, where m is the number of edges of G and $\mu(G)$ is the cyclomatic number of G. Note that the cyclomatic number is the minimum number of edges that must be removed from G in order to transform it to an acyclic graph; it can be calculated using $\mu(G) = m - n + 1$ where n is the number of vertices and $d(i)$ is the sum of distances between vertex i and all other vertices of G, and the summation goes over all edges from the edge set $E(G)$. The Balaban index appears to be a very useful molecular descriptor with attractive properties (Diudea et al. 2006; Trinajstic 1992). The connectivity indices are extensively used as molecular descriptors in predicting the retention indices in chromatographic analysis of various isomeric aliphatic, aromatic and polycyclic hydrocarbons (Ehrman et al. 1981; Ringo 1996; Spieth and Ringo 1983).

In a series of papers, Balaban index of some nanotubes are computed (Iranmanesh and Ashrafi 2007; Yousefi-Azari et al. 2008; Zhou and Trinajsti 2008).

Another topological index is Schultz index. This index was introduced by Harry Schultz in 1989 (Schultz 1989). This molecular topological index studied in many papers (Dobrynin 1999; Gutman 1994; Schultz 2000).

The Schultz index is defined as:

$$S(G) = \sum_{\{u,v\} \subseteq V(G)} (\delta_u + \delta_v)\, d(u,v)$$

Klavžar and Gutman (1997) defined the Modified Schultz index as:

$$MS(G) = \sum_{\{u,v\} \subseteq V(G)} (\delta_u \delta_v)\, d(u,v)$$

In Hosoya (1988), Haruo Hosoya used polynomials to generate distance distributions for graphs.

The Schultz polynomial of G is:

$$H_1(G,x) = \sum_{\{u,v\} \subseteq V(G)} (\delta_u + \delta_v)\, x^{d(u,v)}$$

Also the modified Schultz polynomial of G is defined as:

$$H_2(G,x) = \sum_{\{u,v\} \subseteq V(G)} (\delta_u \delta_v)\, x^{d(u,v)}$$

Observe that the degree of the Schultz polynomial and the modified Schultz polynomial is equal to the diameter of G. Also, notice that

$$H_1'(G,x) = S(G) \tag{5.5}$$

$$H_2'(G,x) = MS(G)$$

In a series of papers, the Schultz and the modified Schultz indices for some fullerene and nanotubes are computed (Alizadeh et al. 2009; Heydari and Taeri 2007a, b; Iranmanesh and Ashrafi 2007; Iranmanesh et al. 2009).

Another topological indices are Zagreb indices. These indices have been introduced more than 30 years ago by Gutman and Trinajestic (1972). They are defined as:

$$M_1(G) = \sum_{v \in V(G)} \deg(v)^2,$$
$$M_2(G) = \sum_{uv \in E(G)} \deg(u)\deg(v).$$

We refer the reader to consult (Braun et al. 2005; Gutman and Das 2004; Nikolic et al. 2003; Zhou 2004; Zhou and Gutman 2004, 2005) for historical background, computational techniques and mathematical properties of Zagreb indices. In a series of papers, these topological indices are computed (Behtoei et al. 2009; Khalifeh et al. 2009; Sun and Chen 2009; Taherkhani et al. 2009).

GAP stands for Groups, Algorithms and Programming (Schonert et al. 1992). The name was chosen to reflect the aim of the system, which is theoretical software for solving computational problems in group theory. The last years, a rapid spread of interest in the understanding, design and even implementation of group theoretical algorithms was recorded. GAP software was built by GAP's team in Aachen. We encourage the reader to consult (Dabirian and Iranmanesh 2005; Trinajstic 1992) for background materials and computational techniques related to applications of GAP in solving some problems in chemistry and biology.

In this chapter, at first, we give an algorithm for computing the Schultz and modified Schultz polynomials for any graph and by this algorithm we compute the Schultz polynomial and Schultz index of C_{60} Fullerene by GAP program. Then we compute the Schultz and the modified Schultz polynomials of C_{80} fullerene by GAP program. In continue, we give a GAP program for computing the Wiener polynomial of any graph. Also we compute the Wiener index, hyper Wiener index and Wiener polynomial of C_{80} fullerene by this program. In Section 5.4, we give an algorithm for computing the Balaban index of any simple connected graph. Also we compute this index for IPR C_{80} fullerene isomers by GAP program. In last section, we give an algorithm that enables us to compute the Zagreb indices of any graph. Also by this algorithm, we compute the Zagreb indices for C_{80} fullerene.

5.2 Computing Schultz Polynomial, Schultz Index of C_{60} Fullerene by GAP Program

According to the Eq. (5.5), we can obtain the Schultz index of the graph by the Schultz polynomial.

In this section, we give an algorithm for obtaining Schultz and modified Schultz indices for any graph. For this purpose, the following algorithm is presented:

At first, we label the vertices of the graph, then we determine all of adjacent vertices set of the vertex u, $u \in V(G)$, and this set is denoted by $N(u)$.

The set of vertices that their distance to vertex u is equal to t, ($t \geq 0$), is denoted by $D_t(u)$ (or $D_{u,t}$).

- $D_{u,t} = D_t(u) = \{v \mid d(u,v) = t\}$
- $D_0(u) = \{u\}$
- $D_1(u) = N(u)$

The distance between each vertex of set $N(v) \backslash (D_t(u) \cup D_{t-1}(u))$ and the vertex u is equal to t + 1, thus we have:

$$D_{i,t+1} = \cup_{j \in D_{i,t}} (N(j) \backslash (D_{i,t} \cup D_{i,t-1}), \ \ t \geq 1.$$

We have the following relations:

- $V(G) = \bigcup_{t \geq 0} D_t(u), \qquad \forall u \in V(G)$
- $$H_1(G,x) = \sum_{\{u,v\} \subseteq V(G)} (\delta_u + \delta_v) \, x^{d(u,v)} = \frac{1}{2} \sum_{i=1}^{n} \sum_{i \neq j=1}^{n} (\delta_i + \delta_j) \, x^{d(i,j)} = \frac{1}{2} \sum_{i=1}^{n} \sum_{j \in D_t(i)} (\delta_i + \delta_j) \, x^{t}$$
- $$H_2(G,x) = \sum_{\{u,v\} \subseteq V(G)} (\delta_u \delta_v) \, x^{d(u,v)} = \frac{1}{2} \sum_{i=1}^{n} \sum_{i \neq j=1}^{n} (\delta_i \delta_j) \, x^{d(i,j)} = \frac{1}{2} \sum_{i=1}^{n} \sum_{j \in D_t(i)} (\delta_i \delta_j) \, x^{t}$$

5.2.1 Schultz Polynomial and Schultz Index of C_{60} Fullerene

In this part,we compute the Schultz polynomial and Schultz index of C_{60} fullerene by GAP program. Fullerenes are cage-like molecules formed as a twenty-sided geometric shape. In order to name fullerenes, letter C is followed by the number of carbon atoms existent in the networks of fullerenes. (e.g. C_{60}). Fullerenes consist of the networks of pentagons and hexagons. To be a closed shape, a fullerene should

Fig. 5.1 C_{60} fullerene

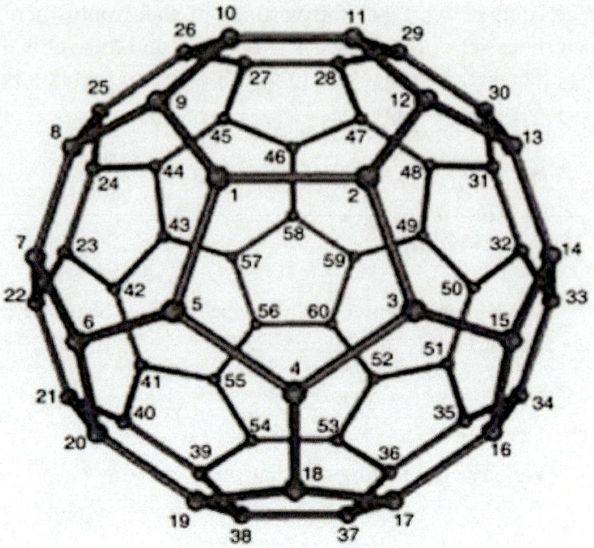

Table 5.1 The schultz polynomial and schultz index of C_{60} fullerene

$H(C_{60}, x)$	$540x + 1080x^2 + 1440x^3 + 1800x^4 + 1800x^5 + 1800x^6 + 1440x^7 + 540x^8 + 180x^9$
$SG(C_{60})$	50040

exactly have 12 pentagon sides, but the number of hexagon sides can be extremely variable. Fullerenes were discovered in 1985 by Robert Curl, Harold Kroto and Richard Smalley at the University of Sussex and Rice University, and are named after Richard Buckminster Fuller (Fig. 5.1).

The following program computes the Schultz polynomial's coefficients of C_{60} fullerene (Table 5.1).

```
n:=60; k:=[];N:=[];
k[1]:=[1..5]; k[2]:=[6..20]; k[3]:=[21..40];
k[4]:=[41..55]; k[5]:=[56..60];
for i in [1..5] do
y:=Size(k[i]);
for j in [1..y] do
x:=k[i][j];
N[x]:=[x-1,x+1];
od;
od;
D1:=[9,12,15,18];
for i in [1..4] do
x:=D1[i]; N[i][3]:=x; N[x][3]:=i;
```

```
od;
D2:=Difference(k[2],Filtered(k[2],i->(i mod 3)=0));
D3:=Filtered(k[3],i->(i mod 4) in [1,2]);
for i in [1..9] do
x:=D2[i]; N[D3[i+1]][3]:=x; N[x][3]:=D3[i+1];
od;
D4:=Difference(k[3],D3);
D5:=Filtered(k[4],i->(i mod 3)<>1);
for i in [1..9] do
x:=D4[i]; N[D5[i+1]][3]:=x; N[x][3]:=D5[i+1];
od;
D6:=Difference(k[4],D5);
for i in [1..4] do
x:=D6[i]; N[k[5][i+1]][3]:=x; N[x][3]:=k[5][i+1];
od;
N[1]:=[2,5,9];N[5]:=[1,4,6];N[6]:=[5,7,20];N[20]:=[19,21,6];
N[21]:=[20,22,40]; N[40]:=[39,41,21]; N[41]:=[40,42,55];
N[55]:=[54,56,41]; N[56]:=[55,57,60];N[60]:=[52,56,59];
md:=1; v:=[]; D:=[];
for i in [1..n] do
D[i]:=[]; u:=[i]; D[i][1]:=N[i]; v[i]:=Size(N[i]);
u:=Union(u,D[i][1]); r:=1; t:=1;
while r<>0 do
D[i][t+1]:=[];
for j in D[i][t] do
for m in Difference (N[j],u) do
AddSet(D[i][t+1],m);
od; od;
u:=Union(u,D[i][t+1]);
if D[i][t+1]=[] then r:=0;fi;
t:=t+1; od;
md:=MaximumList([md,Size(D[i])]);
od;
p:=[];
for i in [1..md] do p[i]:=0; od;
for t in [1..md] do
for i in [1..n] do
x:=0;
for j in D[i][t] do
x:=x+Size(N[j]); od;
p[t]:=p[t]+((Size(D[i][t]))*v[i]+x);
od;od;
p:=p/2;#(p is the set the Schultz polynomial's coefficients
of C60 fullerene)
```

5.2.2 Computing Schultz and the Modified Schultz Polynomials of C_{80} Fullerene

In this part, we compute the Schultz and the modified Schultz polynomials of C_{80} fullerene by GAP program. There are 120 edges in the molecular graph of C_{80} fullerene (Fig. 5.2). There are seven IPR-satisfying isomers for C_{80}. For these seven isomers, the Schultz and the modified Schultz polynomials are computed by the following GAP program in Table 5.2.

The following program computes the Schultz and the modified Schultz polynomials of C_{80} fullerene:

```
n:=80; k:=[]; N:=[];
k[1]:=[1..5]; k[2]:=[6..20]; k[3]:=[21..40];
k[4]:=[41..60]; k[5]:=[61..75]; k[6]:=[76..80];
for i in [1..6] do
y:=Size(k[i]);
for j in [1..y] do
x:=k[i][j];
N[x]:=[x-1,x+1];
od;od;
D1:=[9,12,15,18];
for i in [1..4] do
x:=D1[i]; N[i][3]:=x; N[x][3]:=i;
od;
D2:=Difference(k[2],Filtered(k[2],i->(i mod 3)=0));
D3:=Filtered(k[3],i->(i mod 2)=1);
for i in [1..9] do
x:=D3[i+1]; N[D2[i]][3]:=x; N[x][3]:=D2[i];
od;
D4:=Difference(k[3],D3); D5:= Filtered(k[4],i->i mod 2=1);
for i in [1..9] do
x:=D5[i+1]; N[D4[i]][3]:=x; N[x][3]:=D4[i];
od;
D6:=Difference(k[4],D5);
D7:= Filtered(k[5],i->(i mod 3)<>2);
for i in [1..9] do
x:=D7[i+1]; N[D6[i]][3]:=x; N[x][3]:=D6[i];
od;
D8:=Difference(k[5],D7);
for i in [1..4] do
x:=k[6][i+1]; N[D8[i]][3]:=x; N[x][3]:=D8[i];
od;
```

```
N[1]:=[2,5,9]; N[5]:=[1,4,6]; N[6]:=[5,7,20];
N[20]:=[6,19,21]; N[21]:=[20,22,40];
N[40]:=[21,39,41]; N[41]:=[40,42,60]; N[60]:=[61,59,41];
N[61]:=[60,62,75];
N[74]:=[73,75,76]; N[75]:=[58,61,74]; N[76]:=[74,77,80];
N[80]:=[71,76,79];
v:=[]; D:=[]; md:=1;
for i in [1..n] do
D[i]:=[]; u:=[i]; D[i][1]:=N[i]; v[i]:=Size(N[i]);
u:=Union(u,D[i][1]);
r:=1; t:=1;
while r<>0 do
D[i][t+1]:=[];
for j in D[i][t] do
for m in Difference (N[j],u) do
AddSet(D[i][t+1],m);
od; od;
u:=Union(u,D[i][t+1]);
if D[i][t+1]=[] then r:=0;fi;
t:=t+1;
od;
md:=MaximumList([md,Size(D[i]-1)]);
od;
C:=[];MC:=[];
for k in [1..md] do
C[k]:=0; MC[k]:=0;
od;
Deg:=[];
for i in [1..n] do
Deg[i]:=Size(N[i]);
od;
for i in [1..n] do
for t in [1..Size(D[i])-1] do
for j in D[i][t] do
C[t]:=C[t]+(Deg[i]+Deg[j]);
MC[t]:=MC[t]+(Deg[i]*Deg[j]);
od;od;od;
C:=C/2; #(C is the Schultz polynomial's coefficients of C80
Fullerene)
MC:=MC/2;#(MC is the modified Schultz polynomial's
coefficients of C80 Fullerene)
```

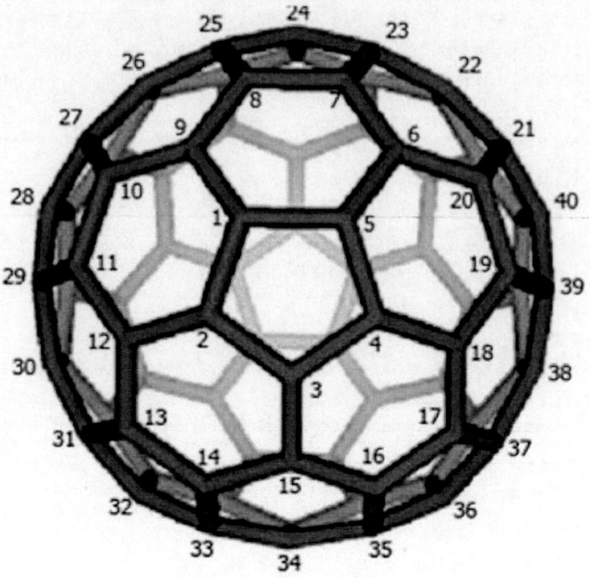

Fig. 5.2 C_{80} fullerene

Table 5.2 Schultz and modified Schultz polynomials of C_{80} fullerene's isomers

Isomers of C_{80} fullerene	Schultz polynomials	Modified Schultz polynomials
N0.1-D5d	$720x + 1440x^2 + 1980x^3 + 2520x^4 + 2700x^5 + 2520x^6 + 1980x^7 + 1440x^8 + 720x^9 + 240x^{10}$	$1080x + 2160x^2 + 2970x^3 + 3780x^4 + 4050x^5 + 4050x^6 + 3780x^7 + 2970x^8 + 2160x^9 + 1080x^{10} + 360x^{11}$
No.2-D2	$720x + 1440x^2 + 1980x^3 + 2520x^4 + 2700x^5 + 2880x^6 + 2640x^7 + 2100x^8 + 1500x^9 + 420x^{10} + 60x^{11}$	$1080x + 2160x^2 + 9720x^3 + 3780x^4 + 4050x^5 + 4320x^6 + 3960x^7 + 3150x^8 + 2250x^9 + 630x^{10} + 90x^{11}$
No.3-C2v	$720x + 1440x^2 + 1980x^3 + 2520x^4 + 2700x^5 + 2844x^6 + 2610x^7 + 2106x^8 + 1494x^9 + 546x^{10}$	$1080x + 2160x^2 + 2970x^3 + 3780x^4 + 4050x^5 + 4266x^6 + 3915x^7 + 3159x^8 + 2241x^9 + 819x^{10}$
No.4-D3	$720x + 1440x^2 + 1980x^3 + 2520x^4 + 2700x^5 + 2796x^6 + 2574x^7 + 2052x^8 + 1470x^9 + 624x^{10} + 84x^{11}$	$1080x + 2160x^2 + 2970x^3 + 3780x^4 + 4050x^5 + 4194x^6 + 3861x^7 + 3078x^8 + 2205x^9 + 936x^{10} + 126x^{11}$
No.5-C2v	$720x + 1440x^2 + 1980x^3 + 2520x^4 + 2700x^5 + 2868x^6 + 2628x^7 + 2088x^8 + 1488x^9 + 516x^{10} + 12x^{11}$	$1080x + 2160x^2 + 2970x^3 + 3780x^4 + 4050x^5 + 4302x^6 + 3942x^7 + 3132x^8 + 2232x^9 + 774x^{10} + 18x^{11}$
No.6-D5h	$720x + 1440x^2 + 1980x^3 + 2520x^4 + 2700x^5 + 2832x^6 + 2598x^7 + 2058x^8 + 1470x^9 + 582x^{10} + 60x^{11}$	$1080x + 2160x^2 + 2970x^3 + 3780x^4 + 4050x^5 + 4248x^6 + 3897x^7 + 3087x^8 + 2205x^9 + 873x^{10} + 90x^{11}$
No.7-lh	$720x + 1440x^2 + 1980x^3 + 2520x^4 + 2700x^5 + 2760x^6 + 2550x^7 + 2040x^8 + 1470x^9 + 660x^{10} + 120x^{11}$	$1080x + 2160x^2 + 2970x^3 + 3780x^4 + 4050x^5 + 4140x^6 + 3825x^7 + 3060x^8 + 2205x^9 + 990x^{10} + 180x^{11}$

5.3 Computing Wiener Polynomial, Wiener Index and Hyper Wiener Index of C_{80} Fullerene by GAP Program

In this section, we give a GAP program for computing the Wiener polynomial of any graph. Also we compute the Wiener index, hyper Wiener index and Wiener polynomial of C_{80} fullerene by this program.

In Section 5.2.1, we introduce some notations. According to these notations and relation (5.1),

we have the following relations:

$$\sum_{v \in V(G)} d(u, v) = \sum_{t \geq 1} t \times |D_{u,t}|, \quad \forall u \in V(G)$$

$$W(G, x) = \frac{1}{2} \sum_{\{u,v\} \subseteq V(G)} x^{d(u,v)} = \frac{1}{2} \sum_{u \in V(G)} \sum_{t \geq 1} |D_{u,t}| x^t$$

According to the above relations, by determining $D_{u,t}$, $t \geq 1$, we can obtain the Wiener polynomial of the graph G.

The following program computes the Wiener polynomial's coefficients of C_{80} fullerene (Table 5.3).

```
n:=80; k:=[]; N:=[];
k[1]:=[1..5]; k[2]:=[6..20]; k[3]:=[21..40];
k[4]:=[41..60]; k[5]:=[61..75]; k[6]:=[76..80];
for i in [1..6] do
y:=Size(k[i]);
for j in [1..y] do
x:=k[i][j];
N[x]:=[x-1,x+1];
od;
od;
D1:=[9,12,15,18];
for i in [1..4] do
x:=D1[i]; N[i][3]:=x; N[x][3]:=i;
od;
D2:=Difference(k[2],Filtered(k[2],i->(i mod 3)=0));
D3:=Filtered(k[3],i->(i mod 2)=1);
for i in [1..9] do
x:=D3[i+1]; N[D2[i]][3]:=x; N[x][3]:=D2[i];
od;
D4:=Difference(k[3],D3);
D5:= Filtered(k[4],i->i mod 2=1);
for i in [1..9] do
x:=D5[i+1]; N[D4[i]][3]:=x; N[x][3]:=D4[i];
od;
D6:=Difference(k[4],D5);
```

```
D7:= Filtered(k[5],i->(i mod 3)<>2);
for i in [1..9] do
x:=D7[i+1]; N[D6[i]][3]:=x; N[x][3]:=D6[i];
od;
D8:=Difference(k[5],D7);
for i in [1..4] do
x:=k[6][i+1]; N[D8[i]][3]:=x; N[x][3]:=D8[i];
od;
N[1]:=[2,5,9]; N[5]:=[1,4,6]; N[6]:=[5,7,20];
N[20]:=[6,19,21];N[21]:=[20,22,40];
N[40]:=[21,39,41]; N[41]:=[40,42,60];N[60]:=[61,59,41];
N[61]:=[60,62,75];
N[74]:=[73,75,76]; N[75]:=[58,61,74]; N[76]:=[74,77,80];
N[80]:=[71,76,79];
md:=1; v:=[]; D:=[];
for i in [1..n] do
D[i]:=[]; u:=[i]; D[i][1]:=N[i]; v[i]:=Size(N[i]);
u:=Union(u,D[i][1]);
r:=1; t:=1;
while r<>0 do
D[i][t+1]:=[];
for j in D[i][t] do
for m in Difference (N[j],u) do
AddSet(D[i][t+1],m);
od; od;
u:=Union(u,D[i][t+1]);
if D[i][t+1]=[] then r:=0;fi;
t:=t+1;
od;
md:=MaximumList([md,Size(D[i])]);
od;
p:=[];
for i in [1..md] do p[i]:=0; od;
for t in [1..md] do
for i in [1..n] do
p[t]:=p[t]+Size(D[i][t]);
od;od;
p:=p/2;#(p is the set the Wiener polynomial's coefficients of
C80 fullerene)
```

Table 5.3 The Wiener polynomial, Wiener index and hyper Wiener index of C_{80} fullerene

$W(C_{80},x)$	$120x + 240x^2 + 330x^3 + 420x^4 + 450x^5 + 450x^6 + 420x^7 + 330x^8 + 240x^9 + 120x^{10} + 40x^{11}$
$W(C_{80})$	17600
$WW(C_{80})$	66900

5.4 Balaban Index of IPR C_{80} Fullerene Isomers

In this section, we give an algorithm for computing the Balaban index of any graph. Also we compute this index for isolated pentagon rule (Hirsch 1994; Kroto 1987; Schmalz et al. 1988) (IPR) C_{80} fullerene isomers by GAP program.

In Section 5.2.1, we introduced some notations. According to these notations, we have the following relations:

- $V(G) = \bigcup_{t \geq 0} D_t(u), \quad \forall u \in V(G)$
- $d(u) = \sum_{v \in V(G)} d(u, v) = \sum_{t \geq 1} t \times |D_t(u)|, \quad \forall u \in V(G)$
- $\delta_u = |N(u)|$

According to the above relations, we can obtain the Balaban index by determining $D_t(u)$ for every vertex u.

For computing this index for IPR C_{80} fullerene isomers, at first we assign to any vertex of the graph one number (Fig. 5.2) and then according to the above algorithm, we write a GAP program to determine $N(i)$ and $D_{i,t}$.

For these seven isomers, Balaban index is computed by the following GAP program in Table 5.4:

```
n:=80; k:=[]; N:=[];
k[1]:=[1..5]; k[2]:=[6..20]; k[3]:=[21..40];
k[4]:=[41..60]; k[5]:=[61..75]; k[6]:=[76..80];
for i in [1..6] do
y:=Size(k[i]);
for j in [1..y] do
x:=k[i][j];
N[x]:=[x-1,x+1];
od;
od;
D1:=[9,12,15,18];
for i in [1..4] do
x:=D1[i]; N[i][3]:=x; N[x][3]:=i;
od;
D2:=Difference(k[2],Filtered(k[2],i->(i mod 3)=0));
D3:=Filtered(k[3],i->(i mod 2)=1);
for i in [1..9] do
x:=D3[i+1]; N[D2[i]][3]:=x; N[x][3]:=D2[i];
od;
D4:=Difference(k[3],D3);
D5:= Filtered(k[4],i->i mod 2=1);
for i in [1..9] do
x:=D5[i+1]; N[D4[i]][3]:=x; N[x][3]:=D4[i];
```

```
od;
D6:=Difference(k[4],D5);
D7:= Filtered(k[5],i->(i mod 3)<>2);
for i in [1..9] do
x:=D7[i+1]; N[D6[i]][3]:=x; N[x][3]:=D6[i];
od;
D8:=Difference(k[5],D7);
for i in [1..4] do
x:=k[6][i+1]; N[D8[i]][3]:=x; N[x][3]:=D8[i];
od;
N[1]:=[2,5,9]; N[5]:=[1,4,6]; N[6]:=[5,7,20];
N[20]:=[6,19,21];N[21]:=[20,22,40];
N[40]:=[21,39,41]; N[41]:=[40,42,60];
N[60]:=[61,59,41];N[61]:=[60,62,75];
N[74]:=[73,75,76]; N[75]:=[58,61,74]; N[76]:=[74,77,80];
N[80]:=[71,76,79];
v:=[]; D:=[];
for i in [1..n] do
D[i]:=[]; u:=[i]; D[i][1]:=N[i]; v[i]:=Size(N[i]);
u:=Union(u,D[i][1]);
r:=1; t:=1;
while r<>0 do
D[i][t+1]:=[];
for j in D[i][t] do
for m in Difference (N[j],u) do
AddSet(D[i][t+1],m);
od; od;
u:=Union(u,D[i][t+1]);
if D[i][t+1]=[] then r:=0;fi;
t:=t+1;
od;od;
m:=(1/2)*Sum(v)-n+1;
d:=[];deg:=[];
for i in [1..n] do
d[i]:=0;
deg[i]:=Size(N[i]);
for t in [1..Size(D[i])] do
d[i]:=d[i]+t*Size(D[i][t]);
od;od;
B:=0;R:=0;
for i in [1..n] do
for j in N[i] do
B:=B+ER(1/((d[i]*d[j])));
Od; od;
B:=m*B/2;#(this value is equal to Balaban index of the graph)
```

Table 5.4 Balaban index of IPR C_{80} fullerene isomers

IPR C_{80} fullerene isomers	Balaban index
No.1-D5d	0.81074
No.2-D2	0.80994
No.3-C2v	0.80701
No.4-D3	0.80915
No.5-C2v	0.80499
No.6-D5h	0.80281
No.7-lh	0.79822

5.5 Computing Zagreb Indices of C_{80} Fullerene

In this section, we give an algorithm that enables us to compute the Zagreb indices of any graph. Also by this algorithm, we compute the Zagreb indices for C_{80} fullerene.

For this purpose, the following algorithm is presented:

- We assign to any vertex one number (Fig. 5.2).
- We determine all of adjacent vertices set of the vertex i, $i \in V(G)$ and this set is denoted by *N(i)*.

The degree of any vertex *i* equal to the number of adjacent vertices to *i*. Therefore, by determining the adjacent vertices of each vertex; its degree can also be obtained.

In the start of program, we set $M1 = 0$, $M2 = 0$, then we perform the following operation for each vertex *i*:

- We add $(deg(i))^2$ to *M1*, then for each vertex *j* in the set of adjacent vertices to vertex *i*, we add the value *(deg(i)). (deg(j))* to *M2*.

At the end of this operation, *M1* and *M2* are equal to the values of the first and the second Zagreb indices respectively. Therefore, by determining the vertices adjacent to the vertex of each graph and the above operation, the Zagreb indices of that graph can be obtained.

Now, we compute the Zagreb indices of C_{80} Fullerene by GAP program.

The following program computes the Zagreb indices of C_{80} Fullerene (Table 5.5).

```
n:=80; k:=[]; N:=[];
k[1]:=[1..5]; k[2]:=[6..20]; k[3]:=[21..40];
k[4]:=[41..60]; k[5]:=[61..75]; k[6]:=[76..80];
for i in [1..6] do y:=Size(k[i]);
for j in [1..y] do x:=k[i][j]; N[x]:=[x-1,x+1];
od;
od;
```

```
D1:=[9,12,15,18];
for i in [1..4] do x:=D1[i]; N[i][3]:=x; N[x][3]:=i;
od;
D2:=Difference(k[2],Filtered(k[2],i->(i mod 3)=0));
D3:=Filtered(k[3],i->(i mod 2)=1);
for i in [1..9] do x:=D3[i+1]; N[D2[i]][3]:=x; N[x][3]:=D2[i];
od;
D4:=Difference(k[3],D3); D5:= Filtered(k[4],i->i mod 2=1);
for i in [1..9] do x:=D5[i+1]; N[D4[i]][3]:=x; N[x][3]:=D4[i];
od;
D6:=Difference(k[4],D5); D7:= Filtered(k[5],i->(i mod 3)<>2);
for i in [1..9] do x:=D7[i+1]; N[D6[i]][3]:=x; N[x][3]:=D6[i];
od;
D8:=Difference(k[5],D7);
for i in [1..4] do x:=k[6][i+1]; N[D8[i]][3]:=x; N[x][3]:=D8[i];
od;
N[1]:=[2,5,9]; N[5]:=[1,4,6]; N[6]:=[5,7,20];
N[20]:=[6,19,21]; N[21]:=[20,22,40];
N[40]:=[21,39,41]; N[41]:=[40,42,60]; N[60]:=[61,59,41];
N[61]:=[60,62,75];
N[74]:=[73,75,76]; N[75]:=[58,61,74]; N[76]:=[74,77,80];
N[80]:=[71,76,79];
deg:=[];
M1:=0;
M2:=0;
for i in [1..n] do
 deg[i]:=Size(N[i]);
od;
for i in [1..n] do
  M1:=M1+(deg[i])^2;
for j in N[i] do
M2:=M2+deg[i]*deg[j];
od;
od;
M1; #(the value of M1 is equal to the first Zagreb index)
M2:=M2/2; #(the value of M2 is equal to the second Zagreb
index)
```

Table 5.5 The Zagreb indices of C_{80} fullerene

The first Zagreb index	The second Zagreb index
720	1080

References

AlizadehY, Iranmanesh A, Mirzaie S (2009) Digest J Nanomater Biostruct 4:7
Balaban AT (1982) Chem Phys Lett 82:404
Balaban AT (1983) Pure Appl Chem 55:199
Behtoei A, Jannesari M, Taeri B (2009) Appl Math Lett 22:1571
Braun J, Kerber A, Meringer M, Rucker C (2005) MATCH Commun Math Comput Chem 54:163
Dabirian M, Iranmanesh A (2005) MATCH Commun Math Comput Chem 54:75
Diudea MV, Florescu MS, Khadikar PV (2006) Molecular topology and its applications. EfiCon Press, Bucarest
Dobrynin AA (1999) Croat Chem Acta 72:869
Ehrman L, Parsons PA (1981) In behaviour genetics and evolutions. McGraw-Hill, New York, NY
Gutman I (1994) J Chem Inf Comput Sci 34:1087
Gutman I (1999) Publikacije Elektrotehnickog Fakulteta, Univerzitet u Beogradu 10:53
Gutman I, Das KC (2004) MATCH Commun Math Comput Chem 50:83
Gutman I, Potgieter JH (1997) J Serb Chem Soc 62:185
Gutman I, Trinajstic N (1972) Chem Phys Lett 17:535
Gutman I, Yeh YN, Lee SL, Luo L (1993) Indian J Chem 32A:651
Hirsch A (1994) The chemistry of the fullerenes. Georg Thieme, Stuttgart
Heydari A, Taeri B (2007a) J Comp Thoer NanoSci 4:158
Heydari A, Taeri B (2007b) MATCH Commun Math Comput Chem 57:665
Hosoya H (1988) Disc Appl Math 19:239
Iranmanesh A, Alizadeh Y (2008) Am J Appl Sci 5:1754
Iranmanesh A, Alizadeh Y, Mirzaei S (2009) Nanotubes Carbon Nanostruct 17:560
Iranmanesh A, Ashrafi AR (2007) J Comput Theor Nanosci 4:514
Iranmanesh A, Mirzaie S Int J Chem Model 3(1–2) In press
Khadikar PV, Karmarkar S (2002) Acta Chim Slov 49:755
Khalifeh MH, Yousefi-Azari H, Ashrafi AR (2009) Discrete Appl Math 157:804
Klavžar S, Gutman I (1997) Disc Appl Math 80:73
Kroto HW (1987) Nature 329:529
Lepovic M, Gutman I (1998) J Chem Inf Comput Sci 38:823
Nikolic S, Trinajstić N, Mihalić Z (1995) Croat Chem Acta 68:105
Nikolic S, Kovacevic G, Milicevic A, Trinajstic N (2003) Croat Chem Acta 76:113
Ringo J (1996) Annu Rev Entomol 41:473
Sagan BE, Yen YN, Zhang P (1996) Int J Quant Chem 60:959
Schmalz TG, Seitz WA, Klein DG, Hite GE (1988) J AmChem Soc 110:1113
Schonert M et al (1992) GAP, groups, algorithms and programming; lehrstuhl de fur mathematik. RWTH Achen
Schultz HP (1989) J Chem Inf Comput Sci 29:227
Schultz HP (2000) J Chem Inf Comput Sci 40:1158
Spieth HT, Ringo JM (1983) In the genetics and biology of drosophila. Academic, London
Sun L, Chen T (2009) Discrete Appl Math 157:1650
Taherkhani B, Iranmanesh A, Alizadeh Y, Mirzaiea S (2009) Digest J Nanomater Biostruct 4:885
Trinajstic N (1992) Chemical graph theory. CRC Press, Boca Raton, FL
Wiener H (1947) J Am Chem Soc 69:17
Yousefi-Azari H, Ashrafi AR, KHalifeh MH (2008) Digest J Nanomater Biostruct 3:251
Zhou B (2004) MATCH Commun Math Comput Chem. 52:113
Zhou B, Gutman I (2004) Chem Phys Lett 394:93
Zhou B, Gutman I (2005) MATCH Commun Math Comput Chem 54:233
Zhou B, Trinajsti N (2008) Croatica Chemica Acta 81:319

Chapter 6
4-Regular and Self-Dual Analogs of Fullerenes

Mathieu Dutour Sikirić and Michel Deza

Abstract An *i-hedrite* is a 4-regular plane graph with faces of size 2, 3 and 4. We do a short survey of their known properties (Deza et al. Proceedings of ICM Satellite Conference On Algebra and Combinatorics, 2003b; Deza et al. J Math Res Expo 22:49,2002; Deza and Shtogrin, Polyhedra in Science and Art 11:27, 2003a) and explain some new algorithms that allow their efficient enumeration. Using this we give the symmetry groups of all *i*-hedrites and the minimal representative for each. We also review the link of 4-hedrites with knot theory and the classification of 4-hedrites with simple central circuits. An *i-self-hedrite* is a self-dual plane graph with faces and vertices of size/degree 2, 3 and 4. We give a new efficient algorithm for enumerating them based on *i*-hedrites. We give a classification of their possible symmetry groups and a classification of 4-self-hedrites of symmetry T, T_d in terms of the *Goldberg-Coxeter construction*. Then we give a method for enumerating 4-self-hedrites with simple zigzags.

6.1 Introduction

A *fullerene* is a 3-regular plane graph whose faces have size 5 or 6. As a consequence of Euler's formula any fullerene has exactly 12 5-gonal faces. For a 3-regular plane graph G and a r-gonal face F of G, the quantity $6-r$ is called *curvature* and Euler's formula is then a statement about the curvature on the sphere. A natural generalization of fullerene is the class of 3-regular plane graphs with faces of size between 3 and 6 (see, for example, Deza et al. (2009)).

Here we consider another generalization, that is a suitable k-regular plane graph. The Euler formula $V-E+F=2$ becomes then

$$\sum_{j=2}^{\infty} p_j(s-j) = \frac{4k}{k-2} \text{ with } s = \frac{2k}{k-2}; \tag{6.1}$$

M. Dutour Sikirić (✉)
Ruđer Bošković Institute, Bijenička 54, 10000 Zagreb, Croatia
e-mail: mdsikir@irb.hr

F. Cataldo et al. (eds.), *The Mathematics and Topology of Fullerenes*,
Carbon Materials: Chemistry and Physics 4, DOI 10.1007/978-94-007-0221-9_6,

we will permit 2-gons (doubled edges) but not 1-gons. The only integral pairs (s, k) are $(6, 3)$, $(4, 4)$ and $(3, 6)$. We will permit only s- and $(s-1)$-gonal faces. So, $p_{s-1} = \frac{4k}{k-2}$ and p_s is not bounded. The number n of vertices is

$$n = \frac{4(k+2)}{(k-2)^2} + p_s \frac{2}{k-2}. \tag{6.2}$$

For $k = 3, 4$ and 6 we get spherical analogs of the regular partition of the Euclidean plane E^2: $\{6^3\}$, $\{4^4\}$ and $\{3^6\}$, respectively, where 12 pentagons, 8 triangles and 6 doubled edges play role of "defects", disclinations needed to increase the curvature to the one of sphere S^2. The graphs with smallest number n of vertices have only $(s-1)$-gons; they are Dodecahedron, Octahedron and $Bundle_6$ (2 vertices connected by 6 edges) for $k = 3, 4, 6$, respectively. The case $k = 3$ gives fullerenes. The case $k = 4$, i.e., of 4-regular plane graphs with faces of size 3 or 4, gives *octahedrites* treated in the foundational paper Deza et al. (2003b). Let us call graphs in the remaining case $k = 6$ (6-regular plane graphs with faces of size 2 or 3) *bundelites*. Thurston's work (Thurston 1998) implies that fullerenes can be parametrized by 10 Eisenstein integers and the number of fullerenes with n vertices grows as n^9; those results can be generalized to octahedrites and bundelites. The ring of definition for octahedrites, respectively bundelites, is the Gaussian, respectively Eisenstein integers. Those cases belong to two of 94 cases enumerated in Thurston (1998).

We present here a short review of known facts about octahedrites as established in Deza et al. (2002, 2003b) and Deza and Shtogrin (2003a) and present a few new facts and applications. We give the possible symmetry groups of octahedrites and the graphs of minimal vertex-sets realizing them. Then we show how octahedrites can be used for the enumeration of all *i-hedrites*, i.e. 4-regular plane graphs with faces of size 2, 3 or 4 and $p_2 + p_3 = i$.

Then we consider *central circuit partition* of the edge-sets of octahedrites and the corresponding knot-theoretic notions, that is *alternating knot*, *Borromean link*, and *equivalence*.

A *i-self-hedrite* is a plane graph with vertices and faces of size 2, 3 or 4 that is isomorphic to its dual with $p_2 + p_3 = i$. Such graphs have $2p_2 + p_3 = 4$ and i-self-hedrites can be enumerated effectively by using $2i$-hedrites with a method detailed below. We determine their possible symmetry groups and we list the minimal representatives for each of them. We characterize the 4-self-hedrites of symmetry T or T_d in terms of the *Goldberg-Coxeter construction* for octahedrites. Then we give a method based on $2i$-hedrites for determining the i-self-hedrites with *simple zigzags*.

The computations of this paper were done using the GAP computer algebra system and the computer packages `polyhedral`, `plangraph` of the first author. The enumeration of octahedrites was done using the ENU program by O. Heidemeier (Heidemeier 1998; Brinkmann et al. 2003) and the program `CaGe` (Brinkmann et al. 1997) was used for making the drawings.

6.2 Structural Properties

A *plane graph* is a graph drawn on the plane with edges intersecting only at vertices. A graph G is *3-connected* if after removing any 2 vertices of G the resulting graph is connected. A *3-polyhedron* is a 3-dimensional polytope, its skeleton defines a 3-connected plane graph and it is known that this characterizes the skeleton of 3-polytopes. Furthermore (Mani, 1971), a 3-connected plane graph G can be represented as a skeleton of a 3-polytope P such that any symmetry of G is realized as a isometry of the polytope P. We refer to Deza et al. (2008) for more details on such questions.

It is proved in Deza et al. (2003b) that any octahedrite is 3-connected which implies that its symmetry groups is realized as isometry of 3-space. Since those group have been classified long ago and are much used in chemistry, we can use the chemical nomenclature here (see, for a possible presentation, Dutour (2004)).

An octahedrite exists for any $n \geq 6$ except $n = 7$ (see Grünbaum 1967, p. 282). For a 4-regular graph with p_j denoting the number of faces of size j, the classical Euler formula $V - E + F = 2$ can be rewritten (see Deza and Dutour-Sikirić 2008, Chapter 1, for the details) as

$$\sum_{j=2}^{\infty}(4-j)p_j = 8. \tag{6.3}$$

For octahedrites this directly implies $p_3 = 8$. Octahedron is the unique octahedrite with $n = 6$.

Theorem 1 *The only symmetry groups of octahedrites are:* C_1, C_s, C_2, C_{2v}, C_i, C_{2h}, S_4, D_2, D_{2d}, D_{2h}, D_3, D_{3d}, D_{3h}, D_4, D_{4d}, D_{4h}, O, O_h. *The minimal possible representative are given in Fig. 6.1.*

The proof that the list of groups is complete is given in Deza et al. (2003b), but the minimal possible representatives were not determined at the time. The method is first to go through the restrictions that vertex degree and face size impose. An m-fold axis of rotation has necessarily $m = 4$ (passing though a face of size 4 or a vertex), $m = 3$ (axis passing through a face of size 3), or $m = 2$ (axis passing though an edge, a vertex of degree 4 or a face of size 4). Then the classification of point groups gives a list of possible candidates. Some candidates are excluded for reason of orbit size and other similar simple arguments. But some groups are excluded for a subtler reason: the existence of a symmetry implies another symmetry. For example a 3-, 4-fold axis of symmetry, i.e. C_3, C_4 implies actually at least D_3, D_4 for possible symmetry groups. See Deza et al. (2003b) for details.

On the other hand, finding the minimal possible representative is done in a very non-clever way: we look at all the generated octahedrites and select the representatives with minimal vertex-sets. The enumeration of octahedrites was done by using

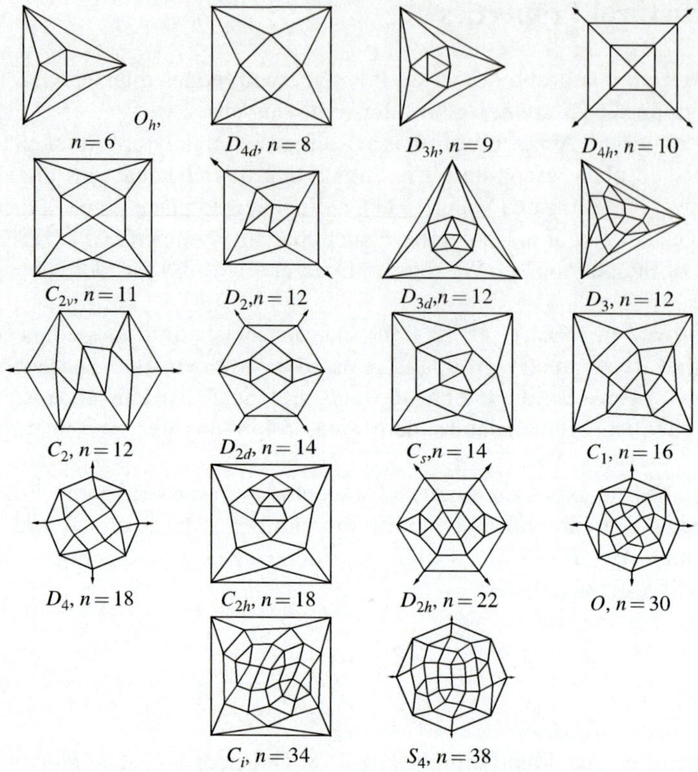

Fig. 6.1 Minimal representatives for each possible symmetry group of an octahedrite

the program ENU (see Heidemeier 1998; Brinkmann et al. 2003) by O. Heidemeier, that enumerates classes of 4-regular graphs with constraint on the size of their faces, fairly efficiently.

6.3 Generation of *i*-Hedrites

Define an *i-hedrite* to be a 4-regular *n*-vertex plane graph, whose faces have size 2, 3 and 4 only and $p_2 + p_3 = i$ (see, for more details, Deza et al. (2003b)). Using Formula 3, we get for an *i*-hedrite $2p_2 + p_3 = 8$ and the only solutions are $i = 4, 5, 6, 7, 8$, which have, respectively, (p_2, p_3)=$(4, 0)$, $(3, 2)$, $(2, 4)$, $(1, 6)$ and $(0, 8)$. So, 8-hedrites are octahedrites. We will be concerned here only about the generation of *i*-hedrites. Actually, 4-hedrites admit a reasonably simple explicit description, see Deza et al. (2003b, 2008), Chapter 2. So, it remains to find efficient methods for the enumeration of 5-, 6- and 7-hedrites. The program ENU cannot deal with faces of size 2; so, we sought a method that allows for reasonable enumeration of such graphs. See Table 6.1 for the number of *i*-hedrites with at most 70 vertices.

Table 6.1 Number of i-hedrites, $4 \leq i \leq 8$, with $2 \leq n \leq 70$

n	4	5	6	7	8	n	4	5	6	7	8	n	4	5	6	7	8
2	1	0	0	0	0	25	0	12	85	107	51	48	21	45	613	1574	2045
3	0	1	0	0	0	26	5	16	119	126	109	49	0	40	614	1751	1554
4	2	0	1	0	0	27	0	21	105	142	78	50	10	54	771	1874	2505
5	0	1	1	0	0	28	8	18	134	179	144	51	0	66	704	1963	1946
6	2	2	2	0	1	29	0	16	135	198	106	52	13	58	771	2247	3008
7	0	3	1	1	0	30	8	24	187	216	218	53	0	48	788	2419	2322
8	4	1	5	1	1	31	0	32	149	257	150	54	12	66	989	2511	3713
9	0	2	5	1	1	32	12	24	189	304	274	55	0	92	849	2735	2829
10	3	3	9	3	2	33	0	18	197	329	212	56	18	68	938	3041	4354
11	0	5	7	4	1	34	6	26	251	382	382	57	0	49	1005	3187	3418
12	5	3	14	5	5	35	0	37	218	431	279	58	9	71	1175	3453	5233
13	0	4	14	7	2	36	13	23	278	483	499	59	0	98	1038	3659	4063
14	3	7	23	9	8	37	0	24	275	547	366	60	22	70	1215	3954	6234
15	0	10	17	12	5	38	6	38	354	601	650	61	0	63	1193	4315	4784
16	7	6	28	18	12	39	0	45	313	643	493	62	9	96	1440	4526	7301
17	0	6	27	22	8	40	15	37	361	764	815	63	0	104	1328	4674	5740
18	5	7	44	25	25	41	0	30	359	838	623	64	21	92	1378	5248	8514
19	0	12	35	36	13	42	10	33	472	889	1083	65	0	74	1440	5600	6631
20	7	9	54	46	30	43	0	52	405	998	800	66	14	80	1751	5741	10103
21	0	8	57	48	23	44	11	44	480	1134	1305	67	0	122	1531	6159	7794
22	4	15	77	62	51	45	0	34	511	1197	1020	68	16	98	1675	6730	11572
23	0	20	59	76	33	46	7	56	609	1324	1653	69	0	72	1792	7005	9097
24	11	11	87	88	76	47	0	69	519	1435	1261	70	14	120	2066	7465	13428

Easy to check that an n-vertex i-hedrite exists for even $n \geq 2$ if $i = 4$, $n \geq 5$ (and $n = 3$) if $i = 5$, $n \geq 4$ if $i = 6$, $n \geq 7$ if $i = 7$, $n \geq 8$ (and $n = 6$) if $i = 8$.

Take an i-hedrite G with $i \in \{5, 6, 7\}$. Then, if F is a face of size 2, we reduce it to a vertex by using the following reduction operation:

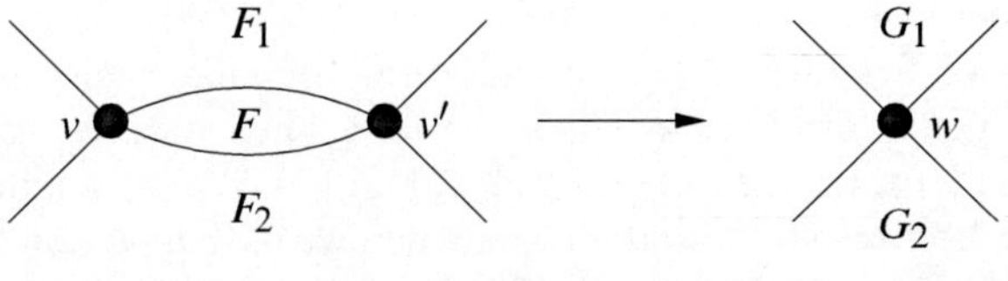

and get a graph denoted by $Red_F(G)$. During this operation the vertices v and v' are merged into one vertex w and the faces F_1 and F_2 are changed into G_1 and G_2 with one edge less. Thus, it is possible that G_1 and/or G_2 are themselves of size 2. We apply the reduction operation whenever, by doing it, the reduced graph is still an i-hedrite. Eventually, since every application of the technique diminish the vertex-set one obtains a graph, denoted by $Red_\infty(G)$ for which we cannot apply the reduction operation anymore.

We call a graph *unreducible* if we cannot apply to it any reduction operation. Let G' be an unreducible graph. If G' has no faces of size 2, then it is an 8-hedrite, i.e.

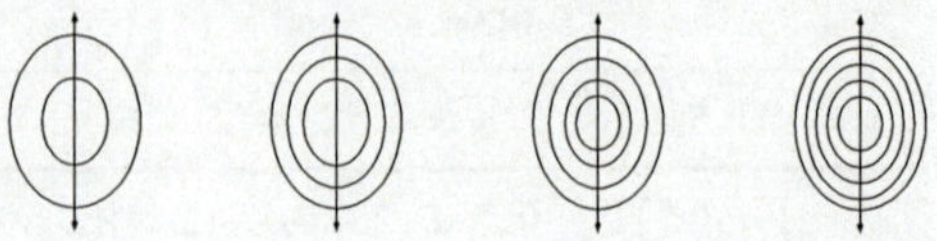

Fig. 6.2 Infinite family of unreducible 4-hedrites

an octahedrite. If G' has a face F of size 2, then denote by e_1, e_2 the two edges of F. Since G' is unreducible, F is adjacent on e_1 or e_2, say e_1, to another face of size 2.

If e_2 is incident to another face of size 2, then G' is actually 2_1, i.e. the unique graph with two vertices, and four faces of size 2, i.e. the 1st one on Fig. 6.2. It is easy to see that e_2 cannot be incident to another face of size 3, but it can be incident to another face of size 4 and in that case G' is not 3-connected and thus (see Deza et al. 2003b) it belongs to the infinite family depicted in Fig. 6.2.

Call *expansion operation* the reverse of the reduction operation. The generation method of i-hedrite is to consider all unreducible i-hedrites and all possible ways of expanding them. For an unreducible graph G denote by $\mathcal{E}xp(G)$ the set of all possible i-hedrites that can be obtained by repeated application of the expansion operation. For the graphs of the infinite family of Fig. 6.2 no expansion operation is possible and thus no i-hedrite is obtained from them. A priori, the set $\mathcal{E}xp(G)$ can be infinite but, as far as we know, for any 8-hedrite G the set $\mathcal{E}xp(G)$ is finite although we have no proof of it. It turns out that $\mathcal{E}xp(2_1)$ is infinite but it has a simple description.

Theorem 2

(i) The only symmetry groups of 4*-hedrites are* D_{4h}, D_4, D_{2h}, D_{2d} *and* D_2.
(ii) The only symmetry groups of 5*-hedrites are:* D_{3h}, D_3, C_{2v}, C_s, C_2 *and* C_1.
(iii) The only symmetry groups of 6*-hedrites are:* D_{2d}, D_{2h}, D_2, C_{2h}, C_{2v}, C_i, C_2, C_s, C_1.
(iv) The only symmetry groups of 7*-hedrites are:* C_{2v}, C_2, C_s *and* C_1.

The theorem is proven in the same way as for octahedrites. Minimal representative for each symmetry groups are given in Figs. 6.3, 6.4, 6.5 and 6.6.

Further generalization of octahedrites are 4-regular plane graphs with 4-, 3-, 2- and 1-gonal faces only. Then, besides i-hedrites, we get graphs with (p_1, p_2, p_3)=(2, 1, 0), (2, 0, 2), (1, 2, 1), (1, 1, 3), (1, 0, 5). The enumeration method is then to use i-hedrites and to add a 1-gon when we have a pair of 2-gon and 3-gon that are adjacent in all possible ways. This is simlar to the strategy of squeezing of 2-gons used for the enumeration of i-hedrites.

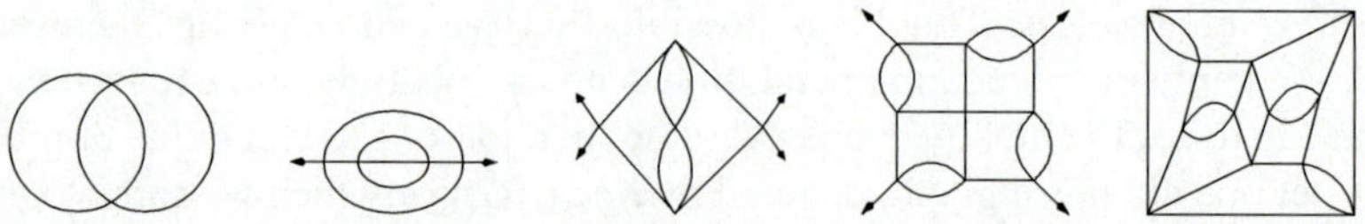

Fig. 6.3 Minimal representatives for each possible symmetry group of a 4-hedrite

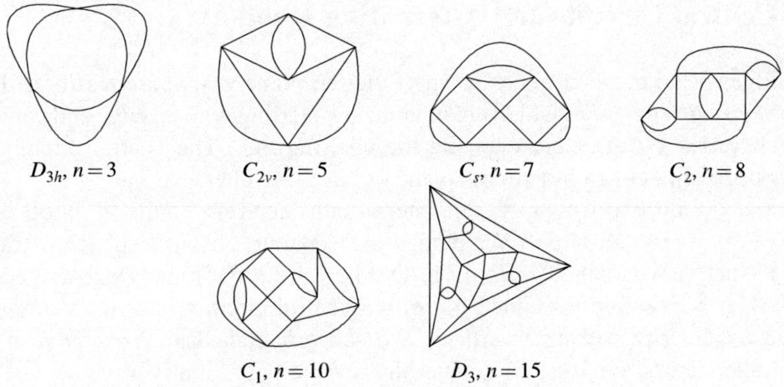

Fig. 6.4 Minimal representatives for each possible symmetry group of a 5-hedrite

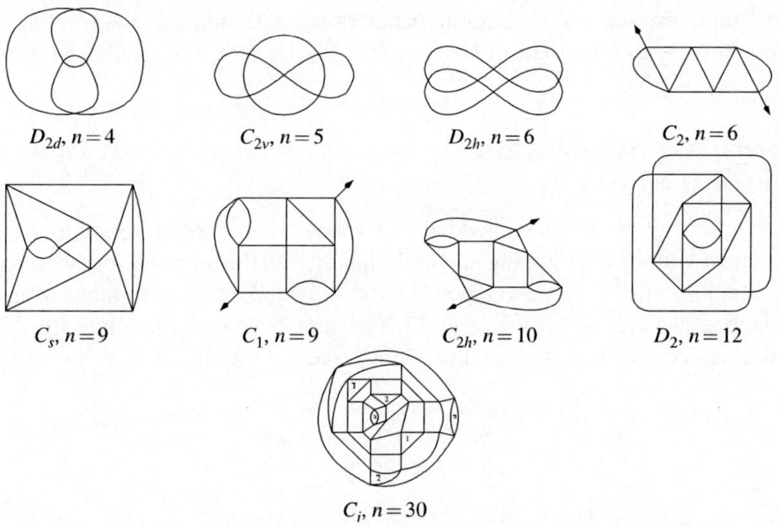

Fig. 6.5 Minimal representatives for each possible symmetry group of a 6-hedrite

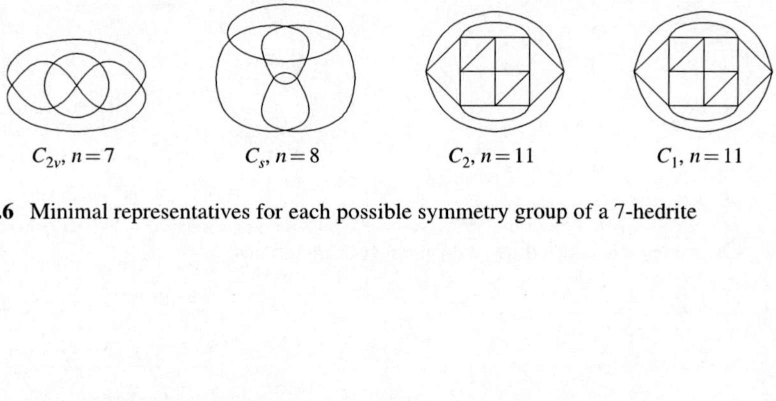

Fig. 6.6 Minimal representatives for each possible symmetry group of a 7-hedrite

6.4 Central Circuits and Alternating Knots

The edges of an octahedrite, as of any Eulerian plane graph, are partitioned by its *central circuits*, i.e. those which are obtained by starting with an edge and continuing at each vertex by the edge opposite the entering one. The central circuits of an octahedrite can define circle in the plane or have self-intersections.

If C_1, C_2 are two (possibly, self-intersecting) central circuits of an octahedrite G, then they are called *parallel* if they are separated by a sequence of faces of size 4 (such pair is called *railroad* in Deza et al. (2003a) and Deza and Shtogrin (2003)). It is possible to reduce those two central circuits into just one and thus get an octahedrite with less vertices. We call an octahedrite *irreducible* if it has no parallel central circuits. Of course, the reverse operation is possible, i.e. split a central circuit into two or more parallel central circuits. In this way every octahedrite is obtained from an irreducible octahedrite.

It is proved in Deza et al. (2003a) that an irreducible octahedrite has at most 6 central circuits and in Deza et al. (2003a) that an irreducible i-hedrite has at most $i-2$ central circuits. All irreducible octahedrites with non self-intersecting central circuits have been classified in Deza et al. (2003a) (see, for another presentation, Deza et al. 2003b).

Theorem 3 *There are exactly eight irreducible octahedrites with simple central circuits (see Fig. 6.7).*

A *link* is a set of circles embedded in 3-space that do not intersect; a link can be represented with its overlapping and underlapping on the plane. A link with only one component is called a *knot* and Knot Theory is concerned with characterizing different plane presentations of links (see Lickorish (1997) for a pleasant introduction). A link is called *alternating* if it admits a plane representation in which overlappings

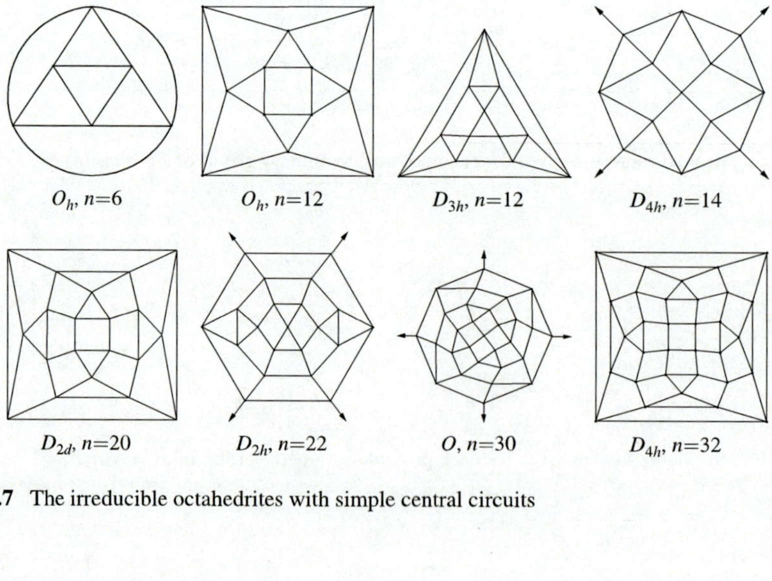

Fig. 6.7 The irreducible octahedrites with simple central circuits

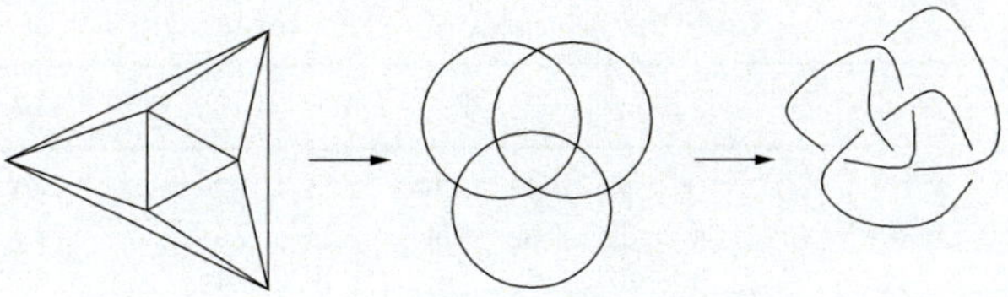

Fig. 6.8 The link corresponding to the octahedron

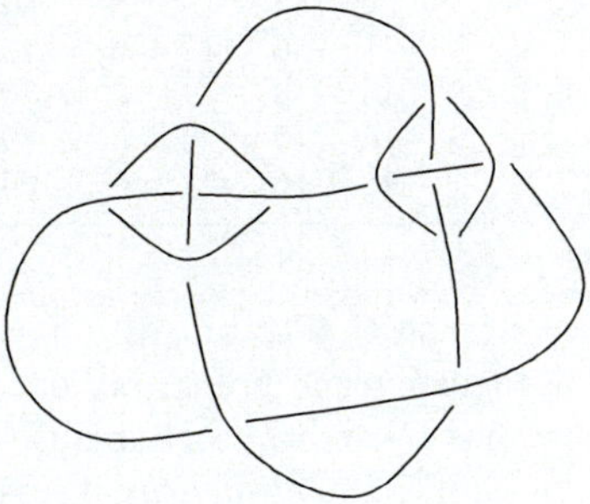

Fig. 6.9 A Borromean link

and underlappings alternate. For a 4-regular plane graph we can define a corresponding alternating link, where the central circuits correspond to the components of the link (see an example on Fig. 6.8). It is interesting that there is no known topological characterization of alternating links.

Since an octahedrite with n vertices is 3-connected, there is no disjointing vertex and thus (Lickorish 1997, Chapter 5) the corresponding alternating link cannot be represented with less than n crossings. But it can happen that two octahedrites that are not equivalent as graphs give rise to equivalent alternating links. A link with m components is called *Borromean* if after removal of any $m-2$ components the remaining two components can be separated one from the other. It is conjectured in Deza et al. (2003a) that an alternating link obtained from a 4-regular 3-connected plane graph is Borromean if and only if for any two central circuits the distance between any two of its consecutive points of its intersection is even. This condition is, of course, sufficient but there are reasons to think that it is not necessary since there exist 4-regular plane graphs (but not 3-connected) which are Borromean without satisfying the specified condition, see Fig. 6.9.

6.5 Self-Dual Graphs

A graph G is called *self-dual* if it is isomorphic to its dual G^*. The *medial graph* $Med(G)$ of a plane graph G is the plane graph obtained by putting a vertex on any edge with two edges adjacent if they share a common vertex and are contained in a common face. One has $Med(G) = Med(G^*)$. The graph $G' = Med(G)$ is always 4-regular and its dual $(Med(G))^*$ is bipartite, that is the face-set $\mathcal{F}$ of $Med(G)$ is split into two sets $\mathcal{F}_1(G')$ and $\mathcal{F}_2(G')$, which correspond to the vertices and faces of

Table 6.2 Number of i-self-hedrites with $4 \le n \le 40$ and $2 \le i \le 4$

n	2	3	4	n	2	3	4	n	2	3	4	n	2	3	4
2	1	0	0	12	4	29	24	22	10	90	191	32	9	239	584
3	1	1	0	13	6	30	33	23	7	119	198	33	9	256	631
4	2	1	1	14	5	42	40	24	7	131	234	34	14	232	748
5	2	4	1	15	5	47	48	25	10	124	276	35	10	290	760
6	3	6	2	16	8	48	69	26	10	162	304	36	14	308	857
7	3	7	4	17	5	64	73	27	8	170	332	37	16	286	956
8	3	11	6	18	6	72	92	28	12	158	407	38	11	342	1002
9	3	16	8	19	8	70	114	29	10	190	421	39	11	359	1070
10	5	16	15	20	6	89	130	30	9	210	476	40	16	332	1239
11	4	26	16	21	8	104	148	31	14	202	550				

the graph G. The bipartition $\mathcal{F}_1(G')$, $\mathcal{F}_2(G')$ can be computed easily from a given 4-regular plane graph, i.e. one can compute easily from a graph G' the two dual graphs G_1 and G_2 such that $G' = Med(G_1) = Med(G_2)$.

Call G a *i-self-hedrite* if it is a self-dual plane graph with vertices of degree 2, 3 or 4 with $v_2 + v_3 = i$ and, consequently, faces of size 2, 3 or 4. If G is a i-self-hedrite then $Med(G)$ is a $2i$-hedrite.

The Euler formula $V - E + F = 2$ for a self-dual plane graph is, clearly:

$$\sum_{j=2}^{\infty} p_j(4-j) = 4; \tag{6.4}$$

we again permit 2-gons but not 1-gons. Define an *i-self-hedrite* to be such a graph with faces of size 2, 3, 4 only and $p_2+p_3 = i$. So, $2p_2+p_3 = 4$ and p_4 is not bounded; also $n = p_4+\frac{p_3}{2}+2 = p_4-p_2+4$. Clearly, an i-self-hedrite can have $i = 2, 3, 4$ only with $(p_2, p_3) = (2,0), (1,2), (0,4)$, respectively. The i-self-hedrites with smallest number n of vertices have no 4-gons; they are $Bundle_2$ (2 vertices connected by 2 edges), triangle with one doubled edge and Tetrahedron, respectively. Easy to check that n-vertex an i-self-hedrite exists if $n \ge i$.

Thus our enumeration method for i-self-hedrites is to consider all $2i$-hedrites G', determine for them the graphs G_1, G_2 such that $G' = Med(G_1) = Med(G_2)$ and keep the ones that have G_1 isomorphic to G_2. We denote by $Med^{-1}(G') = G_1 \simeq G_2$ the obtained plane graph if it exists. Using the enumeration of $2i$-hedrites, we can derive the i-self-hedrite, see Table 6.2. Another method would be possible with the results of Archdeacon and Richter (1992) (but it would require more hard programming work and the speed gain is uncertain).

Theorem 4

(i) The possible symmetry groups of 2-self-hedrites graphs are C_2, C_{2v}, C_{2h}, D_2 and D_{2h}. Minimal representatives are given in Fig. 6.11.

(ii) The possible symmetry groups of 3-self-hedrites graphs are C_1, C_2, C_s and C_{2v}. Minimal representatives are given in Fig. 6.12.

(iii) *The possible symmetry groups of 4-self-hedrites graphs are* C_1, C_2, C_{2h}, C_{2v}, C_3, C_{3v}, C_4, C_{4v}, C_i, C_s, D_2, D_{2d}, D_{2h}, S_4, T, T_d. *Minimal representatives are given in Fig.* 6.13.

Proof If G is a 4-self-hedrite then $G' = Med(G)$ is an octahedrite. If Γ, Γ' are the symmetry groups of G, G', then the self-duality of G becomes a symmetry in G' that exchanges $\mathcal{F}_1(G')$ and $\mathcal{F}_2(G')$. Thus Γ is identified with the subgroup of Γ' formed by the transformations preserving the bipartition. Obviously, the order of Γ is half the one of Γ'. The possible groups of G' are known (see Theorem 1). So, we set out to enumerate the index 2 subgroups of each of the 18 groups and found, besides the groups in the statement, the groups D_3, D_4, C_{4h}, C_{3h}, S_6, S_8 and T_h.

The graph G has 4 vertices of degree 3 and 4 faces of size 3; both should be partitioned in the same number of orbits and this excludes D_4, D_3, C_{3h}, S_6 and T_h. Suppose G has symmetry C_{4h}. Due to the plane of symmetry, the 4-fold axis pass through, either two vertices of degree 4, or through two faces of size 4. But self-duality requires that it passes through a vertex and a face. The same argument excludes S_8.

For 2-self hedrites, using the known groups for 4-hedrites gives candidates C_2, C_{2h}, C_{2v}, C_4, C_{4h}, C_{4v}, D_2, D_{2d}, D_{2h}, D_4, S_4. Same kind of orbit reasons exclude C_4, C_{4h}, C_{4v}, D_{2d}, D_4, S_4. A 3-self-hedrite has only one vertex of degree 2 that has to be preserved by any symmetry. So, the symmetry is a subgroup of C_{2v} and all possible subgroups do occur. □

It is known (Deza et al. 2003b; Dutour and Deza 2004) that all octahedrites of symmetry O or O_h are obtained from the Goldberg-Coxeter construction, i.e. they are of the form $GC_{k,l}(Octahedron)$ for some integer $0 \le l \le k$. The pairs (k, l) correspond to the relative position of the triangles; see Fig. 6.10 for the smallest such graphs and Dutour and Deza (2004) for more details on the construction itself.

Theorem 5 *All* 4-*self-hedrites of symmetry* T *or* T_d *are obtained by the Goldberg Coxeter construction as* $Med^{-1}(GC_{k,l}(Octahedron))$ *with* $k + l$ *odd.*

Proof If G is a 4-self-hedrite of symmetry T or T_d then its medial $G' = Med(G)$ is an octahedrite of symmetry O or O_h. So, $G' = GC_{k,l}(Octahedron)$ for some (k, l). The

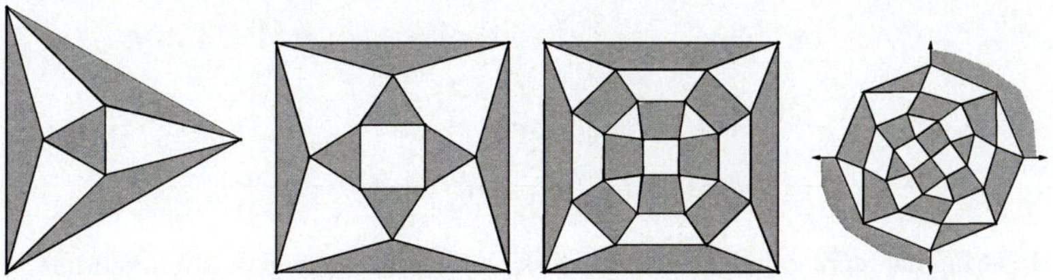

$(k, l) = (1, 0)$ $(k, l) = (1, 1)$ $(k, l) = (2, 0)$ $(k, l) = (2, 1)$

Fig. 6.10 First examples of octahedrites of symmetry O or O_h expressed as $GC_{k,l}(octahedron)$

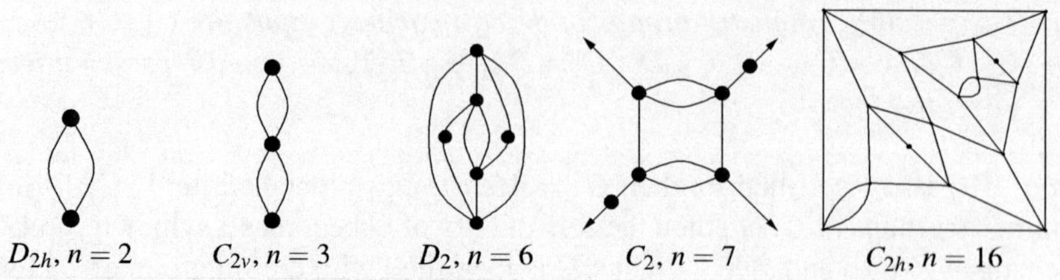

Fig. 6.11 Minimal representatives for each possible symmetry group of 2-self-hedrites

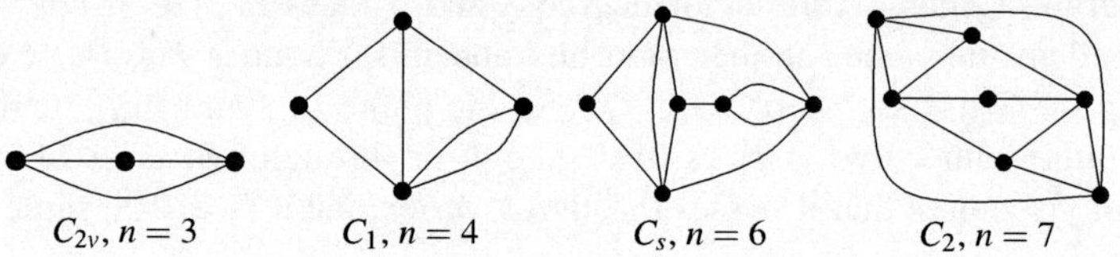

Fig. 6.12 Minimal representatives for each possible symmetry group of 3-self-hedrites

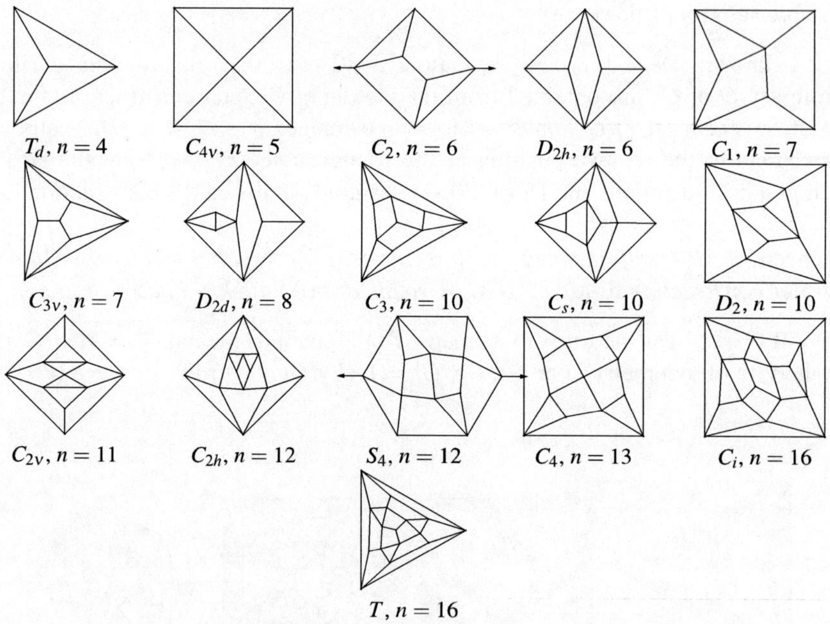

Fig. 6.13 Minimal representatives for each possible symmetry group of 4-self-hedrites

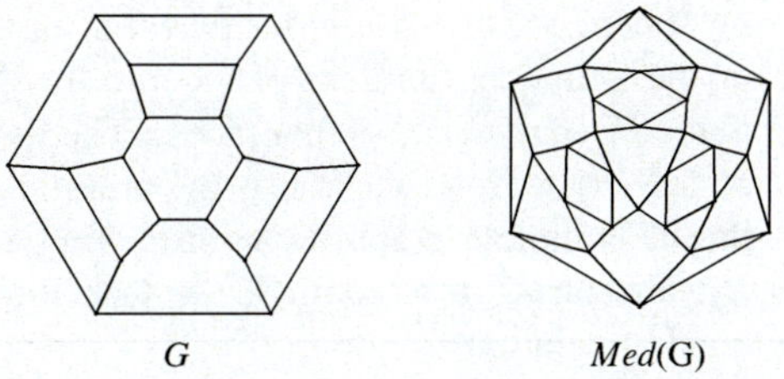

Fig. 6.14 Example of a zigzag in a plane graph G and the corresponding central circuit in $Med(G)$

automorphism group of the plane graph $GC_{k,l}(Octahedron)$ is transitive on triangles; so, we only need to determine when the triangles are not all in $\mathcal{F}_1(G')$ or $\mathcal{F}_2(G')$. Clearly, this correspond to $k+l$ odd. □

For a plane graph G a *zigzag* is a circuit of edges, such that any two but no three, consecutive edges belong to the same face. Zigzags of G correspond to central circuits of $Med(G)$, see an example on Fig. 6.14. So, if G is a 4-self-hedrite with simple zigzags, then $Med(G)$ is an octahedrite with simple central circuits. By Section 6.4, octahedrites G' with simple central circuits are obtained by taking the ones of Fig. 6.7 and splitting each central circuit C_i into m_i parallel central circuits. Then we have to determine for which $m = (m_i)$ the triangles are in two parts $\mathcal{F}_1(G')$ and $\mathcal{F}_2(G')$ which are equivalent under an automorphism of G'. This requires a detailed analysis of the automorphism and a search of the necessary relations between m_i and parity conditions. The details are very cumbersome but in principle we can get a classification of the 4-self-hedrites with simple zigzags.

In particular, 1, 3, 4, 5, 6, 7th irreducible octahedrites in Fig. 6.7 are the medial graphs of 1, 6, 7, 11, 13, 16th 4-self-hedrites in Fig. 6.13, respectively; they are all *irreducible* 4-self-hedrites with simple zigzags.

6.6 Going on Surfaces

In Deza et al. (2000) was considered a generalization of plane fullerenes on any irreducible surface. Similarly, it is easy to see that any *generalized octahedrite*, i.e., a 4-regular map on an irreducible surface, having only 3- and 4-gonal faces, is either an octahedrite on sphere S^2, or a partition of torus T^2 (or Klein bottle K^2) into 4-gons, or the antipodal quotient of a centrally symmetric octahedrite on the projective plane P^2 (having 4 3-gonal faces). Maps on surfaces of high genus can be very complicated. Actually, there are examples with the genus being about half the number of vertices. Here, for the sake of simplicity and the search of more complex examples, we limit ourselves to graphs with no loops or multiple edges. The *minimal*, i.e. with minimal number of vertices, generalized octahedrite on S^2 is Octahedron $K_{2,2,2}$; on P^2 it is the antipodal quotient of Cube with two opposite faces triangulated in their center, that is K_5. On T^2 it is K_5, and on K^2 it is again $K_{2,2,2}$ (but embedded as a quadrangulation); see Fig. 6 in Nakamoto (2001).

Finally, it is easy to check that any *generalized 4-self-hedrite*, i.e., self-dual map on an irreducible surface, having only 3- and 4-gonal faces, is either a 4-self-hedrite on sphere S^2, or a 4-regular partition of torus T^2 (or Klein bottle K^2) into 4-gons, or the antipodal quotient of a centrally symmetric 4-self-hedrite on the projective plane P^2 (having 2 3-gonal faces). The minimal generalized 4-self-hedrite graph on S^2 is Tetrahedron; on P^2 it is the antipodal quotient of the 12th graph on Fig. 6.13, that is complete graph K_6 with disjoint 2- and 4-vertex paths deleted. On T^2 it is K_5, and on K^2 it is again $K_{2,2,2}$ (see Fig. 6 in Nakamoto (2001)) embedded as a quadrangulation.

Similar results hold for generalization of i-hedrites and i-self-hedrites from sphere on any irreducible surface. On T^2 and K^2 it gives the 4-regular quadrangulations. On P^2 they are the antipodal quotients of such centrally symmetric graphs on S^2. So, $2p_2 + p_3$ becomes 4 for i-hedrites and 2 for i-self-hedrites on P^2.

Acknowledgments First author has been supported by the Croatian Ministry of Science, Education and Sport under contract 098-0982705-2707. The authors thank G. Brinkmann for help with the ENU program.

References

Archdeacon D, Richter RB (1992) J Combin Theory *B* 54:37
Brinkmann G, Delgado-Friedrichs O, Dress A, Harmuth T (1997) MATCH 36:233
Brinkmann G, Harmuth T, Heidemeier O (2003) Disc Appl Math 128:541
Deza M, Dutour-Sikirić M (2008) Geometry of chemical graphs. Encyclopedia of mathematics and its applications 119. Cambridge University Press, Cambridge
Deza M, Dutour Sikirić M, Fowler PW (2009) MATCH 61:589
Deza M, Dutour M, Shtogrin MI (2003b). Proceedings of ICM satellite conference on algebra and combinatorics. World Scientific, Singapore, p 73
Deza M, Fowler PW, Rassat A, Rogers KM (2000) J Chem Inf Comput Sci 40:550
Deza M, Huang T, Lih K-W (2002) J Math Res Expo 22:49
Deza M, Shtogrin MI (2003a) Symmetry: Special Issue Polyhedra in Science and Art 11:27, Budapest
Dutour M (2004) Available via DIALOG http://www.liga.ens.fr/~dutour/PointGroups/
Dutour M, Deza M (2004) Elec J Combin 11:R20
Grünbaum B (1967) Convex Polytopes. Wiley New York, NY
Heidemeier O (1998) Die Erzeugung von 4-regulären, planaren, simplen, zusammenhängenden Graphen mit vorgegebenen Flächentypen. Diploma thesis, Bielefeld University, Germany
Mani P (1971) Math Annal 192:279
Nakamoto A (2001) Interdis Inf Sci 7: 77
Raymond Lickorish WB (1997) An introduction to knot theory. Springer, Berlin
Thurston WP (1998) Geometry and Topology Monographs 1, The Epstein Birthday Schrift. Geom Topol Publ., Coventry, p 511

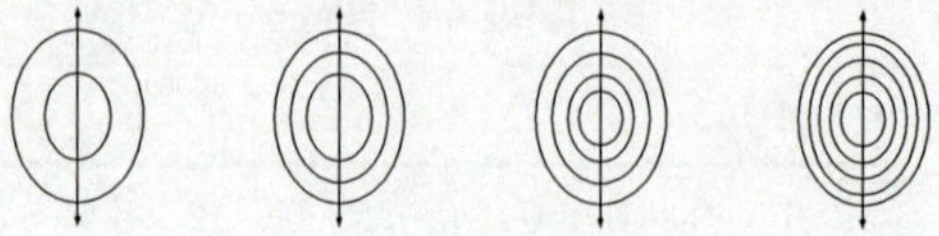

Fig. 6.2 Infinite family of unreducible 4-hedrites

an octahedrite. If G' has a face F of size 2, then denote by e_1, e_2 the two edges of F. Since G' is unreducible, F is adjacent on e_1 or e_2, say e_1, to another face of size 2.

If e_2 is incident to another face of size 2, then G' is actually 2_1, i.e. the unique graph with two vertices, and four faces of size 2, i.e. the 1st one on Fig. 6.2. It is easy to see that e_2 cannot be incident to another face of size 3, but it can be incident to another face of size 4 and in that case G' is not 3-connected and thus (see Deza et al. 2003b) it belongs to the infinite family depicted in Fig. 6.2.

Call *expansion operation* the reverse of the reduction operation. The generation method of i-hedrite is to consider all unreducible i-hedrites and all possible ways of expanding them. For an unreducible graph G denote by $\mathcal{E}xp(G)$ the set of all possible i-hedrites that can be obtained by repeated application of the expansion operation. For the graphs of the infinite family of Fig. 6.2 no expansion operation is possible and thus no i-hedrite is obtained from them. A priori, the set $\mathcal{E}xp(G)$ can be infinite but, as far as we know, for any 8-hedrite G the set $\mathcal{E}xp(G)$ is finite although we have no proof of it. It turns out that $\mathcal{E}xp(2_1)$ is infinite but it has a simple description.

Theorem 2

(i) The only symmetry groups of 4*-hedrites are* D_{4h}, D_4, D_{2h}, D_{2d} *and* D_2.

(ii) The only symmetry groups of 5*-hedrites are:* D_{3h}, D_3, C_{2v}, C_s, C_2 *and* C_1.

(iii) The only symmetry groups of 6*-hedrites are:* D_{2d}, D_{2h}, D_2, C_{2h}, C_{2v}, C_i, C_2, C_s, C_1.

(iv) The only symmetry groups of 7*-hedrites are:* C_{2v}, C_2, C_s *and* C_1.

The theorem is proven in the same way as for octahedrites. Minimal representative for each symmetry groups are given in Figs. 6.3, 6.4, 6.5 and 6.6.

Further generalization of octahedrites are 4-regular plane graphs with 4-, 3-, 2- and 1-gonal faces only. Then, besides i-hedrites, we get graphs with (p_1, p_2, p_3)=$(2,1,0)$, $(2,0,2)$, $(1,2,1)$, $(1,1,3)$, $(1,0,5)$. The enumeration method is then to use i-hedrites and to add a 1-gon when we have a pair of 2-gon and 3-gon that are adjacent in all possible ways. This is simlar to the strategy of squeezing of 2-gons used for the enumeration of i-hedrites.

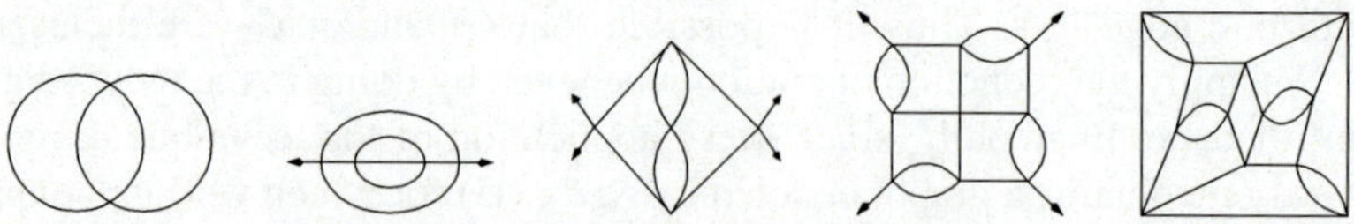

Fig. 6.3 Minimal representatives for each possible symmetry group of a 4-hedrite

Chapter 7
Endohedral Fullerene Complexes and In-Out Isomerism in Perhydrogenated Fullerenes

Why the Carbon Cages Cannot Be Used as the Hydrogen Containers?

Helena Dodziuk

Abstract As shown by our group (Dodziuk and Nowinksi, 1998; Dodziuk et al., manuscript in preparation, 2011) endohedral fullerene complexes are objects of nontrivial topology. An insertion of a guest (atom, ion, or molecule(s)) inside the fullerene cage usually changes properties of both host and guest. Due to their size and complexity, predicting properties of the complexes is a difficult task. For instance, the numbers of endohedral hydrogen molecules in C_{60} and C_{70} have only recently been correctly determined by calculations in agreement with experimental results on the existence of only one hydrogen molecule inside the former molecule and one or two of them (in 96:4 proportion) in the C_{70} cage (Korona et al. 2009). Fullerenes are studied because of their exciting structure and prospects of applications. However, their use for hydrogen storage seems highly improbable.

7.1 Serendipidous Development of Topological Chemistry

The development of topological chemistry is full of peculiarities and surprises. For more than 100 years chemists thought that topology is of no use in chemistry and the first paper by van Gulick considering the possibility of the synthesis of molecules with distinct topological properties was not accepted by a respectable chemical journal *Tetrahedron* in 1960. The paper circulated widely as cited preprint and has been published with preface 33 years later in *New Journal of Chemistry* (Gulick 1993). Interestingly, it is still stimulating today 50 years after it has been written. Even more exciting is the fact that the first synthesis of a molecule with distinct topological properties, catenane **1**, has been described in the same year 1960 (Wasserman 1960) as the rejected van Gulick paper has been submitted. The syntheses of other molecules with distinct topological properties followed. In particular, molecules like **2** (Walba 1985) mimicking

H. Dodziuk (✉)
Institute of Physical Chemistry, Polish Academy of Sciences, Kasprzaka 44/52, 01-224 Warsaw, Poland
e-mail: dodziuk@ichf.edu.pl

F. Cataldo et al. (eds.), *The Mathematics and Topology of Fullerenes*,
Carbon Materials: Chemistry and Physics 4, DOI 10.1007/978-94-007-0221-9_7,

what is called Möbious strip have been obtained. Such objects bear this name although it was not Möbius who first proposed them judging either by the publication date or the date of the first discovery; precedence goes to Johann Benedict Listing (http://www.gap-system.org/~history/Biographies/Mobius.html). Numerous molecules with nontrivial topological properties have been synthesized since then and obtaining lanthanum inside the C_{60} cage (Heath et al. 1985) (just after fullerene discovery in 1985) marked the beginning of endohedral fullerene chemistry which for long time has not been recognized as a part of topological chemistry.

1

2

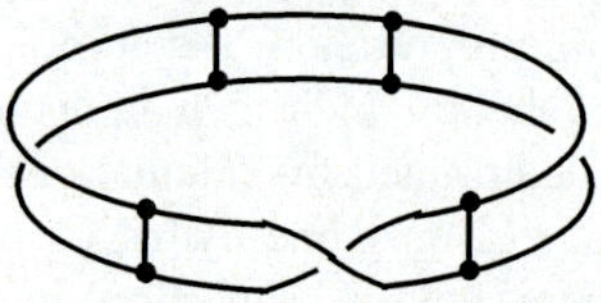

3

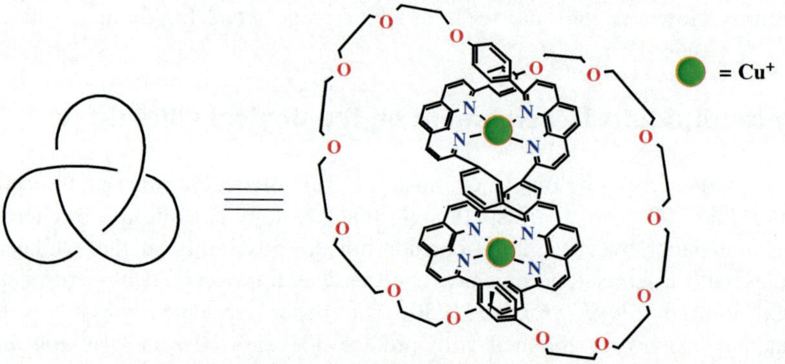

4

5

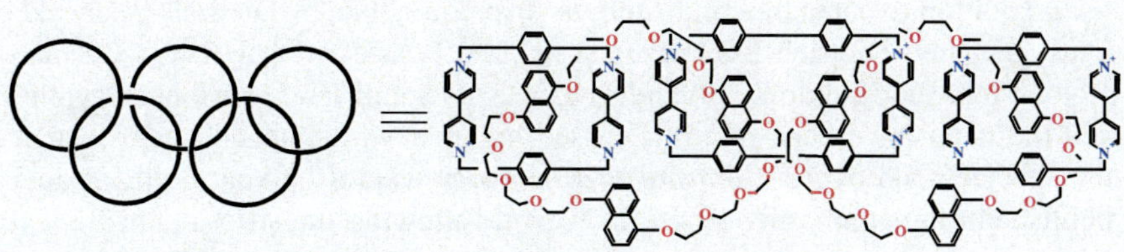

6

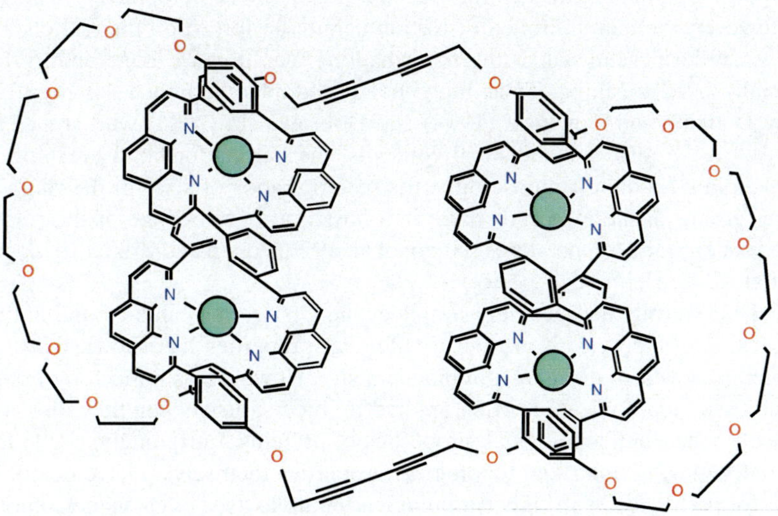

7

8

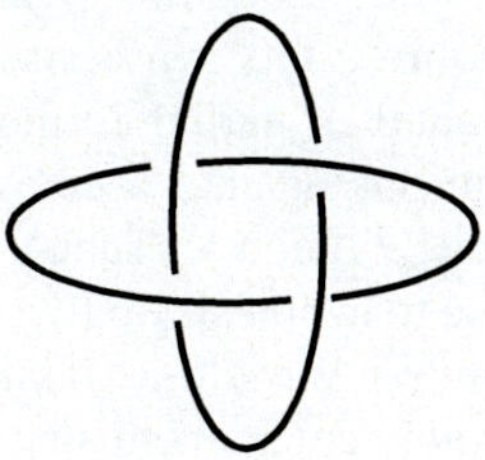

In addition to catenanes and Möbius strip, knots like **3** (Dietrich-Buchecker and Sauvage 1989; Dietrich-Buchecker et al. 2005; Ashton et al. 1997; Carina et al. 1996; Vögtle and Lukin 2005) and rotaxanes **4** (Schill 1971) have been synthesized and the latter have been often discussed together with catenanes (for instance, in the special issue of *New Journal of Chemistry* where the van Gullick paper was published (thematic issue, 1993). However, following the strict mathematical definition they have been treated as objects having no distinct topological properties since a separation of the two fragments forming **4** can be achieved by a considerable enlargement of the central rotaxane ring, allowed in topology for ideal objects. However, such an unlimited distortion cannot be forced on the real chemical entities, molecules, in which the deformations would cause huge energy losses and finally bond breaking. Thus, the generally adopted approach has been questioned by Dodziuk and Nowiński (1998) and Dobrowolski (2003) who argued that, contrary to the ideal mathematical concepts, the barrier for bond breaking has to be taken into account. (Interestingly, for specific ratios of sizes of the ring and blocking groups at the axle ends there is a possibility of slippage of the ring from the axis at elevated temperatures (Agam et al. 1976; Agam and Zilkha 1976; Amabilino et al. 1998; Heim et al. 1999).)

Discovering that relatively simple cyclic DNAs, that can be found in living creatures, are objects with distinct topological properties has shown that topological chemistry is also of interest in biochemistry. However, as noticed by Francl (2009) the term "topological" is often applied by biochemists when they discuss geometrical, rather than topological, properties of proteins. Intriguingly, cyclic DNAs (the objects having nontrivial topological properties themselves) have been established to form structures of higher topological complexity, i.e. catenanes, knots and the figure-of-eight, which role in nature is still to be cleared. The existence of special enzymes, topoisomerases, (Champoux 2001; Corbett and Berger 2004) involved in the syntheses of such molecules should be mentioned here. A polycatenated net built of the cyclic DNAs mimicking hauberk, an idealized form of which is presented in Fig. 7.1 seems to be one of the most complicated existing chemical structures with distinct topological properties (Chen et al. 1995).

At the very beginning, molecules belonging to topological chemistry were synthesized using the so-called statistical approach. In his first catenane synthesis, Wasserman (1960) carried out cyclization of long chain molecules having reactive groups at their ends. During this synthesis, sometimes a ring was created by a chain threaded through another, earlier formed ring. The statistical probability of such an event is low but, taking into account the multitude of molecules in a reaction vessel, not negligible. Of course, such a method (bearing the name "statistical approach") enabled one to synthesize only very simple molecules with distinct topological properties in quite low yields. Making use of the preorganization phenomenon, typical of supramolecular chemistry, drastically changed the situation (Amabilino and Stoddart 1995; Dodziuk, 2002a; Reymo and Stoddart 1999). This phenomenon consists in the orientation of reagents in an appropriate way before the ring(s) closure by making use of complexation of reagents with an appropriate cation (Fig. 7.2), (Carina et al. 1996) π-stacking interactions, (Claessens and

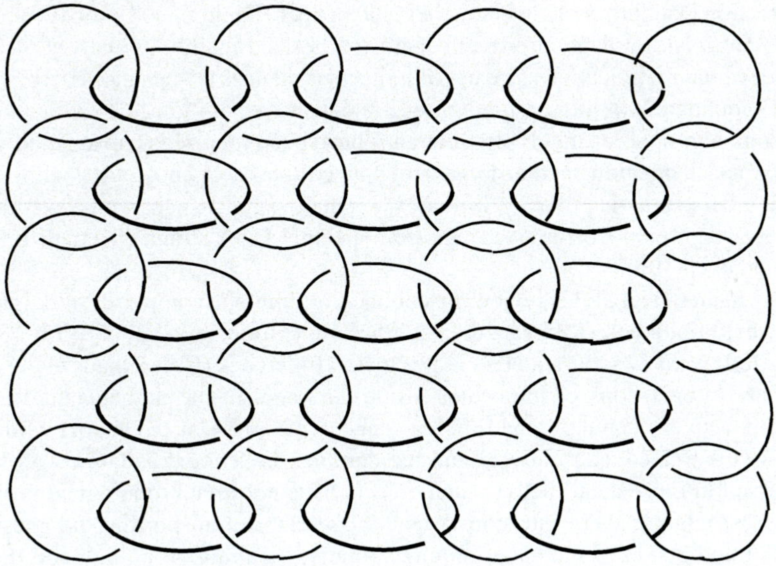

Fig. 7.1 DNA net mimicking hauberk

Stoddart 1997) hydrogen bonds (Schalley et al. 2004) or weak but numerous non-bonding interactions (Yamaguchi et al. 2006). Such a procedure not only allowed one to obtain highly complicated molecules having distinct topological properties but it also leads to considerably higher reaction yields. At present, the synthetic methods applied to obtain these systems have been extended to include dynamic covalent chemistry (Haussmann and Stoddart 2009; Rowan et al. 2002; Stoddart 2009). Using one of these approaches olympiadane **5**, (Amabilino et al. 1994) double knot **6**, (Carina et al. 1996) Borromean rings **7**, (Cantrill et al. 2005) Solomon link **8** (Nierengarten et al. 1994; Peinador et al. 2009; Pentecost et al. 2007) (dubbed "Solomon knot" long before establishment of topology) and other more complicated systems (Pentecost et al. 2006; Williams et al. 2006) have been synthesized. Although there are still several relatively simple structures with nontrivial topology that await their syntheses (some of which are presented in a manuscript prepared for

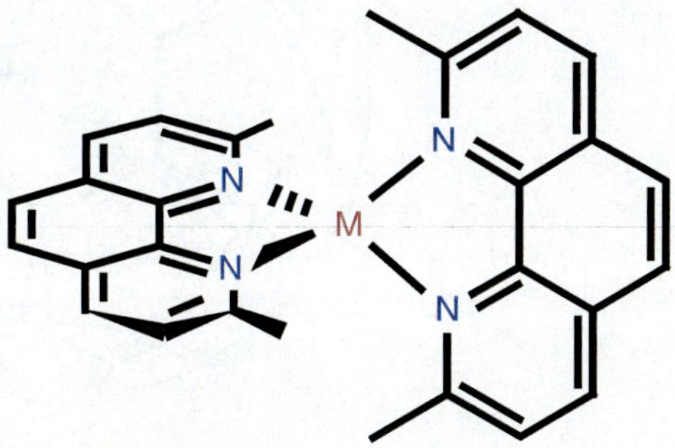

Fig. 7.2 Metal complexation forcing the perpendicular arrangement of the phenanthroline units

publication Dodziuk et al. 2011 and in Fenlon paper (Fenlon 2008)) today molecules with nontrivial topological properties are synthesized mainly to analyze the possibilities of their applications as parts of molecular devices (Fang et al. 2010; Stoddart and Colquhoun 2008). In particular, several derivatives of molecules with nontrivial topological properties involving fullerenes have been reported (Gibson et al. 2009).

A special position in the domain of the syntheses of molecules with distinct topological properties occupy works by Seeman group which synthesized a knot, figure-of-eight and Borromean rings from a single DNA strand (Seeman 1998a, b; Seeman et al. 1998).

The main object of this review are endohedral fullerene complexes, (Akasaka and Nagase 2002) like **9**, (Komatsu et al. 2005b) **10**, (Murata et al. 2008b) **11**, (Goedde et al. 2001) and **12** (Murata et al. 2006) that is fullerenes (Dresselhaus et al. 1995a) that have atoms, ions or molecules inside the cage. In the nomenclature used in supramolecular chemistry, the fullerene cage plays a role of the host H while what is inside is treated as a guest G and the complex is denoted as G@H. As will be discussed in more detail below, such objects have nontrivial topological properties since dissociating them into constituent parts (that is transporting the guest from inside into outside of the cage) requires bond(s) breaking. Topology and the relation of chemistry to this branch of mathematics is discussed in the paper now in preparation (Dodziuk et al. 2011).

Sometimes, the term "chemical topology" is used incorrectly when chemistry of molecules with nontrivial topology is discussed (Fenlon 2008; Frisch and Wasserman 1961; McArdle et al. 2000; Siegel 2004). To be precise, I believe that the term chemical topology is appropriate when one discusses nontrivial topological properties of molecules. Thus, all questions related to topological indices as well as analysis of molecular graphs belong to chemical topology whereas syntheses of molecules with distinct topological properties and analysis of their physicochemical properties are the subjects of topological chemistry. As always happens with closely related sciences, sometimes these domains overlap.

9

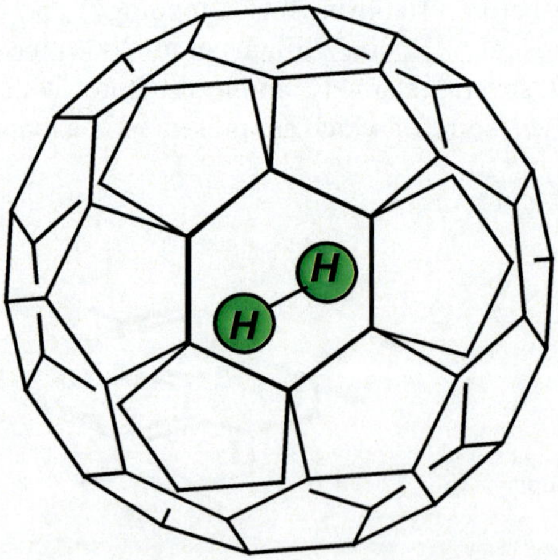

10

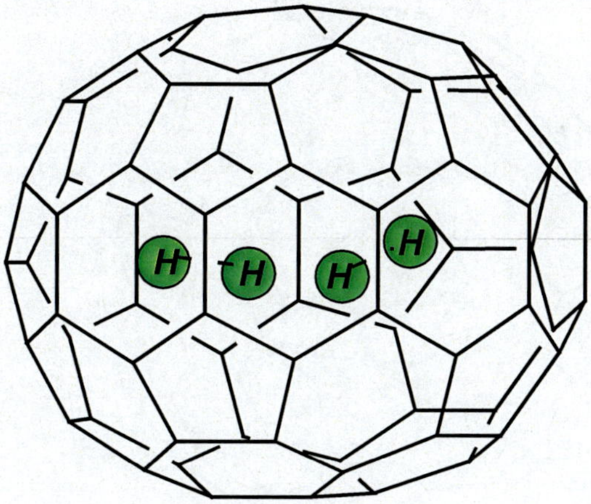

11

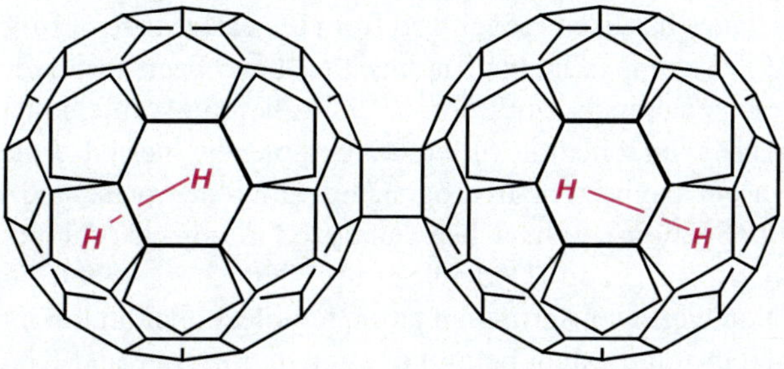

12

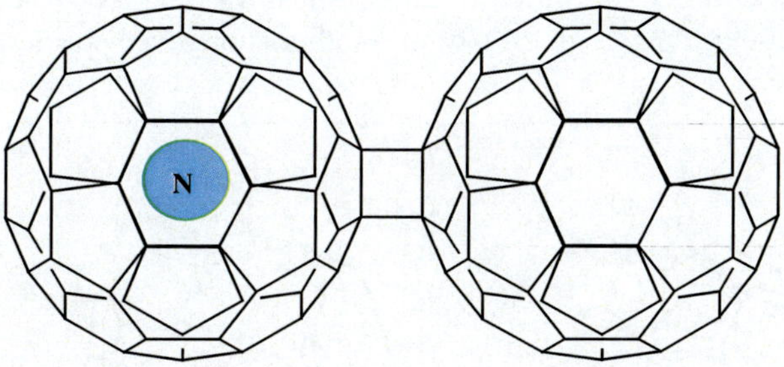

13

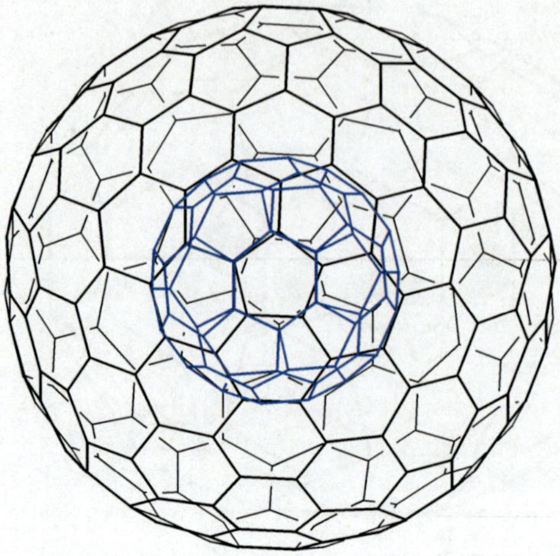

The fascinating idea of an empty space inside the C_{60} cage that could be filled with atoms or molecules has been recognized from the early stage of fullerene study (Heath et al. 1985) (a compendium of atoms that have been inserted until 2007 is given in http://homepage.mac.com/jschrier/endofullerenes_table.html). However, similarly to rotaxanes endohedral fullerene complexes, nested fullerenes like $C_{60}C_{240}$ **13** and in-out isomers of hydrogenated fullerenes presented in Fig. 7.3 (briefly discussed in Section 7.5) have been analyzed as objects of nontrivial topology only in works of our group (Dodziuk et al. 2000, 2001, 1999; Dodziuk and Nowinski 1996). Noteworthy, inserting an atom, a molecule or an ion into a molecular cage (discussed in some extent below) or aggregation (Lee and Kim 2008) can lead to spectacular changes of the host and/or guest properties. As briefly discussed below, it can even stabilize a non-IPR fullerene isomer (Campanera et al. 2002; Cao et al. 2004; Rapta et al. 2008; Shi et al. 2006; Shustova et al. 2007; Takata et al. 2003; Tan et al. 2009; Yang et al. 2007b) or another short-lived species (Cai et al.

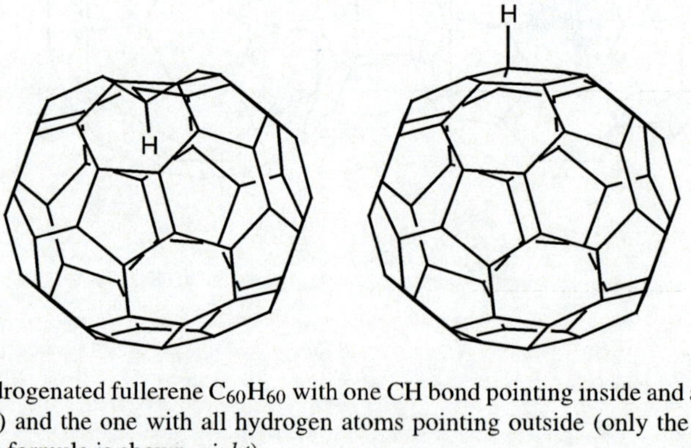

Fig. 7.3 Perhydrogenated fullerene $C_{60}H_{60}$ with one CH bond pointing inside and all other outside (not shown, *left*) and the one with all hydrogen atoms pointing outside (only the one that points inside in the left formula is shown, *right*)

2007; Jimenez-Vazquez et al. 1994) or, as shown for instance in case of Y@C_{82}, change the fullerene symmetry upon the guest inclusion (Takata et al. 1995). The isolation of atomic nitrogen (Lips et al. 2000) or diatomic van der Waals molecules of noble gases He_2 in C_{60} and C_{70} cages as well as of atomic nitrogen inside the C_{60} dimer **11** and Ne_2 in the C_{70} cage (Sternfeld et al. 2002) illustrates radical changes in the guests properties due to its encapsulation into the fullerene cage. Other examples of short-lived species stabilized in "molecular flasks" are briefly reviewed by Dodziuk (2002c).

It should be stressed that although C_{60} is the most common fullerene, it does not form numerous endohedral complexes because of its small internal cavity. As pointed out by Dodziuk et al. (2001) to obtain endohedral complexes with fairly large guest molecules suitable for applications one has to master the synthesis and purification of much larger fullerenes.

Today, endohedral fullerene complexes are studied by several groups not only because they are exciting objects but also in view of their future applications. With a metal guest, so called endohedral metallofullerenes, EMFs, can be metal, small-gap semiconductors or insulators depending upon the fullerene size and the kind and number of encapsulated metal atoms. MEFs applications in medicine among others as radiotracers, (Cagle et al. 1999) MRI contrast agents, (Bolskar 2008; MacFarland et al. 2008), drugs (Bakry et al. 2007; Wilson et al. 1999) (in particular, HIV-1 protease inhibitors (Ibrahim et al. 2010)) in electronics, (Kobayashi et al. 2003; Ross et al. 2009; Shibata et al. 2004; Yasutake et al. 2005) in particular, as single-molecule transistor for quantum computing, (Twamley 2003) and in solar cells (Zhang et al. 2008b) have been proposed. Noteworthy, the application of so-called peapods that is carbon nanotubes filled with endohedral fullerenes encapsulating metal atoms in the latter devices seems promising (Kurokawa et al. 2005).

7.2 Endohedral Fullerene Complexes and In, Out Isomers of Fullerene Derivatives as a Specific Domain of Topological Chemistry

Endohedral fullerene complexes and hypothetical in, out isomers of hydrogenated fullerenes occupy a special position in topological chemistry from both points of view of chemistry and topology. As mentioned earlier, concerning chemistry (contrary to organic molecules with distinct topological properties such as **1–8**) endohedral fullerene complexes **9–12** are not obtained by taking advantage of the preorganization (Dodziuk 2002a, b) or dynamic covalent chemistry (Haussmann and Stoddart 2009; Rowan et al. 2002; Stoddart 2009). They are manufactured by applying completely different procedures. Those having metal cations inside are obtained in an arc reactor or via laser evaporation during the process of fullerene formation (Nishibori et al. 2006) while the ones with noble gas guests are produced by heating fullerenes in the nobel gas atmosphere at high pressure using Krätschmer-Huffmann procedure (Krätschmer et al. 1990; Saunders et al. 1993). Endohedral fullerene complexes of few simple molecules inside the fullerene cage were obtained by so-called molecular surgery approach consisting in making a hole in the cage chemically, inserting a guest inside the cage (for instance, one (Komatsu et al. 2005b) or two (Murata et al. 2008a) hydrogen molecules) and closing the hole chemically. On the other hand, different nitride cluster fullerenes, like $Sc_3N@C_{80}$, $Sc_{3-x}Er_xN@C_{80}$ ($x = 1, 2, 3$), $Sc_3N@C_{78}$, $Y_3N@C_{80}$, $Ho_3N@C_{80}$ and $Tb_3N@C_{80}$, were obtained by a reactive gas addition to the cooling gas of the arc burning process (Yang and Dunsch 2006). It should be stressed that the main effort in studying endohedral fullerene complexes seems to be focused on group 2 and 3 metallofullerenes involving Sc, Y, La, Ca, Sr, Ba and lanthanide (Ce – Lu) metallofullerenes (Stevenson et al. 2000) for which exciting applications are expected. Considerable computational effort centered on hydrogen molecules inserted into the C_{60} cage summarized in Dodziuk (2007) has been triggered by the hope to use fullerenes for hydrogen storage. This idea (presented in few details later) in its pure shape is unreasonable since the guest can be released from C_{60} only as a result of cage breaking. As will be discussed later, the synthetic efforts resulted in putting hydrogen molecule inside C_{60} cage, (Komatsu et al. 2005b) one or two of them inside C_{70} one (Murata et al. 2008a) as well as two hydrogen molecules inside C_{120} (Murata et al. 2006).

As mentioned before, the topological nontriviality of endohedral fullerene complexes has been largely overseen although we studied this phenomenon for several years (Dodziuk 2007; Dodziuk and Dolgonos 2002; Dodziuk et al. 2001; Dodziuk and Nowinski 1998). The hypothetical at present "in", "out" isomers of hydrogenated fullerenes (Fig. 7.3) or, more generally of fullerene derivatives, are also topological isomers (Dodziuk et al. 1999; Dodziuk and Nowinski 1996). From topological point of view both endohedral fullerene complexes and the isomers differ from aforementioned catenanes (links), rotaxanes, möbiusanes and knots **1–8** since the fullerene cage separates 3D space into the "in" and "out" regions. For them, contrary to the latter molecules, it is not sufficient to break and then restore one bond to obtain the separated systems: much larger two-dimensional holes have to be created. Similarly to the rotaxanes case, a condition of bond(s) breaking should be included for these complexes since an included atom, ion or molecule cannot, in general, escape the C_{60} cage without breaking bond(s) which cannot be extended over a certain, very narrow limit.

Interestingly, few, but revolutionary proposals to apply the endohedral complexes have been published soon after the C_{60} discovery (Stoddart 1991) (one of them consisted in using an endohedral fullerene with a door enabling the guest drug escape in appropriately controlled conditions). Some of them have been mentioned above. Until now very few fullerenes applications have been commercialized. The idea of using fullerenes for hydrogen storage also looked promising. However, as mentioned above placing H_2 molecules inside the cage seems impractical for such a purpose since, even if we do not bother for a moment how to put hydrogen inside, one has to destroy the cage irreversibly to release it. Some other possibilities, e.g., enhancing exohedral hydrogen binding electrochemically, or by forming composites either by placing a metal cation inside the fullerene cavity, or by coating fullerene with Ca have been proposed (Arai et al. 2009; Lan et al. 2009; Liu et al. 2009; Yoon et al. 2007, 2009) but their feasibility for practical uses is still unclear.

A special group of endohedral fullerene complexes – nested or onion-like fullerenes like **13** is built of fullerene cages buried inside one another like Russian dolls. Discovered earlier than C_{60} itself, (Iijima 1980) they may consist of two, three (Mordkovich 2000) or several hundred layers (Ugarte 1992). They can be not only spheroidal but also assume different shapes (Terrones and Terrones 1997). Obtaining nested fullerene noncarbon analogues has also been reported (Drummond et al. 1999). Dodziuk et al. have shown that formation of nested fullerenes (Dodziuk et al. 2000) is driven by weak but very numerous nonbonding attractions.

As mentioned before, the topological nontriviality of endohedral fullerene complexes has been overseen although we studied this phenomenon for several years (Dodziuk 2007; Dodziuk et al. 2001; Dodziuk and Nowinski 1998). It is obvious that hypothetical at present "in" isomers of hydrogenated fullerenes (Fig. 7.3) are also topological isomers of the "out" ones (Dodziuk et al. 1999; Dodziuk and Nowinski 1996). Endohedral fullerene complexes form a specific domain of topological chemistry since they not only are obtained in a different way than catenanes, knots and other organic molecules of nontrivial topology but also are 2D objects while catenanes, rotaxanes, knots, Möbious strips, etc. are objects of a lower dimensionality.

7.3 Types of Endohedral Fullerene Complexes

The possibility to include something inside the fullerene cage was first proposed as early as in 1985 (Heath et al. 1985). Depending on the guest character, endohedral fullerenes can be divided into those having an atom(s) or molecule(s) inside their cages. The guest can be metal or nonmetal (or even tritium) (Jimenez-Vazquez et al. 1994; Khong et al. 2000) or muonium (Kiefl et al. 1992; Tan et al. 2006). As mentioned before, multishell so-called nested fullerenes can have two (Mordkovich 2000) or more fullerene cages buried inside one another (Iijima 1980; Ugarte 1992). As mentioned above, various types of endohedral fullerenes are also produced using different procedures.

Endohedral fullerene complexes are often so stable that several their derivatives have been synthesized (Cai et al. 2007, 2008; Cardona et al. 2005; Chaur et al. 2009; Komatsu and Murata 2004; Yamada et al. 2010). Interestingly, the guest can influence the shape of the host cage. To our best knowledge, no endohedral fullerene complex with an isolated hydrogen atom as the guest has been detected. However, as mentioned earlier Cross et al. reported a tritium guest inside C_{60} (Jimenez-Vazquez et al. 1994; Khong et al. 2000).

7.3.1 Endohedral Metallofullerenes

Endohedral metallofullerenes (Akasaka and Nagase 2002; Chaur et al. 2008b, 2009; Kato et al. 2003; Yamada et al. 2010) are probably the most extensively studied today since they are thought to bring marketable applications as e.g advanced materials (Kobayashi et al. 2003; Ross et al. 2009; Yasutake et al. 2005) and in medicine (Cagle et al. 1999; Ji et al. 2006; Murthy and Geckeler 2006; Watanabe et al. 2005). The most abundant classical EMFs are $M@C_{82}$ (Akasaka and Nagase 2002; Kitaura and Shinohara 2007; Liu and Sun 2000; Shinohara 2000a, b) and the most abundant metal guests are lantanides. If one or more atoms of metal reside inside a fullerene then electrons are transferred from the guest to the host cage. However, the amount of the charge transfer is not always simple to be determined. For $La_2@C_{80}$ it is, as in most cases, between 2 and 3. However, the number reaches 6 in case of $Sc_3N@C_{80}$, which should better be denoted as $[Sc_3N]^{+6}@[C_{80}]^{-6}$ (Iezzi et al. 2002). It should be mentioned that endohedral metallofullerenes represent untypical salts since they cannot be dissociated into the cation and anion without the fullerene cage breaking. $Sc_3N@C_{78}$ should also be mentioned since, contrary to typical endohedral complex $Sc_3N@C_{80}$ (in which the guest rotates freely inside the host), the Sc_3N cluster occupies a defined position inside the host cavity (Campanera et al. 2002). It should be stressed that, contrary to expectations in both structures the cluster is planar. Both structures are also noteworthy in this respect that the former one is stable although non-IPR C_{78} is not while in the latter the C_{80} cage isolated host is of I_h symmetry which is the least stable for the empty fullerene.

As mentioned earlier, fullerenes doped with metals are obtained by laser (Wilson et al. 2002) or arc vaporization (Afanas'ev et al. 1997) of metal-graphite composites in a rare gas atmosphere followed by tedious purification (Komatsu 2009; Komatsu et al. 2007; Nagata et al. 2005). As proven by mass spectroscopy, the larger fullerene cages can contain up to 4 metal ions (Shinohara 2000b). The largest clusters inside a fullerene cage reported are probably $Sc_4(\mu_3\text{-}O)_2$ inside $I_h\text{-}C_{80}\cdot Ni^{II}(OEP)\cdot 2C_6H_6$ fullerene adduct (Stevenson et al. 2008b) and $Sc_4(\mu_3\text{-}O)_3$ inside $I_h\text{-}C_{80}$ (Mercado et al. 2010). Today methods (as of 2010) allow one to obtain at best no more than 10% of fullerenes (empty and endohedral) in the raw soot. Therefore, strenuous purification procedures involving extraction with organic solvents followed by multi-step HPLC chromatographic separations are applied to obtain the total yield of purified metallofullerenes generally lower than 1% (Yamada et al. 2010).

As concerns classical EMFs, sometimes, the guest influences considerably the host cavity. For instance, C_{66} does not have IPR isomers. However, $Sc_2@C_{66}$ obtained by Shinohara's group has been found to be a non-IPR stable fullerene (Wang et al. 2000). Later it was shown that even those fullerenes that can form IPR structures are sometimes stabilized as non-IPR ones, *e.q.* $La_2@C_{72}$, (Kato et al. 2003) $Sc_2C_2@C_{68}$, (Shi et al. 2006) and others. Today non-IPR isomers have been isolated for many cages built of less than 60 carbon atoms (like U@C28 (Guo et al. 1992)), and those containing 60, (Löffler et al. 2009; Zettergren and Martin 2008) 66, (Wang et al. 2000) 68, (Park et al. 2005; Rapta et al. 2008; Stevenson et al. 2000) 70, (Yang et al. 2007b; Zettergren and Martin 2008) 72, (Dunsch et al. 2001; Kato et al. 2003; Wakahara et al. 2006) 74, (Rappoport and Furche 2009) 76, (Yang et al. 2007a) 78, (Beavers et al. 2009; Park et al. 2005; Popov et al. 2007) 82, (Fu et al. 2009; Mercado et al. 2008) 84 (Beavers et al. 2006; Fu et al. 2009; Zuo et al. 2008) carbon atoms with one to three ions or an anionic cluster, (Shinohara 2000b) metal nitrides (M_3N) (Chaur et al. 2009) or metal carbides (M_2C_2) (Shi et al. 2006) serving as guests. A guest can also stabilize a less stable IPR isomer (Olmstead et al. 2003). On the basis of semiempirical quantum chemical and DFT calculations for a large number of IPR and non-IPR (having less than four adjacent pentagons), Popov and Dunsch (2007) have found that for the cages smaller than C_{84}, non-IPR isomers of hexaanions C_{2n}^{-6} are comparable or even more stable than the neutral species. They have also predicted the most stable such complexes for C_{68} – C_{98} fullerenes. One of the largest identified EMF seems to be $Gd_2@C_{106}$ (Yang et al. 2008). On the basis of DFT calculations, Scuseria group (Infante et al. 2008) have obtained a stable $U_2@C_{60}$ complex. Contrary to the experimental finding of $U_2@C_{58}$ (also coauthored by Scuseria (Guo et al. 1992)), they claimed that it is an artefact since C_{60} cavity is too small to host this guest.

The EMF family is constantly increasing; the complex having Sc_3CH in the C_{80} cage (Krause et al. 2007) has been recently obtained as well as EMF in the C_{106} cage (Chaur et al. 2008b). Echegoyen and coworkers (Chaur et al. 2009) divided recently endohedral metallofullerenes into four groups:

1. Classical EMFs ($M@C_{2n}$ and $M_2@C_{2n}$, M = metal and $60 \leq 2n \geq 88$). They include mainly alkali metal atom(s) ($K@C_{44}$ and $Cs@C_{48}$, (Weiss et al. 1988) $K@C_{60}$, $Rb@C_{60}$, $Cs@C_{60}$, (Curl 1992) $Cs@C_{82}$, (Anderson et al. 2000)) alkali earth metal Ca, (Wan et al. 1998; Zhang et al. 2006) transition metals, (Bohme 2008; Chaur et al. 2009; Suzuki et al. 1996; Wang et al. 2000; Yamada et al. 2008b) lantanides, (Lu et al. 2008; Wang et al. 1997; Xu et al. 2006a; Yamada et al. 2008a, 2010) iron, (Pradeep et al. 1992) cobalt (Bethune et al. 1993) and uranium (Guo et al. 1992). Not mentioned in (Chaur et al. 2009), there are also such complexes involving three metallic guests $M_3@C_{2n}$ (Yang and Dunsch 2006; Yannoni et al. 1992).
   The last three groups, not classical EMFs, are listed below.
2. Metallic carbides EMFs ($M_2C_2@C_{2n}$, and $M_3C_2@C_{2n}$, M = metal and $68 \leq 2n \geq 92$) (Iiduka et al. 2005; Wakahara et al. 2004; Wang et al. 2001). In their review published in 2006, Dunsch and Yang (2006) asked why metallic carbides EMFs are formed only with scandium or ittrium metals (Iiduka et al. 2006; Wang et al. 2001). Recently, a carbide structure have been proven by X-ray also for $Gd_2C_2@D_3(85)$-C_{92} (Yang et al. 2008). As with numerous other endohedral fullerenes, also in this case the guest can stabilize a non-IPR structure (Wang et al. 2001).
3. Metallic nitrides EMFs ($M_3N@C_{2n}$, M = metal and $68 \leq 2n \geq 96$) (Chaur et al. 2008a, b, 2009; Yang et al. 2009) They include, among other, mixed $Er_xSc_{3-x}N@C_{80}$ (x = 0 – 3) (Campanera et al. 2002; Stevenson et al. 1999), unusual mixed lutetium/yttrium nitride metallofullerenes (Tarabek et al. 2009) and $CeSc_2N@C_{80}$, (Wang et al. 2006), $Gd_2ScN@C_{80}$, and $TbSc_2N@C_{80}$, (Stevenson et al. 2008a) $Tb_3N@C_{80}$, $Tb_3N@C_{86}$ and $Tb_3N@C_{88}$, (Zuo et al. 2007) $Gd_3N@C_{78}$, (Beavers et al. 2009) and $Sc_3N@C_{68}$. (Olmstead et al. 2003) have been proven to have non-IPR structures. On the basis of quantum chemical calculations, Poblet, Echegoyen et al. (Chaur et al. 2008c) have found that redox

properties of metallic nitride endohedral fullerenes do not depend on the cage size supporting the ionic model of binding in these complexes.

4. Metallic oxide EMFs ($M_4O_2@C_{80}$) or $M_4O_3@C_{80}$ (Mercado et al. 2010).

The EMF structures were extensively studied using numerous physicochemical methods, (Chaur et al. 2009; Guha and Nakamoto 2005; Popov 2009; Popov and Dunsch 2009) among others X-ray, (Campanera et al. 2002; Mercado et al. 2010; Wang et al. 2000; Zuo et al. 2007) NMR, (Heine et al. 2004; Iiduka et al. 2006; Koltover 2003; Koltover et al. 2003; Reveles et al. 2005) EPR, (Dresselhaus et al. 1995b; Koltover et al. 2000, 2003; Tagmatarchis et al. 2002) XAFS, (Kubozono et al. 2001) electrochemical, (Chaur et al. 2009) IR and Raman methods, (Guha and Nakamoto 2005; Popov 2009) as well as computational approaches (Campanera et al. 2002; Guha and Nakamoto 2005; Iiduka et al. 2007; Popov and Dunsch 2009). Of particular interest were free movement of guest(s) (Akasaka et al. 1997) or its (their) specific position in the cavity and the nature of host-guest bonding in EMF (Popov 2009). A covalent character of the bond between the C_{66} cage and the covalently bound scandium dimer yielding the trivial topological structure are especially interesting with this respect (Takata et al. 2003).

7.3.2 Group V Endohedral Fullerenes

To our best knowledge, only N, N_2 and P of this group have been found to form the complexes. Interestingly, although in Nature nitrogen is found only in the form of a diatomic molecule, it has been encapsulated as monoatomic radical in C_{60} and C_{70} as well as in four C_{60} adducts. $N@C_{60}$ (Jakes et al. 2002; Mauser et al. 1997; Naydenov et al. 2006; Pietzak et al. 1998; Suetsuna et al. 2002) $N@C_{70}$, (Lips et al. 2000) $N_2@C_{60}$, (Suetsuna et al. 2002) $N@C_{70}$ (Cao et al. 2006) and $P@C_{60}$ (Knapp et al. 1998; Naydenov et al. 2006) together with its two adducts (Scheloske et al. 2006) and a $N@C_{60}$–C_{60} dimer **11** in which only one of the connected C_{60} cages is filled, (Goedde et al. 2001) have been reported.

The $N@C_{60}$ complex is so stable at ambient conditions that it survives exohedral addition reactions. As such it produces a very characteristic, very clear hyperfine split EPR signal with sharp lines even in the solid state and has been proposed as ideal probe for monitoring chemical reactions of C_{60} via changes of the signal (Pietzak et al. 1998). In particular, the effects of cage variation in the series $N@C_{60}$, $N@C_{61}(COOC_2H_5)_2$, $N@C_{66}(COOC_2H_5)_{12}$, $N@C_{66}(COOC_2D_5)_{12}$, $N@C_{61}(COOC_2D_5)_2$ and $N@C_{70}$ have been examined with this respect (Dietel et al. 1999).

The system built of an isolated nitrogen atom and fullerene represents a nanomagnet manipulation of which may enable novel devices for high-density information storage and quantum-state control (Grose et al. 2008). Due to the spin to charge conversion in thin $N@C_{60}$ films at room temperature, the possibility of application of these endohedral fullerene (N@C-60) spin qubits has been proposed (Scheloske et al. 2006).

7.3.3 Nested Fullerenes and Analogous Structures

As mentioned earlier, nested or onion-like fullerenes are built of two or more fullerene cages buried inside one another. The number of fullerene cages in such complexes can be as high as several hundreds (Ugarte 1992). As mentioned before, the latter have been discovered long before C_{60} itself (Iijima 1980). Non-spheroidal nested fullerenes has also been reported (Terrones and Terrones 1997). Dodziuk et al. have shown that the driving force for the formation of nested fullerenes is due to small but very numerous nonbonding attractions (Dodziuk et al. 2000). Mordkovitch reported obtaining two- and three-layered $C_{60}@C_{240}$, $C_{240}@C_{560}$ and $C_{80}@C_{240}@C_{560}$ (Mordkovich 2000). We believe that what he really had observed were complexes involving the C_{540} cage not the C_{560} one (Dodziuk et al. 2000).

An interesting type of clusters in which fullerene forms the core are multi-layered metal (Ca, Sr, Ba) structures grown over a fullerene such as $C_{60}@M_{32}$, $C_{60}@M_{32}@M_{72}$, $C_{60}@M_{32}@M_{72}@M_{132}$, and $C_{60}@M_{32}@M_{72}@M_{132}@M_{212}$ (Zimmermann et al. 1994). As evidenced by $C_{60}@M_{32}$ and $C_{70}@Ba_{37}$, in the first layer one metal atom is placed over each of the 12 pentagonal faces and over each of 20 (or 25 for the larger cage) hexagonal faces.

7.3.4 Fullerenes with Noble Gas Atom(s) or Molecule(s) as Guest(s)

Fullerenes (C_{60} to C_{400}) have not only been found in nature but also their endohedral complexes with noble gases have been detected in the Allende and Murchison meteorites and some sediment samples in deposites associated with two separate events involving the impact of a large bolide (asteroid or comet) with the Earth (Becker et al. 2000). Similarly to fullerene itself (which happily could not be patented because it has been found in Nature), endohedral fullerenes with nobel gas guests are not only man-made.

Similarly to fullerenes themselves, endohedral fullerenes are obtained by heating in the noble gas atmosphere at high pressure using Krätschmer-Huffmann procedure (Krätschmer et al. 1990; Saunders et al. 1993). The C_{60} cage has been shown to be large enough to enclose all the noble gases helium, neon, argon, krypton, and xenon (Rubin et al. 2001; Saunders et al. 1994) and the barrier for the helium atom for the penetration of six-membered fullerene ring has been calculated to be larger than 200 kcal/mol. Although model calculations (Event et al. 2005) indicated that four He atoms inside C_{60} are close to the packing limit, only very small amount of $2He@C_{60}$ have been detected (Sternfeld et al. 2002). Slightly larger amount of $2He@C_{70}$ (Khong et al. 1998) and $2Ne@C_{70}$ (Laskin et al. 1998) have been found. This triggered a discussion whether the noble atoms in the last complexes are present in atomic or molecular form. Krapp and Frenking carried out quantum chemical calculations using DFT (BP86) and ab initio methods (MP2, SCS-MP2) for the endohedral fullerenes $Ng_2@C_{60}$ (Ng = He - Xe) to elucidate the nature and mere existence of the Ng-Ng bond (Krapp and Frenking 2007). The authors state that the Ng-Ng distances in Ng_2C_{60} are much shorter than in free 2Ng. However, they claim that Ng-Ng bond is present only in $Xe_2@C_{60}$ while $He_2@C_{60}$ and $Ne_2@C_{60}$ are weakly bonded van der Waals complexes. In the former case of the heavy noble gas, there is a considerable charge transfer between the guest and the host cavity. This effect, which is thought to be resonsible for the bonding in endohedral metal complexes, is considered to be responsible for the Xe-Xe bonding in the complex with C_{60}. In view of a relatively close distance between He and Ne atoms inside C_{60} and changes in properties of the host and guests after complexation one often speaks about the $He_2@C_{60}$ (Sternfeld et al. 2002) and $Ne_2@C_{70}$ (Laskin et al. 1998) complexes in which the noble gases form weakly bound van der Waals molecules. In any case, the latter complexes ilustrate the impact the encapsulation can have on a guest.

As stated above, endohedral complexes with nobel gas guests are obtained using a modified Krätschmer-Huffmann procedure (Krätschmer et al. 1990; Saunders et al. 1993). Obtaining argon and krypton in opened fullerene cage (Stanisky et al. 2009) has to be acknowledged here since it may pave the way for a synthesis of $Xe_2@C_{60}$ allowing one to check the Krapp and Freking prediction (Krapp and Frenking 2007) of a real Xe_2 molecule inside the C_{60} cage.

NMR studies of endohedral fullerenes labeled with ^{3}He showed that the magnetic field inside the cage is altered by aromatic ring current effects (Ruttimann et al. 1997) each chemical derivative of a fullerene has given a distinct ^{3}He NMR peak (Cross et al. 1996). Thus, this technique, using this and other noble gases, plays an important role in fullerene studies.

7.3.5 Fullerenes with Neutral or Slightly Polar Molecule(s) as Guest(s)

Only few endohedral fullerenes with neutral or slightly polar molecules (other than noble gases) as guests have been reported. They encompass $H_2@C_{60}$, (Komatsu et al. 2005b, 2007) $D_2@C_{60}$, (Tanabe et al. 2006) $H_2@C_{70}$, and $2H_2@C_{70}$, (Murata et al. 2008b) $^{13}CO@C_{60}$, (Peres et al. 2001), and nitrogen molecule buried in C_{60} (Suetsuna et al. 2002) and nitrogen atom in C_{70} (Jakes et al. 2002) as well as two hydrogen molecules and atomic nitrogen inside the C_{60} dimer **11** (Murata et al. 2006) and **12** (Goedde et al. 2001), respectively. One hydrogen molecule put separately into one of the C_{60} cages of the dimer, N_2 inserted into one of the C_{60} cages of the dimer as well as two N atoms or two nitrogen molecules inserted into two C_{60} cages of C_{120} still await their synthesis. CO and N_2 have been inserted into the fullerene cages by more standard procedures during the fullerene manufacturing while hydrogen molecules have been inserted into the fullerene cages by so-called “molecular surgery”, that is by chemical opening the cage, carrying out the guest insertion and chemical closing the cage. As mentioned before, this method of obtaining fullerene complexes with molecular guests has been proposed earlier independently by Patchkovskii and Thiel (1996) and Dodziuk et al. (2001).

H_2, (Rubin et al. 2001) D_2, (Tanabe et al. 2006) H_2O, (Iwamatsu and Murata 2004; Iwamatsu et al. 2004; Xiao et al. 2007) CO, (Iwamatsu et al. 2006; Stanisky et al. 2009) N_2, (Stanisky et al. 2009) NH_3, (Whitener Jr. et al. 2008) and CH_4

(Whitener Jr. et al. 2009) have been inserted inside an "opened" fullerene, mainly C_{60}. Whether those involving water, carbon monoxide, ammonia and methane can be closed without losing the guest remains to be seen. Our recent SAPT calculations indicate that these complexes should be stable (Korona and Dodziuk 2011).

7.4 Unusual Properties of Endohedral Fullerenes

Formation of a topologically nontrivial molecule is known to change its properties as compared with the trivial one. These changes are less pronounced for catenanes, rotaxanes, knots, etc. than for molecules that are endohedral complexes of cage compounds such as endohedral fullerene complexes. The former ones are characterized by a considerable restriction of molecular dynamics. Worm-like movement of knots (Sauvage and Dietrich-Buchecker 1999) or catenanes and rotaxanes (Watanabe et al. 2001) has to be mentioned with this respect. Dynamic properties of mechanically interlocked molecules and their manifestations in NMR spectra are presented in a very detailed review, mainly based on Stoddart group works (Vignon and Stoddart 2005).

One of the most interesting problems in the structure of fullerene complexes is whether there is a bonding of included atom(s) or molecule(s) to the cage. A covalent bond between the C_{66} cage and the endohedral scandium dimer creates a molecule with exciting but topologically trivial structure (Takata et al. 2003). In $La_2@I_h{-}C_{80}$ (Akasaka et al. 1997) and $Ce_2@I_h{-}C_{80}$ complexes, the guest(s) do not occupy a specific position but circulate three-dimentionally executing large amplitude movements (Yamada et al. 2010). Such a movement was also documented in $Sc_3N@C_{80}$ complex in which the guest rotates freely inside the host while dynamic motion of ytterbium ion in the complex $Yb@C_{74}$ has been reported by Xu et al. (2006b).

Encapsulating one or more atoms or molecules in a fullerene can significantly change properties of both host and guest. For instance, in Nature noble gases exist as monoatomic species but the weakly bonded van der Waals diatomic molecules of noble gases He_2 in C_{60} (Sternfeld et al. 2002) and of Ne_2 in C_{70} (Laskin et al. 1998) have been reported. On the other hand, nitrogen is usually present in form of diatomic molecules. However, monoatomic nitrogen radical inside C_{60}, C_{70} and C_{120} dimers (Dietel et al. 1999; Goedde et al. 2001; Naydenov et al. 2006; Pietzak et al. 1998; Suetsuna et al. 2002; Zhang et al. 2008a) and in the $C_{61}(COOC_2H_5)_2$, $C_{66}(COOC_2H_5)_{12}$, $C_{66}(COOC_2D_5)_{12}$, $C_{61}(COOC_2D_5)_2$ cages (Dietel et al. 1999).

"Empty" C_{2n} fullerenes are known to obey the Isolated Pentagon Rule, IPR, stating that the most stable are those in which pentagons do share at least one common atom. This rule holds for empty fullerene cages larger than C_{60} since there are no IPR structures for fullerenes smaller than this cage. There is only one IPR isomer for C_{60} and C_{70}, and the number of isomers increases rapidly with the increase of the cage. It can be shown that there are only 450 IPR isomers of C_{100} while the number of all possible isomers (enantiomers excluded) for this molecule equals 285 913 (Fowler and Manolopoulos 1995). However, a guest can invalidate this rule and several stable non-IPR endohedral fullerene isomers (Rapta et al. 2008; Shi et al. 2006; Shustova et al. 2007; Stevenson et al. 2000; Takata et al. 2003; Tan et al. 2009) have been observed. Interestingly, as shown for instance in case of $Y@C_{82}$ the cage symmetry can change upon the guest inclusion (Takata et al. 1995).

In agreement with the "induced fit" mechanism, (Koshland 1994) not only are the non IPR isomers $Sc_2@C_{66}$ (Takata et al. 2003) and $Sc_3N@C_{68}$ (Stevenson et al. 2000) stable but they also were isolated in spite of the fact that neither the C_{66} and C_{68} hosts nor the latter guests are stable molecules. Noteworthy, $Sc_3N@C_{78}$ does not represent a "real" endohedral complex since there is a considerable bonding between three Sc and cage carbon atoms (Campanera et al. 2002) whereas in the $Sc_3N@C_{80}$ complex the nonbonded guest rotates freely inside the cage.

Fullerenes act as Faraday cage isolating the guest from the outside and/or suppressing the influence of the guest on the whole system (Delaney and Greer 2004; Lips et al. 2000; Zope 2008) For instance, the DFT calculations of the electronic structure and static dipole polarizability of $C_{60}@C_{240}$ **13** have shown practically identical values of dipole polarizability of the complex and that of the isolated C_{240} (449 $Å^3$ vs. 441 $Å^3$). This means that the outer shell almost completely shields the inner one (Zope 2008). At low temperatures, hydrogen molecules can exist in form of *para*- and *ortho*-allotropes which interconvert when H_2 is immersed inside C_{60} (Turro et al. 2008).

Another interesting property of endohedral fullerenes is a very strong sensivity of the NMR signals of the host and encapsulated guest to the charge distribution enabling isomers identification, (Peera et al. 2003) probing fullerene reactivity, (Rosenthal et al. 2006) and internal magnetic field in the cage (Sternfeld et al. 2002) or its aromaticity (Sternfeld et al. 2003).

7.5 "In"-"Out" Isomerism of Hydrogenated Fullerenes

Saunders was the first to notice that not only there is enough space inside fully hydrogenated $C_{60}H_{60}$ to allow for the inward orientation of some CH bonds (Saunders 1991) but also that in this case the molecule becomes much less strained. By applying a simple molecular mechanics, MM, model Saunders has found that the most stable is a unsymmetrical isomer with 10 bonds pointing inside. He also proposed heating as a method which, by breaking and restoring CC bonds, should lead to "in" isomers of hydrogenated fullerenes. To our best knowledge, the most hydrogenated fullerene up-to-date is $C_{60}H_{52}$ (Darwish et al. 2000). $C_{60}H_{60}$ is a hypothetical molecule. As shown by model calculations by Saunders (1991) and Dodziuk and Nowiński (1996) it is highly strained and this explains difficulties in its obtaining. Interestingly, perfluorinated $C_{60}F_{60}$ has been synthesized (Taylor et al. 1992). However, it proved highly unstable producing by decomposition on air the strongest acid HF. Thus, the prospects of its application as an ideal lubricant that should revolutionized whole industry (Stoddart 1991) have to be abandoned. Dodziuk and Nowiński MM calculations (Dodziuk and Nowinski

1996) carried out using a similar procedure as that applied by Saunders (1991) also found the isomer with ten CH bonds pointing inside to be the most stable but in this case it was a symmetrical isomer. Similar but less definite conclusion on the larger stability of “in” isomers of $C_{60}H_{60}$ having several CH bonds pointing inside the cage have been drawn by Dunlap et al. (1991) and Yoshida with coworkers (Yoshida et al. 1993). Taking into account that $C_{60}H_{60}$ is a hypothetical molecule and that the MM calculations applied by Saunders and Dodziuk and Nowiński can yield semiquantitative results at best, one can conclude that an isomer with 10 “in” CH bonds is most probably the most stable. However, the crude methods used today do not allow one for any conclusion concerning the symmetry of such isomer. The dependence of the calculated steric energy on the number of “in” CH bonds is shown in Fig. 7.4. Not only 10 CH bonds were found to be the most stable but also the methyl or ethyl groups pointing inside were calculated to be more stable than when they pointed outside. Moreover, $C_{60}H_{58}(CH_3)_2$ with the methyl groups situated at opposite sides of the cage and pointing inside were predicted to be more stable than when one of them pointed outside and, in turn, the latter isomer was more stable than the one with both Me groups pointing outside. Contrary to Saunders opinion on heating as the method of obtaining the “in” isomers, Patchkovski and Thiele (1996) and independently Dodziuk et al. (2001) advocated that only aimed synthesis can lead to such topologically nontrivial systems. Buchachenko and Breslavskaya (2007) recently reported the B3LYP/6-31G* DFT calculations for two paramagnetic $C_{59}X$ ($X = B, P$) heterofullerenes and found that the systems with CH bonds pointing inside are characterized by the energy lower by 40–50 kcal/mol as compared to the values for the structures with CH bonds pointing outside.

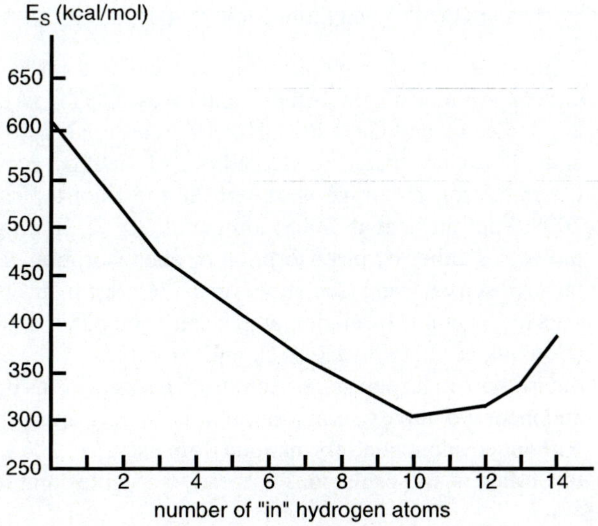

Fig. 7.4 Dependence of the steric energy of $C_{60}H_{60}$ on the number of “in” CH bonds

7.6 Hydrogen Molecule Inside C_{60}: One or More? Can Fullerenes Be Applied for Hydrogen Storage?

Numerous experimental and theoretical papers have been devoted to studies of endohedral fullerene complexes involving guest hydrogen molecule(s). These studies can be divided into syntheses and physicochemical studies on the experimental side and molecular mechanics, MM, and quantum chemical, QC, calculations on the other, with few dynamic simulations in between. In addition to the "in" isomers of hydrogenated fullerenes discussed in Section 7.5, hydrogen can also be present inside the fullerene cage bonded to it, as is the case of $Sc_3CH@C_{80}$ mentioned earlier. Of course, with hydrogen covalently connected to the cage the latter compound is not an object with distinct topological properties.

7.6.1 Hydrogen Storage

It should be stressed that hydrogen as fuel has some important advantages:

a. It is an alternative to coal, oil and gas present in huge amount mainly in H_2O and clathrate hydrates (Dodziuk 2002d).
b. It can easily and effectively be transformed into other forms of energy.
c. It does not create pollution since the only product of its burning is H_2O.

However, hydrogen as fuel has also several disadvantages limiting its applications:

a. As gas, it requires very heavy containers.
b. As liquid, it is very expensive and its use is characterized by huge losses.
c. As metal hydrides, it has a small energy content and is sensitive to impurities (Liu et al. 2010; Sakintuna et al. 2007).

In view of the above arguments, there is a massive search for hydrogen containers. Among others carbon nanotubes, CNTs, (Arai et al. 2009; Becher et al. 2003; Geetha and Gayathri 2010; Hirscher et al. 2001), and fullerenes (Pupysheva et al. 2008) have been considered one of the most promising targets and fullerenes use in hydrogen storage is one of the most hot topics in their applications (Denis 2008; Pupysheva et al. 2008; Shin et al. 2006). Hydrogen can be captured by these molecules either by physisorption or chemisorption. The latter process, which will be not discussed here (see, however, an interesting proposal, (Zhao et al. 2007)) consists in CH bonds formation which can point either outside or inside the carbon cage (Dodziuk et al. 1999; Dodziuk and Nowinski 1996; Saunders 1991) (so called in-out isomerism discussed in Section 7.5). As concerns physisorption, it consists in the endohedral complex formation, that is placing an H_2 molecule inside the C_{60} cage without covalent bond formation (Komatsu et al. 2005a) (the external complexation to a fullerene molecule does not lead to a stable complex). The first report about the nanotubes application for hydrogen storage by Dillon et al. (1997, 1999) seemed very promising but turned out erroneous since absorbance of hydrogen by single wall nanotube bundles was measured without purifying the sample which contained metal catalyst and/or several other impurities. Thus, in spite of several attempts no one could repeat the results by Dillon et al. (Becher et al. 2003; Hirscher et al. 2001) and, to our best knowledge, materials other than carbon nanotubes for hydrogen storage seem more promising. Some nontrivial materials, other than CNTs and fullerenes (such as graphene, (Du et al. 2010) neon hydrate, (Hakim et al. 2010) and nanoporous carbon (Gao et al. 2010)), are also considered for hydrogen storage.

Moreover as mentioned earlier, even if hydrogen would be put inside a fullerene cage it could not be released without the cage breaking. Thus, fullerenes are not suitable for repeated usage as hydrogen container. However, numerous, especially computational works have been carried out with such proclaimed purpose. A recent idea, to store hydrogen in an chemically opened fullerene cages (Hu and Ruckenstein 2005) could be a route to overcome this obstacle. Other approaches to the fullerenes usage for hydrogen storage explored by Yoon (2007) and Kuc et al. (2007) and Patchkovskii and Heine (2007) have consisted in taking advantage of exohedral complex formation involving hydrogen molecules, for instance in the solid or porous state. In spite of the fact that fullerenes cannot serve as hydrogen containers, numerous calculations researchers have been carried out with the aim to determine how many hydrogen molecules can be hosted by C_{60} and C_{70} cages.

7.6.2 Modeling Fullerene Endohedral Complexes with Hydrogen Molecule(s) Guests

Interest in endohedral fullerene complexes involving hydrogen guests arouse long before their experimental observations. Therefore, molecular modeling has been applied in these studies. However, few words on the limitations of fullerenes modeling should be said before discussing them. The molecules are large and difficult to study both experimentally and theoretically. A very revealing assessment of the difficulties encountered by the experimental determination of bond lengths in **14** and **15** and their comparison with the calculations has been given some time ago by Andreoni (1998). Her statements "Many aspects of the physics and chemistry of fullerenes are understood by now but many others still constitute an open question. In a number of cases, we seem to have contrasting but still coexisting views." seem valid today. This opinion is even more valid for the number of hydrogen molecules that can be hosted by a fullerene C_{2n} cage which will be discussed in detail later in this section. In spite of a low reliability of quantum chemical calculations for fullerenes discussed in some detail below, fullerene cages attract numerous theoreticians which often publish calculations using unreliable and/or unchecked methods, e.g., on relative stabilities of fullerene C_{2n} isomers which have not yet been detected thus without a possibility of any check against experiment.

13

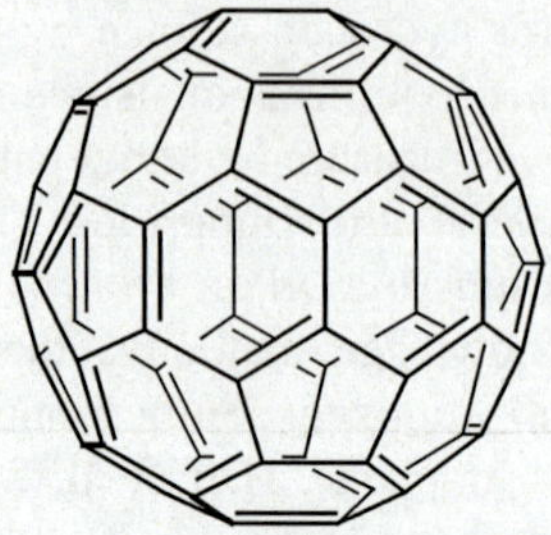

14

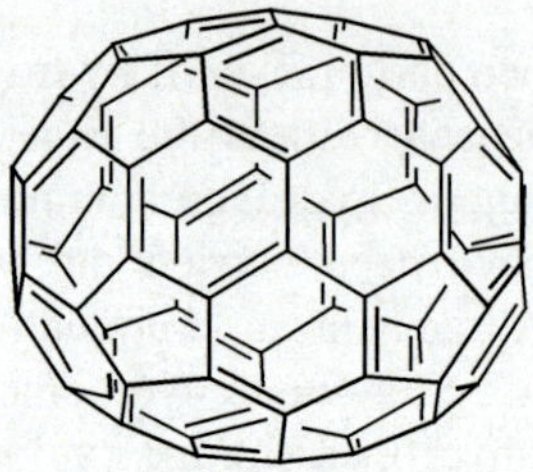

The title of this section suggests that it deals only with hydrogen molecules inside a fullerene cage. However, few words should be said about a possibility of the hydrogen atom or ion enclosed in C_{60} since in few theoretical papers stability of such complexes have been analyzed. Such systems are unknown and it is not clear how they could be obtained experimentally. $H_2@C_{60}$ has been obtained. Thus, if there is enough room for H_2 inside the cage then it is too much room for the atomic hydrogen and the fullerene complex with atomic or ionic hydrogen would not be energetically favored. This very simple reasoning indicates that it will be very difficult, if possible, to obtain H, H^+ or H^- inside C_{60}. Of course, once such a system has been formed (how?) it would be stable. This argument refers to all, also to quantum chemical, theoretical papers, discussed in Section 7.6.2.2 dealing with such complexes.

7.6.2.1 Molecular Mechanics Calculations

As early as in 1993 Williams et al. have analyzed stability and dynamics of some isomers of C_{60} and C_{70} with H_2, F_2 and other diatomic, triatomic and larger guests using a very simple model of atom-atom potential energies (Williams et al. 1993). They have found that not only diatomic molecules H_2, F_2, etc. but even the pentaatomic CH_4 molecule is stabilized inside the host cage.

In 2001, Dodziuk and coworkers (2001) carried out MM calculations of stabilization energies (defined as a difference between the steric energies of the complex minus the sum of the energies of its constituent parts) of several endohedral fullerene complexes using three different force fields (CFF91, (Maple et al. 1988) ESFF (Barlow et al. 1996) and CVFF (Dauber-Osguthorpe et al. 1988)) with the aim to qualitatively analyze which molecules could be hosted by C_{60} (I_h), C_{70} (D_{5h}), C_{76} (D_2, T_d), C_{80} (C_{2v}(3)), C_{80} ($C_{2v'}$(5), D_2(2), D_3(4), D_{5d}(1), D_{5h}(6), I_h(7) (Fig. 7.5) and C_{82} (C_{2v}(9), C_{3v}(7), C_{3v}(8), C_2(1), C_2(3), C_2(5), C_s(6), C_s(2), C_s(4)) cages (the numbers in parentheses give the symmetry and isomers

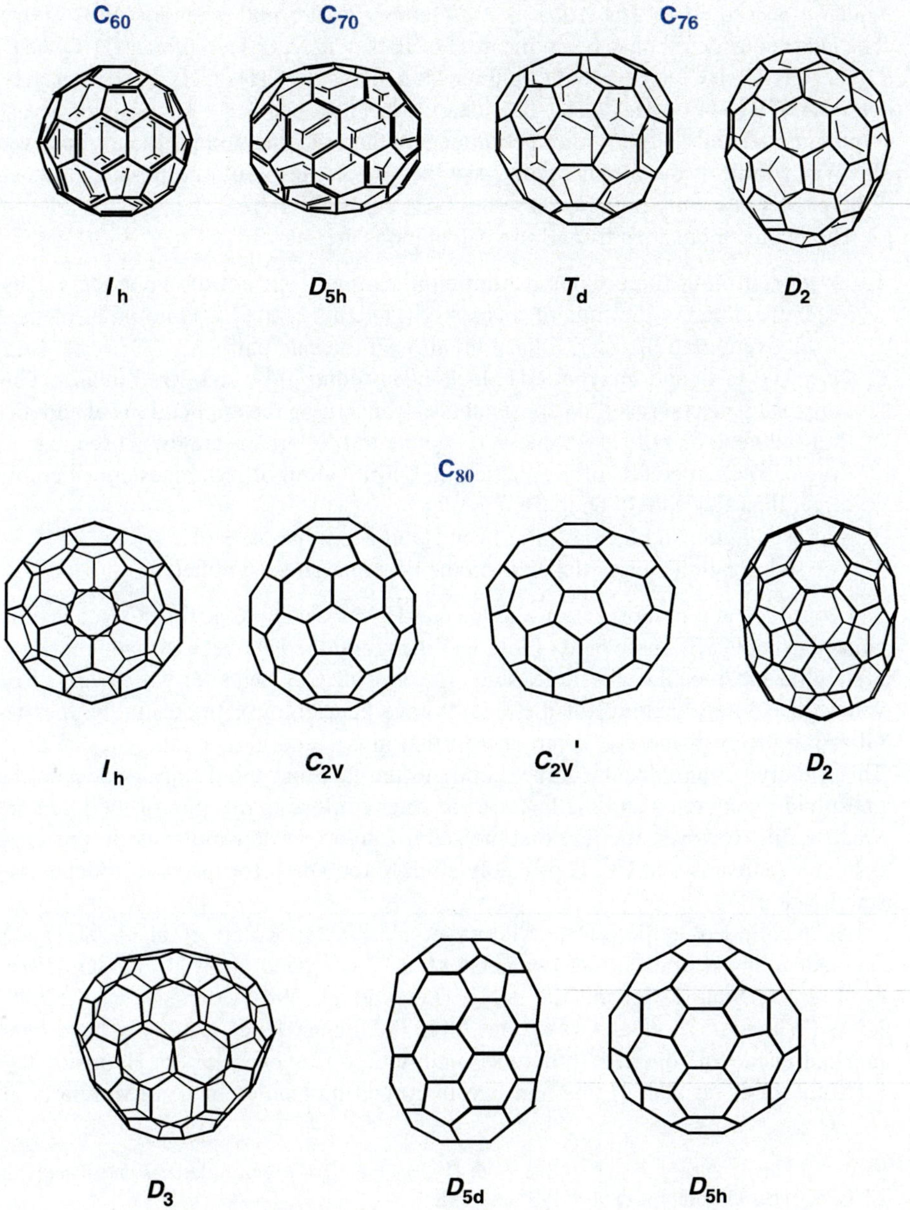

Fig. 7.5 IPR isomers of C_{60}, C_{70}, C_{76} and C_{80} fullerenes and their symmetries

notation according to *The Atlas of Fullerenes* (Fowler and Manolopoulos 1995). The guest considered have been linear (H_2, HCN, ClCN, C_2H_2), planar (H_2O, CO_2, C_2H_4, CH_2O, $HCONH_2$, H_2S) or three dimensional (NH_3, CH_4, CH_3Cl, C_2H_6, CH_3OH, CH_3NH_2, CH_3COH). The idea has been to check on the basis of a very simple model, only slightly more complicated than the one taking into account van der Waals radii of the atoms involved, which molecules could be hosted in which fullerenes. Thus, only physisorption has been analyzed there.

The main conclusions from those calculations are:

1. In general, both force fields exhibit similar trends but not the concrete values. We believe that at the time of this study (Dodziuk et al. 2001) no other method could give more reliable results at the time of the calculations.
2. Only H_2, H_2O and, maybe, NH_3 have been found to be stabilized in C_{60}. This conclusion has far reaching consequences concerning the applications of endohedral fullerene complexes, namely, if endohedral fullerenes are to be used, e.g. as drug carriers, methods of production and purification of fullerenes considerably larger than C_{60} should be mastered.
3. Almost all guest molecules have been stabilized in isomers of C_{82}.
4. The stabilization energy depends on the host and guest symmetry.

Let us look at the stabilization energies for H_2 inside C_{60}, C_{70}, C_{76}, C_{80} cages calculated using CVFF force field (Table 7.1). As mentioned above, hydrogen molecule is stabilized inside all cages under study (also inside C_{82} cages for which very close values of 6.3–6.6 kcal/mol of the energy have been found). Interestingly, the stabilization of H_2 inside C_{60} is larger than that in C_{70} and higher fullerenes studied. This, purely supramolecular effect is due to the fact that the distance between the guest hydrogen atoms and the host carbon ones is close to the sum of their van der Waals radii. However, the fact that the stabilization of water molecule in this cage is higher indicates that C_{60} is probably slightly too small for the most efficient H_2 stabilization.

As mentioned earlier, H_2, (Murata et al. 2008a; Rubin et al. 2001) H_2O, (Iwamatsu and Murata 2004; Iwamatsu et al. 2004; Xiao et al. 2007) CO, (Peres et al. 2001; Stanisky et al. 2009) N_2, (Peres et al. 2001; Stanisky et al. 2009) NH_3, (Whitener Jr. et al. 2008) and CH_4 (Whitener Jr. et al. 2009) have been inserted inside an "opened" fullerene, mainly C_{60}. However, except H_2 inside C_{60} **9** (Komatsu et al. 2005a) one and two hydrogen molecules in C_{70} (Murata et al.

Table 7.1 Steric energy of the complexes of H_2, H_2O and NH_3 with fullerenes calculated with CVFF (Dauber-Osguthorpe et al. 1988) force field

	C_{60}	C_{70}	C_{76}	C_{76}	C_{80}	C_{80}	C_{80}	C_{80}	C_{80}	C_{80}	C_{80}
	I_h	D_{5h}	D_2	T_d	$C_{2v}(3)$	$C_{2v'}(5)$	$D_2(2)$	$D_3(4)$	$D_{5d}(1)$	$D_{5h}(6)$	$I_h(7)$
H_2	−7.7	−7.2	−6.8	−6.6	−6.5	−6.4	−6.7	−6.5	−6.9	−6.2	−6.4
H_2O	−10.9	−11.2	−10.6	−11.1	−10.6	−10.5	−10.5	−10.2	−10.5	−10.4	−10.7
NH_3	−0.3	−8.0	−8.0	−11.7	−11.5	−12.0	−9.1	−9.0	−7.5	−12.2	−12.3

2008a) and in the C_{60} dimer (Murata et al. 2006) as well as nitrogen atom in C_{60} (Suetsuna et al. 2002) and in the C_{60} dimer **11**, (Goedde et al. 2001) N_2 inside C_{60} (Suetsuna et al. 2002) and C_{70} (Peres et al. 2001) other attempts to close the cage have proven unsuccessful until now.

After learning about observation of two H_2 molecules inserted into an opened C_{70} cage at a conference, (Komatsu 2005a) Dodziuk has carried out the MM calculations for one to four hydrogen molecules inside the IPR isomers of C_{60}, C_{70}, C_{76}, C_{78}, and C_{80} (Dodziuk 2005). This modeling has shown that both one and two hydrogen molecules could be stabilized inside a closed C_{70} cage. However, the absolute value of the stabilization energy for one H_2 guest is larger than the corresponding value for two guests (–4.8 kcal/mol vs −3.0 kcal/mol, respectively). Thus, on the basis of this modeling Dodziuk concluded that it will be very difficult to close the C_{70} cage with both guests inside. Later results by Murata et al. (2008a) confirmed these predictions. Namely, the nonseparated mixture of $H_2@C_{70}$ and $2H_2@C_{70}$ was obtained in the proportion 96:4 somewhat accidentally exactly corresponding to the energy difference of 1.8 kcal/mol calculated by Dodziuk using the MM method (Dodziuk 2005). In addition, the calculations indicated that of the fullerenes studied C_{76} of D_2 symmetry seems to be the smallest for which the endohedral complex with two hydrogen molecules should be the most stable.

Two general conclusions have been drawn from the calculations by Dodziuk et al. (2001). First, to be able to apply endohedral fullerene complexes in reality chemists have to master obtaining and purification of larger fullerenes. Secondly, knowing that most endohedral fullerene complexes have been obtained during the syntheses of the parent cage, we foresaw that the most prospective method to obtain $H_2@C_{60}$ would be a chemical opening of the cage, followed by the guest insertion with a subsequent chemical closure of the opening (Dodziuk et al. 2001). This conclusion was independent on the earlier one published by Patchkovskii and Thiele (1996) Insertion of H_2 into the cage has been realized in this way (there is only one report using another approach (Oxengorn 2003) that has never been followed). Rubin and coworkers (Rubin et al. 2001) have reported opening the C_{60} cage and inserting helium atom or H_2 into it. Analogous procedure has been developed in the Komatsu group (Murata et al. 2003; Sawa et al. 2005) (summarized in Hara et al. 2009) who succeeded in closing the cage with hydrogen molecule remaining inside thus generating $H_2@C_{60}$ **9** (Komatsu et al. 2005a) using this method. As mentioned above, Komatsu, Murata et al. group succeeded in obtaining not only the latter complex but also $H_2@C_{70}$ (Komatsu et al. 2005a, b) $2H_2@C_{70}$ **10** (Murata et al. 2008b) and $2H_2@C_{120}$ **12** (Murata et al. 2006) while Goedde et al. obtained $N@C_{120}$ **11** (Goedde et al. 2001). The method was dubbed "molecular surgery".

The results of MM modeling, as well as simple arguments on the basis of van der Waals radii limiting the closest approach of the hydrogen and carbon atoms, cast doubts in the results of papers claiming, on the basis of semiempirical, ab initio, or DFT calculations, that more than one molecule of hydrogen, (Barajas-Barraza and Guirado-Lopez 2002; Ganji 2008; Shigeta and Saito 2003; Shigeta and Takatsuka 2005; Turker and Erkoc 2003) water, (Ramachandran and Sathyamurthy 2005) or ammonia (Erkoc and Turker 2003) can be hosted by the C_{60} cage.

7.6.2.2 Quantum Chemical Calculations of Fullerene Complexes

Putting hydrogen inside a fullerene was first studied by quantum chemical calculations as early as in 1991 by Cioslowski (1991) who carried out the *ab initio* quantum chemical calculations for the frozen geometry of an empty C_{60} cage using 4–31G basis set for the cage and DZP basis set for the optimized included guest (H_2, N_2, CO, HF, LiH and LiF). As usually, stabilization energies of the complexes under investigation have been defined as the differences between the energies of a complex and its free constituent parts. The results obtained (for two different symmetries of the C_{60} cage for H_2 and HF guests) shown in Table 7.2 indicate that the purely qualitatively the method yielded stabilization of the complexes with strongly polar guests and destabilization in case of nonpolar H_2 and N_2 and slightly polar CO included molecules. H_2@C_{60}, (Komatsu et al. 2005b) N_2 (Peres et al. 2001; Suetsuna et al. 2002) and CO@C_{60} (Peres et al. 2001) have been obtained later proving the inadequacy of the pure Hartree-Fock approach for description of dispersion interactions.

The largest endohedral content of 29 hydrogen molecules has been reported by Yang (2007) and Pupysheva et al. (2008) on the basis of DFT calculations which are known to inadequately describe nonbonding interactions (Bartlett et al. 2005; Kamiya et al. 2002). If, according to Yang, Pupysheva et al. the hydrogen content is high, then in addition to physisorption, chemisorption, that is formation of CH bonds pointing inside, takes place. Taking into account that "C_{60} has a relative large and robust cage structure", Yang concluded that "as many as 29 hydrogen molecules can be stored without rupturing the cage" although it should lead to unphysically close location of hydrogen and carbon nonbonded atoms. The latter value is even higher than 24 H_2 molecules inside the cage claimed by Turker and Erkoc (2003) discussed earlier. The situation analyzed in both studies is not the same since in the latter work only physisorption has been analyzed while in Yang work some CH "in" bonds are formed. However, the hydrogen atoms involved in these bonds occupy a considerable part of the inside volume. Then, according to simple arguments similar to those given in the former section when discussing Turker and Erkoc (2003) work there is no space even for one physisorped hydrogen molecule inside the hydrogenated fullerene with some CH bonds buried inside the cage. The critique of Yang paper and its conclusions has been published by Dolgonos (2008) who has repeated his (Dolgonos 2005) and Dodziuk (2006) earlier arguments on the unreliability of Turker and Erkoc paper (Turker and Erkoc 2003) (in which, as discussed in section II.1, "only" 24 hydrogen molecules had been found to be hosted by the C_{60} cage) on the basis of the pure geometrical considerations but rather disclaimed the

Table 7.2 Calculated stabilization energies of endohedral complexes

Molecule	ΔE (kcal/mol)	Molecule	ΔE (kcal/mol)
H_2 (D_{5d})	1.22	HF (C_{5v})	−1.94
H_2 (D_{3d})	1.22	HF (C_{3v})	−1.96
N_2 (D_{5d})	9.60	LiH	−2.39
CO	11.20	LiF	−14.94

methodology applied by Yang. The reply (Yang 2008) to the latter commentary has been unsubstantial.

Similarly, contrary to the results of DFT calculations, (Jin et al. 2008) there is no sufficient room inside the C_{60} cage to host even one acetylene, ethylene or ethane molecule. Since MP2 calculations are known to overestimate the attractive dispersion interactions (Sponer et al. 1996).

To overcome the limitations of the classical DFT method, either special functionals like modified Perdew-Wang and Becke functional, MPWB1K, (Zhao and Truhlar 2004) or explicit inclusion of London dispersion term into the model (DFT-D method) (Grimme 2004) have been developed. The former method has been applied to $H_2@C_{60}$, $He@C_{60}$, and $N_2@C_{60}$ by Slanina et al. (2006) to check the applicability of the method to such systems. The latter authors have found that their values of stabilization energy are close to the MP2 or SCS-MP2 (spin-component scaled MP2). Taking into account that MP2 method is known to overestimate the values of nonbonding interactions, (Hobza et al. 1996) a real check of the method would be showing that $2H_2@C_{60}$ is not stable. The latter conclusion is corroborated not only by simple model considerations on the basis of van der Waals radii and the results of MM calculations by Dodziuk et al. (2001) but also by recent quantum chemical calculations by Symmetry-Adapted Perturbation Theory, SAPT, by Korona et al. (2009). Kruse and Grimme (2009) came to the same conclusion on the $2H_2@C_{60}$ instability on the basis of quantum chemical calculations involving double-hybrid density functionals. However, their conclusion (and that by Pupyshva et al. 2008) on the larger stability of $2H_2@C_{70}$ than that of $H_2@C_{70}$ seems incompatible with the larger content of the former molecule than the latter one in the mixture obtained by Murata et al. (2008a). Interestingly and somewhat accidentally, the 96:4 relative content of the complexes in the mixture corresponds to the energy difference of 1.8 kcal/mol estimated by Dodziuk using the MM calculations (Dodziuk 2005).

On the basis of a generalized gradient approximation of the DFT method, Bai and coworkers (Bai et al. 2008) correctly found stabilization of one hydrogen molecule in C_{60} and its dimer and a free guest rotation inside the cage. As mentioned earlier, the erroneous conclusion on the possibility of several clusters of molecular hydrogen inside C_{60}, C_{82} and some nanotubes drawn by Barajas-Barraza and Guirado-Lopez (2002) who used semiempirical MNDO and DFT calculations is based on the deficiencies of the applied methods. Similarly, semiempirical PM3 and DFT calculations by Ren et al. (2006) on the configurations of hydrogen molecules inside C_{60} could not bring reliable results. Taking into account that no more than one hydrogen molecule can be housed by the C_{60} cage, we strongly believe that studying catalytic activity of C_{60} on enclosed two to five hydrogen molecules (Lee and McKee 2008) is a purely in silico experiment on the highly strained hypothetical systems the results of which have no chances to be compared with experiment.

7.7 Conclusions

The studies of endohedral fullerene complexes are booming due to their unusual properties and prospective applications. Larger fullerenes are obtained and more guests are inserted into the fullerene cages producing exciting complexes. The practical applications are still *in statu nascendi*. They will most probably include uses in medicine, molecular devices (e.g. as single-molecule transistor for quantum computing, (Twamley 2003) in solar cells, (Zhang et al. 2008b) etc. One of the most exciting is a prospective application of so-called peapods that is carbon nanotubes filled with endohedral fullerenes encapsulating metal atoms (Kurokawa et al. 2005). However, using hydrogen filled fullerenes for the gas storage seems at the moment highly unlikely. Further development of theoretical methods is needed for a reliable description of endohedral fullerene complexes.

References

Afanas'ev DV, Bogdanov AA, Dyuzhev GA, Kruglikov AA (1997) Zh Tekh Fiz 67:125–128

Agam G, Graiver D, Zilkha A (1976) J Am Chem Soc 98:5206–5214

Agam G Zilkha A (1976) J Am Chem Soc 98:5214–5216

Akasaka T, Nagase S (eds) (2002) Review endofullerenes: a new family of carbon clusters. Kluwer, Dordrecht

Akasaka T, Nagase S, Kobayashi K, Waelchli M, Yamamoto K, Funasaka H, Kako M, Hoshino T, Erata T (1997) Angew Chem Int Ed Engl 36:1643

Amabilino DB, Asakawa M, Ashton PR, Ballardini R, Balzani V, Belohradsky M, Credi A, Higuchi M, Raymo FM, Shimizu T, Stoddart JF, Venturi M, Yase K (1998) New J Chem 22:959–972

Amabilino DB, Ashton PR, Reder AS, Spencer JF, Stoddart JF (1994) Angew Chem Int Ed Engl 33:1286

Amabilino DB, Stoddart JF (1995) Chem Rev 95:2725–2828

Anderson MR, Dorn HC, Stevenson SA (2000) Carbon 38:1663–1670

Andreoni W (1998) Annu Rev Phys Chem 49:405–439

Arai M, Utsumi S, Kanamaru M, Urita K, Fujimori T, Yoshizawa N, Noguchi D, Nishiyama K, Hattori Y, Okino F, Ohba T, Tanaka H, Kanoh H, Kaneko K (2009) Nano Lett 9:3694–3698

Ashton PR, Matthews OA, Menzer S, Raymo FM, Spencer R, Stoddart JF, Williams DJ (1997) Liebigs Ann 2485–2494

Bai YJ, Fu SY, Deng KM, Tang CM, Chen X, Tan WS, Lin YZ, Huang DC (2008) Acta Phys Sin 57:3684–3689

Bakry R, Vallant RM, Najam-ul-Haq M, Rainer M, Szabo Z, Huck CW, Bonn GK (2007) Int J Nanomed 2:639–649

Barajas-Barraza RE, Guirado-Lopez RA (2002) Phys Rev B 66:155426

Barlow S, Rohl AL, Shi SG, Freeman CM, O'Hare D (1996) J Am Chem Soc 118:7578

Bartlett RJ, Lotrich VF Schweigertm IV (2005) J Chem Phys 123:062205

Beavers CM, Chaur MN, Olmstead MM, Echegoyen L, Balch AL (2009) J Am Chem Soc 131:11519–11524

Beavers CM, Zuo TM, Duchamp JC, Harich K, Dorn HC, Olmstead MM, Balch AL (2006) J Am Chem Soc 128:11352–11353

Becher M, Haluska M, Hirscher M, Quintel A, Skakalova V, Dettlaff-Weglikovska U, Chen X, Hulman M, Choi Y, Roth S, Meregalli mV, Parrinello M, Strobel R, Jorissen L, Kappes MM, Fink J, Zuttel A, Stepanek I, Bernier P (2003) Comp Rend Phys 4:1055–1062

Becker L, Poreda RJ, Bunch TE (2000) Proc Natl Acad Sci USA 97:2979–2983

Bethune DS, Kiang CH, de Vries MS, Gorman G, Savoy R, Vazquez J, Beyers R (1993) Nature 363:605–607

Bohme DK (2008) Can J Chem 86:86

Bolskar RD (2008) Nanomedicine 3:201–213

Buchachenko AL Breslavskaya NN (2007) Russ Chem Bull 56:1283–1288

Cagle DW, Kennel SJ, Mirzadeh S, Alford JM, Wilson LJ (1999) Proc Natl Acad Sci USA 96:5182–5187

Cai T, Xu L, Gibson HW, Dorn HC, Chancellor CJ, Olmstead MM, Balch AL (2007) J Am Chem Soc 129:10795–10800
Cai T, Xu L, Shu C, Champion HA, Reid JE, Anklin C, Anderson MR, Gibson HW, Dorn HC (2008) J Am Chem Soc 130:2136–2137
Campanera JM, Bo C, Olmstead MM, Balch AL, Poblet JM (2002) J Phys Chem A 106:12356–12364
Cantrill SJ, Chichak KS, Peters AJ, Stoddart JF (2005) Acc Chem Res 38:1–9
Cao BP, Peres T, Lifshitz C, Cross RJ, Saunders M (2006) Chem Eur J 12:2113–2221
Cao BP, Wakahara T, Tsuchiya T, Kondo M, Maeda Y, Rahman GMA, Akasaka T, Kobayashi K, Nagase S, Yamamoto K (2004) J Am Chem Soc 126:9164–9165
Cardona CM, Kitaygorodskiy A, Ortiz AL, Herranz MA, Echegoyen L (2005) J Org Chem 70:5092–5097
Carina RF, Dietrich-Buchecker C-O, Sauvage J-P (1996) J Am Chem Soc 118:9110–9116
Champoux JJ (2001) Annu Rev Biochem 70:369–413
Chaur MN, Anthans AJ, Echegoyen L (2008a) Tetrahedron 64:11387–11393
Chaur MN, Melin F, Elliott B, Kumbhar A, Athans AJ, Echegoyen L (2008b) Chem Eur J 14:4594–4599
Chaur MN, Melin F, Ortiz AL, Echegoyen L (2009) Angew Chem Int Ed 48:7514–7538
Chaur MN, Valencia R, Rodríguez-Fortea A, Poblet JM, Echegoyen L (2008c) Angew Chem Int Ed 48:1425–1428
Chen J, Rauch CA, White JK, Englung PT, Cozarelli NR (1995) Cell 80:61–69
Cioslowski J (1991) J Am Chem Soc 113:4139–4141
Claessens CG Stoddart JF (1997) J Phys Org Chem 10:254–272
Corbett KD, Berger JM (2004) Ann Rev Biochem 33:95–118
Cross RJ, Jiménez-Vázquez HA, Lu Q, Saunders M, Schuster DI, Wilson SR, Zhao H (1996) J Am Chem Soc 118:11454–11459
Curl RF (1992) Carbon 30:1149–1151
Darwish AD, Taylor R, Loutfy R (2000) Electrochem Soc Proc 11:179–185
Dauber-Osguthorpe P, Roberts VA, Osguthorpe DJ, Wolff J, Genest M, Hagler AT (1988) Proteins 4:31–47
Delaney P, Greer JC (2004) Appl Phys Lett 84:431–433
Denis PA (2008) J Phys Chem C 112:2791–2796
Dietrich-Buchecker CO, Sauvage JP (1989) Chem Rev 89:795–810
Dietrich-Buchecker CO, Colasson B, Sauvage JP (2005) Top Curr Chem 249:261–283
Dillon AC, Gennett T, Jones KM, Alleman JL, Parilla PA, Heben MJ (1999) Adv Mat 11:1354–1358
Dillon AC, Jones KM, Bekkedahl TA, Kiang CH, Bethune MJ, Heben MJ (1997) Nature 386:377–379
Dobrowolski JC (2003) 76:145–152
Dodziuk H (2002a) In: Introduction to supramolecular chemistry. Kluwer, Dordrecht, pp 275–284
Dodziuk H (2002b) In: Introduction to supramolecular chemistry. Kluwer, Dordrecht, pp 27–39
Dodziuk H (2002c) Int J Molec Sci 3:814–821
Dodziuk H (2002d) In: Introduction to supramolecular chemistry. Kluwer, Dordrecht, pp 294–298
Dodziuk H (2005) Chem Phys Lett 410:39–41
Dodziuk H (2006) Chem Phys Lett 426:224–225
Dodziuk H (2007) Properties J Nanosci Nanotechnol 7:1102–1110
Dodziuk H, Dolgonos G (2002) Chem Phys Lett 356:79–83
Dodziuk H, Dolgonos G, Lukin O (2000) Chem Phys Lett 329:351–356
Dodziuk H, Dolgonos G, Lukin O (2001) Carbon 39:1907–1911
Dodziuk H, Lukin O, Nowinski KS (1999) Pol J Chem 73:299–306
Dodziuk H, Nowinski KS (1996) Chem Phys Lett 249:406–412
Dodziuk H, Nowinski, KS (1998) Tetrahedron 54:2917–2930
Dodziuk H, Balaban A, Nowinski KS, Dobrowolski JC (2011) manuscript in preparation

Dolgonos G (2005) J Mol Struct (Theochem) 732:239–241
Dolgonos G (2008) Carbon 46:704–705
Dresselhaus MS, Dresselhau G, Eklund PC (1995a) Science of fullerenes and carbon nanotubes: their properties and applications. Elsevier, Oxford
Dresselhaus MS, Dresselhau G, Eklund PC (1995b) In: Science of fullerenes and carbon nanotubes: their properties and applications. Elsevier, Oxford, pp 670–673
Drummond C, Homyonfer M, Feldman Y, Tenne R, Israelachvili J (1999) Adv Mater 11:934–937
Du AJ, Zhu Z, Smith SC (2010) J Am Chem Soc 132:2876–2877
Dunlap BI, Brenner DW, Mintmire JW, Mowrey RC, White CT (1991) J Phys Chem 95:5763–5768
Dunsch L, Bartl A, Georgi P, Kuran P (2001) Synth Met 121:1113–1114
Dunsch L, Yang S (2006) Electrochem Soc Interface 15:34–39
Erkoc S, Turker L (2003) J Mol Struct (THEOCHEM) 640:57–61
Event W, Smith J, Roth MW (2005) Mol Simul 31:207–213
Fang L, Olson MA, Benitez D, Tkatchouk E, Goddard WA, Stoddart JF (2010) Chem Soc Rev 39:17–29
Fenlon EE (2008) Eur J Org Chem:5023–5035
Fowler PW, Manolopoulos DE (1995) An atlas of fullerenes. Clarendon, Oxford
Francl M (2009) Nature Chem 1:334–335
Frisch HL, Wasserman E (1961) J Am Chem Soc 83:3789–3795
Fu WJ, Xu LS, Azurmendi H, Ge JC, Fuhrer T, Zuo TM, Reid J, Shu CY, Harich K, Dorn HC (2009) 131:11762–11769
Ganji MD (2008) Mol Simul 34:821–828
Gao JB, Adelhelm P, Verkuijlen MHW, Rongeat C, Herrich M, van Bentum PJM, Gutfleisch O, Kentgens APM, de Jong KP, de Jongh PE (2010) J Phys Chem C 114:4675–4682
Geetha R, Gayathri V (2010) Current Nanosci 6:131–136
Gibson HW, Ge ZX, Jones JW, Harich K, Pederson A, Dorn HC (2009) J Polym Sci A 47:6472–6495
Goedde B, Waiblinger M, Jakes P, Weiden N, Dinse KP, Weidinger A (2001) Chem Phys Lett 334:12–17
Grimme S (2004) J Comput Chem 25:1463–1473
Grose JE, Tam ES, Timm C, Scheloske M, Ulgut B, Parks JJ, Abruna HD, Harneit W, Ralph DC (2008) Nat Mat 7:884–889
Guha S, Nakamoto K (2005) Coord Chem Rev 249:1111–1132
Gulick v (1993) New J Chem 17:619–625
Guo T, Diener MD, Chai Y, Alford JM, Haufler RE, McClure SM, Ohno TR, Weaver JH, Scuseria GE, Smalley RE (1992) Science 257:1661–1664
Hakim L, Koga K, Tanaka H (2010) Physica A 389:1834–1838
Hara T, Konno T, Nakamura Y, Nishimura Y (2009) In: H Dodziuk (ed) Strained hydrocarbons. Wiley-VCH, Weinheim
Haussmann PC, Stoddart JF (2009) Chem Rec 9:136–154
Heath JR, O'Brien SC, Zhang Q, Liu Y, Curl RF, Tittel FK, Smalley RE (1985) J Am Chem Soc 107:7779–7780
Heim C, Affeld A, Nieger M, Vögtle F (1999) Helv Chim Acta 82:746–759
Heine T, Vietze K, Seifert G (2004) Magn Res Chem 42:S199–S201 Sp. Iss.
Hirscher M, Becher M, Haluska M, Dettlaff-Weglikovska U, Quintel A, Duesberg GS, Choi Y, Downes P, Hulman M, Roth S, Stepanek I, Bernier P (2001) Appl Phys A 72:129–132
Hobza P, Selzle HL, Schlag EW (1996) J Phys Chem 100:18790–18794
Hu YH, Ruckenstein E (2005) J Chem Phys 123:art 144303
Ibrahim M, Saleh NA, Elshemey WM, Elsayed (2010) J Comput Theor Nanos 7:224–227
Iezzi EB, Duchamp JC, Harich K, Glass TE, Lee HM, Olmstead MM, Balch AL, Dorn HC (2002) J Am Chem Soc 124:524–525
Iiduka Y, Wakahara T, Nakahodo T, Tsuchiya T, Sakuraba A, Maeda Y, Akasaka T, Yoza K, Horn E, Kato T, Liu MTH, Mizorogi N, Kobayashi K (2005) J Am Chem Soc 127:12500–12505

Iiduka Y, Wakahara T, Nakajima K, Nakahodo T, Tsuchiya T, Nakahodo T, Maeda Y, Akasaka T, Yoza K, Liu MTH, Mizorogi N, Nagase S (2007) Angew Chem Int Ed 46:5562–5564
Iiduka Y, Wakahara T, Nakajima K, Tsuchiya T, Nakahodo T, Maeda Y, Akasaka T, Mizorogi N, Nagase S (2006) Chem Commun 2057–2059
Iijima S (1980) J Cryst Growth 50:675
Infante I, Gagliardi L, Scuseria GE (2008) J Am Chem Soc 130:7459–7465
Iwamatsu S, Murata S (2004) Tetrah Lett 45:6391
Iwamatsu SI, Stanisky CM, Cross RJ, Saunders M, Mizorogi N, Nagase S, Murata S (2006) Angew Chem Int Ed 45:5337–5340
Iwamatsu S-i, Uozaki T, Kobayashi K, Re S, Nagase S, Murata S (2004) J Am Chem Soc 126:2668–2669
Jakes P, Weiden N, Eichel RA, Gembus A, Dinse KP, Meyer C, Harneit W, Weidinger A (2002) J Magn Res 156:303–308
Ji ZQ, Sun HF, Wang HF, Xie QY, Liu YF, Wang Z (2006) J Nanoparticle Res 8:53–63
Jimenez-Vazquez HA, Cross RJ, Saunders M, Poreda RJ (1994) Chem Phys Lett 229:111–114
Jin LJ, Zhang M, Su ZM, Shi LL (2008) J Theor Comput Chem 7:1–11
Kamiya M, Tsuneda T, Hirao K (2002) J Chem Phys 117:6010–6015
Kato H, Taninaka A, Sugai T, Shinohara H (2003) J Am Chem Soc 125:7782–7783
Khong A, Cross RJ, Saunders M (2000) J Phys Chem A 104:3940–3943
Khong A, Jiménez-Vázquez HA, Saunders M, Cross RJ, Laskin J, Peres T, Lifshitz C, Strongen R, Smith AB (1998) J Am Chem Soc 120:6380–6383
Kiefl RF, Duty TL, Schneider JW, MacFarlane A, Chow K, Elzey J, Mendels P, Morris GD, Brewer JH, Ansaldo EJ, Niedermayer C, Noakes DR, Stronach CE, Hitti B, Fischer JE (1992) Phys Rev Lett 69:2005–2008
Kitaura R, Shinohara H (2007) Jap J Appl Phys Part 1 46:881–891
Knapp C, Weiden N, Kass K, Dinse KP, Pietzak B, Waiblinger M, Weidinger A (1998) Mol Phys 95:999–1004
Kobayashi S, Mori S, Iida S, Ando H, Takenobu T, Taguchi Y, Fujiwara A, Taninaka A, Shinohara H, Yoshihiro I (2003) J Am Chem Soc 125:8116–8117
Koltover VK (2003) Carbon 42:1179–1183
Koltover VK, Bubnov VP, Estrin YI, Lodygina VP, Davydov RM, Subramoni M, Manoharan PT (2003) Phys Chem Chem Phys 5:2774–2777
Koltover VK, Estrin YI, Bubnov VP, Laukhina EE (2000) Russ Chem Bull 49:1745–1748
Komatsu K, Murata M, Murata Y (2005a) In: XIX international winterschool on electronic properties of novel materials (ed Kuzmany H, Mehring M, Fink J, Roth S) Am. Inst. Phys. Kirchberg, Tirol
Komatsu K, Murata M, Murata Y (2005b) Science 307:238–240
Komatsu K, Murata Y (2004) J Synth Org Chem Jpn 62:1138–1147
Komatsu N (2009) J Jpn Petrol Inst 52:73–80
Komatsu N, Kadota N, Kimura T, Kikuchi Y, Arikawa M (2007) Full Nanotubes Carb Nanostr 15:217–226
Korona T, Hesselmann M, Dodziuk H (2009) J Chem Theory Comp 5:1585–1596
Korona T, Dodziuk H (2011) manuscript in preparation
Koshland DE, Jr (1994) Angew Chem Int Ed 33:2375–2378
Krapp A, Frenking G (2007) Chem Eur J 13:8256–8270
Krätschmer W, FostiropoulosK, Huffmann DR (1990) Chem Phys Lett 170:167
Krause M, Ziegs F, Popov AA, Dunsch L (2007) Chem Phys Chem 8:537–540
Kruse H, Grimme S (2009) J Phys Chem C 113:17006–17010
Kubozono Y, Inoue T, Takabayashi Y, Fujiki S, Kashino S, Akasaka T, Wakahara T, Inakuma M, Kato H, Sugai T, Shinohara H, Emura S (2001) J Synchrotr Rad 8:551–553
Kuc A, Zhechkov L, Patchkovskii S, Seifert G, Heine T (2007) Nano Lett 7:1–5
Kurokawa Y, Ohno Y, Shimada T, Ishida M, Kishimoto S, Okazaki T, Shinohara H, Mizutani T (2005) Jap J Appl Phys Part 2 44:L1341–L1343

Lan JH, Cao DP, Wang WC (2009) ACS NANO 3:3294–3300

Laskin J, Peres T, Lifshitz C, Saunders M, Cross RJ, Khong A (1998) Chem Phys Lett 285:7

Lee J, Kim JH (2008) Environ Sci Technol 42:1552–1557

Lee TB, McKee ML (2008) J Am Chem Soc 130:17610–17619

Lips K, Waiblinger M, Pietzak B, Weidinger A (2000) Mol Mater 13:217–224

Liu C, Li F, Ma LP, Cheng HM (2010) Adv Mat 22:E28–E62

Liu S, Sun S (2000) J Organomet Chem 599:74–86

Liu W, Zhao YH, Li Y, Lavernia EJ, Jiang Q (2009) Phys Chem Chem Phys 11:9233–9240

Löffler D, Bajales N, Cudaj M, Weis P, Lebedkin S, Bihlmeier A, Tew DP, Klopper W, Böttcher A, Kappes MM (2009) J Chem Phys 130:art 164705

Lu X, Nikawa H, Nakahodo T, Tsuchiya T, Ishitsuka MO, Maeda Y, Akasaka T, Toki M, Sawa H, Slanina Z, Mizorogi N, Nagase S (2008) J Am Chem Soc 130:9129–9136

MacFarland DK, Walker KL, Lenk RP, Wilson SR, Kumar K, Kepley CL, Garbow JR (2008) J Med Chem 51:3681–3683

Maple JR, Dinur U, Hagler AT (1988) Proc Natl Acad Sci USA 85:5350–5354

Mauser H, Hommes NJRV, Clark T, Hirsch A, Pietzak B, Weidinger A, Dunsch L (1997) Angew Chem Int Ed 36:2835–2838

McArdle CP, Vittal JJ, Puddephat RJ (2000) Angew Chem Int Ed 39:3819–3822

Mercado BQ, Beavers CM, Olmstead MM, Chaur MN, Walker K, Holloway BC, Echegoyen L, Balch AL (2008) J Am Chem Soc 130:7854–7855

Mercado BQ, Olmstead MM, Beavers CM, Easterling ML, Stevenson S, Mackey MA, Coumbe CE, Phillips JD, Phillips JP, Poblet JM, Balch AL (2010) Chem Commun 279–281

Mordkovich VZ (2000) Chem Mater 12:2813

Murata M, Maeda S, Morinaka Y, Murata Y, Komatsu K (2008a) J Am Chem Soc 130:15800–15801

Murata M, Murata Y, Komatsu K (2006) J Am Chem Soc 128:8024–8033

Murata Y, Maeda S, Murata M, Komatsu K (2008b) J Am Chem Soc 130:6702

Murata Y, Murata M, Komatsu K (2003) J Am Chem Soc 125:7152–7153

Murthy CN, Geckeler KE (2006) Curr Org Synth 3:1–7

Nagata K, Dejima E, Kikuchi Y, Hashiguchi M (2005) Chem Lett 34:178–179

Naydenov B, Spudat C, Harneit W, Süss HI, Hulliger J, Nuss J, Jansen M (2006) Chem Phys Lett 424:327–332

Nierengarten J-F, Dietrich-Buchecker C-O, Sauvage J-P (1994) JACS 116:375–376

Nishibori E, Narioka S, Takata M, Sakata M, Inoue T, Shinohara H (2006) Chem Phys Chem 7:345–348

Olmstead MM, Lee HM, Duchamp JC, Stevenson S, Marciu D, Dorn HC, Balch AL (2003) Angew Chem Int Ed Engl 42:900–903

Oxengorn B (2003) C R Chimie 6:467–472

Park SS, Liu D, Hagelberg F (2005) J Phys Chem A 109:8865–8873

Patchkovskii S, Heine T (2007) Phys Chem Chem Phys 9:2697–2705

Patchkovskii S, Thiel W (1996) J Am Chem Soc 118:7164

Peera A, Saini RK, Alemany LB, Billups WE, Saunders M, Khong A, Syamala MS, Cross RJ (2003) Eur J Org Chem 21:4140–4145

Peinador C, Blanco V, Quintela JM (2009) J Am Chem Soc 131:920–921

Pentecost CD, Chichak KS, Peters AJ, Cave GWV, Cantrill SJ, Stoddart JF (2007) Angew Chem Int Ed 46:218–222

Pentecost CD, Peters AJ, Chichak KS, Cave GWV, Cantrill SJ, Stoddart JF (2006) Angew Chem Int Ed Engl 45:4099–4104

Peres T, Cao BP, Cui WD, Lifshitz C, Khong A, Cross RJ, Saunders M (2001) Int J Mass Spectr 210:241–247

Pietzak B, Waiblinger M, Murphy TA, Weidinger A, Hohne M, Dietel E, Hirsch A (1998) Carbon 36:613–615

Popov AA (2009) J Comput Theor Nanosci 6:292–317 Sp. Iss. SI

Popov AA, Dunsch L (2007) J Am Chem Soc 129:11835–11849
Popov AA, Dunsch L (2009) Chem Eur J 15:9707–9729
Popov AA, Krause M, Yang SF, Wong J, Dunsch L (2007) J Phys Chem B 111:3363–3369
Pradeep T, Kulkarni GU, Kannan KR, Row TNG, Rao CN (1992) J Am Chem Soc 114:2272
Pupysheva OV, Farajian AA, Yakobson BI (2008) Nano Lett 8:767–774
Ramachandran CN, Sathyamurthy N (2005) Chem Phys Lett 410:348–351
Rappoport D, Furche F (2009) Phys Chem Chem Phys 11:6353–6358
Rapta P, Popov AA, Yang S, Dunsch L (2008) J Phys Chem A 112:5858–5865
Ren YX, Ng TY, Liew KM (2006) Carbon 44:397–406
Reveles JU, Heine T, Koster AM (2005) J Phys Chem A 109:7068–7072
Reymo FM, Stoddart JF (1999) Chem Rev 99:1643–1663
Rosenthal J, Schuster DI, Cross RJ, Khong A (2006) J Org Chem 71:1191–1199
Ross RB, Cardona CM, Swain FB, Guldi DM, Sankaranarayanan SG, Van Keuren E, Holloway BC, Drees M (2009) Adv Funct Mat 19:2332–2337
Rowan SJ, Cantrill SJ, Cousins GRL, Sanders JKM, Stoddart JF (2002) Angew Chem Int Ed Engl 41:898–952
Rubin Y, Jarrosson T, Wang W, Bartberger MD, Houk KN, Schick G, Saunders M, Cross RJ (2001) Angew Chem Int Ed 40:1543–1546
Ruttimann M, Haldimann RF, Isaacs L, Diederich F, Khong A, Jimenez-Vazquez H, Cross RJ, Saunders M (1997) Chem Eur J 3:1071–1076
Sakintuna B, Lamari-Darkrim F, Hirscher M (2007) Int J Hydrogen En 32:1121–1140
Saunders M (1991) Science 253:330
Saunders M, JimenezVazquez HA, Cross RJ, Mroczkowski S, Gross ML, Giblin DE, Poreda RJ (1994) J Am Chem Soc 116:2193–2194
Saunders M, Jimenez-Vazquez HA, Cross RJ, Poreda RJ (1993) Science 259:1428
Sauvage J-P, Dietrich-Buchecker CO (eds) (1999) In: Molecular Catenanes, Rotaxanes and Knots. A Journay through the World of Molecular Topology. Wiley-VCH, Weinheim, pp 107–142
Sawa H, Wakabayashi Y, Murata Y, Murata M, Komatsu K (2005) Angew Chem Int Ed 44: 1981–1983
Schalley CA, Weilandt T, Bruggemann J, Vogtle F (2004) Top Curr Chem 248:141–200
Scheloske M, Naydenov B, Meyer C, Harneit W (2006) Isr J Chem 46:407–412
Schill G (1971) Catenanes. Rotaxanes and knots. Academic, New York, NY
Seeman NC (1998a) Biophys J 74:A5–A5 Part 2
Seeman NC (1998b) Ann Rev Biophys Biomol Struct 27:225–248
Seeman NC, Mao CD, Sun WQ (1998) Math Intell 20:3–3 SUM
Shi ZQ, Wu X, Wang CR, Lu X, Shinohara H (2006) Angew Chem Int Ed 45:2107–2111
Shibata K, Kubozono Y, Kanbara T, Hosokawa T, Fujiwara A, Ito Y, Shinohara H (2004) Appl Phys Lett 84:2004
Shigeta Y, Saito H (2003) Synth Met 135:765–766
Shigeta Y, Takatsuka K (2005) J Chem Phys 123:131101
Shin WH, Yang SH, Goddard WA, Kang JK (2006) Appl Phys Lett 88:art. 053111
Shinohara H (2000a) Rep Progr Phys 63:843–892
Shinohara H (2000b) In: Kadish KM, Ruoff RS (eds) Fullerenes: chemistry, physics and technology. Wiley, New York, NY, pp 357–394
Shustova NB, Popov AA, Newell BS, Miller SM, Anderson OP, Seppelt K, Bolskar RD, Boltalina OV, Strauss SH (2007) Angew Chem Int Ed 46:4111–4114
Siegel JS (2004) Science 5675:1256–1257
Slanina Z, Pulay P, Nagase S (2006) J Chem Theory Comp 2:782–785
Sponer J, Leszczynski J, Hobza P (1996) J Comput Chem 17:841–850
Stanisky CM, Cross RJ, Saunders M (2009) J Am Chem Soc 131:3392–3395
Sternfeld T, Hoffman RE, Saunders M, Cross RJ, Syamala MS, Rabinovitz M (2002) J Am Chem Soc 124:8786–8787
Sternfeld T, Saunders M, Cross RJ, Rabinovitz M (2003) Angew Chem Int Ed 42:3136–3139

Stevenson S, Chancellor CJ, Lee HM, Olmstead MM, Balch AL (2008a) Inorg Chem 47: 1420–1427

Stevenson S, Fowler PW, Heine T, Duchamp JC, Rice G, Glass T, Harich K, Hajdu E, Bible R, Dorn HC (2000) Nature 408:427–428

Stevenson S, Mackey MA, Stuart MA, Phillips JP, Easterling ML, Chancellor CJ, Olmstead MM, Balch AL (2008b) J Am Chem Soc 130:11844–11845

Stevenson S, Rice G, Glass T, Harich K, Cromer F, Jordan MR, Craft J, Hadju E, Bible R, Olmstead MM, Maitra K, Balch AL, Dorn HC (1999) Nature 401:55–57

Stoddart JF (1991) Angew Chem Int Ed Engl 30:70–71

Stoddart JF (2009) Chem Soc Rev 38:1802–1820

Stoddart JF, Colquhoun HM (2008) Tetrahedron 64:8231–8263

Suetsuna T, Dragoe N, Harneit W, Weidinger A, Shimotani S, Ito S, Takagi H, Kitazawa K (2002) Chem Eur J 8:5079–5083

Suzuki T, Kikuchi K, Oguri F, Nakao Y, Suzuki S, Achiba Y, Yamamoto K, Funasaka H, Takahashi T (1996) Tetrahedron 52:4973–4082

Tagmatarchis N, Taninaka A, Shinohara H (2002) Chem Phys Lett 355:226–232

Takata M, Nishibori E, Sakata M, Wang CR, Shinohara H (2003) Chem Phys Lett 372:512–518

Takata M, Umeda B, Nishibori E, Sakata M, Saito Z, Ohno M, Shinohara H (1995) Nature 377:46

Tan Y-Z, Lu X, Wang CR (2006) J Phys Chem B 110:11098–11102

Tan Y-Z, Xie ST, Huang RB, Zheng LS (2009) Nature Chem 1:450–460

Tanabe F, Murata M, Murata Y, Komatsu K (2006) Nippon Kagakkai Koen Yokoshu 86:1282

Tarabek J, Yang S, Dunsch L (2009) Chem Phys Chem 10:1037–1043

Taylor R, Avent AG, Dennis TJ, Hare JP, Kroto HW, Holloway JH, Hope EG, Langley GJ (1992) Nature 355:27

Terrones H, Terrones M (1997) J Phys Chem Solids 38:1789–1796 thematic issue (1993) New J Chem 17

Turker L, Erkoc S (2003) J Mol Struct (THEOCHEM) 638:37–40

Turro NJ, Mart AA, Chen JY-C, Jockusch S, Lawler RG, Ruzzi M, Sartori E, Chuang S-C, Komatsu K, Murata Y (2008) J Am Chem Soc 130:10506

Twamley J (2003) Phys Rev A 67:art. no. 052318

Ugarte D (1992) Nature 359:707

Vignon SA, Stoddart JF (2005) Collect Czech Chem Commun 70:1493–1576

Vögtle F, Lukin O (2005) Angew Chem Int Ed 44:1456–1477

Wakahara T, Nikagawa H, Kikuchi T, Nakahodo T, Rahman A, Tsuchiya T, Maeda Y, Akasaka T, Yoza K, Horn E, Yamamoto K, Mizorogi N, Slanina Z, Nagase S (2006) J Am Chem Soc 128:14228–14229

Wakahara T, Sakuraba A, Iiduka Y, Okamura M, Tsuchiya T, Maeda Y, Akasaka T, Okubo S, Kato T, Kobayashi K, Nagase S, Kadish KM (2004) Chem Phys Lett 398:553–556

Walba DM (1985) Tetrahedron 41:3161

Wan TSM, Zhang H-W, Nakane T, Xu Z, Inakuma M, Shinohara H, Kobayashi K, Nagase S (1998) J Am Chem Soc 120:6806–6807

Wang CR, Kai T, Tomiyama T, Yoshida T, Kobayashi Y, Nishibori E, Takata M, Sakata M, Shinohara H (2000) Nature 408:426–427

Wang CR, Kai T, Tomiyama T, Yoshida T, Kobayashi Y, Nishibori E, Takata M, Sakata M, Shinohara H (2001) Angew Chem Int Ed 40:397–399

Wang CR, Zuo TM, Olmstead MM, Duchamp JC, Glass TE, Cromer F, Balch AL, Dorn HC (2006) J Am Chem Soc 128:art. no. JA061434I

Wang W, Ding J, Yang S, Li X-Y (1997) Proc Electrochem Soc 97–14:417–428

Wasserman E (1960) J Am Chem Soc 82:4433–4434

Watanabe K, Ishioka NS, Sekine T, Kudo H, Shimomura H, Muratsu H, Kume T (2005) J Radioanal Nucl Chem 266:499–502

Watanabe N, Furusho Y, Kihara N, Takata T, Kinbara K, Saigo K (2001) Bull Chem Soc Japan 74:149–155

Weiss FD, O'Brien SC, Elkind PC, Curl RF, Smalley RE (1988) J Am Chem Soc 110:4464–4465

Whitener Jr KE, Cross RJ, Saunders M, Iwamatsu S-i, Murata S, Mizorogi N, Nagase S (2009) J Am Chem Soc 131:6338–6339
Whitener Jr KE, Frunzi M, Iwamatsu S-i, Murata S, Cross RJ, Saunders M (2008) J Am Chem Soc 130:13996–13999
Williams AR, Northrop BH, Chang T, Stoddart JF, White AJP, Williams DJ (2006) Angew Chem Int Ed Engl 45:6665–6669
Williams CI, Whitehead MA, Pang I (1993) J Phys Chem 97:11652–11656
Wilson LJ, Cagle DW, Thrash TP, Kennel SJ, Mirzadeh S, Alford JM, Ehrhardt GJ (1999) Coord Chem Rev 192:199–207
Wilson M, Kannangara K, Smith G, Simmons M, Raguse B. (2002) Nanotechnology: basic science and emerging technologies. Chapman & Hall/CRC, Virginia Beach, VA
Xiao Z, Yao JY, Yang DZ, Wang FD, Huang SH, Gan LB, Jia ZS, Jiang ZP, Yang XB, Zheng B, Yuan G, Zhang SW, Wang ZM (2007) J Am Chem Soc 129:16149–16162
Xu J, Li M, Shi Z, Gu Z (2006a) Chem Eur J 12:562–567
Xu J, Tsuchiya T, Hao C, Wakahara T, Mi W, Gu Z, Akasaka T (2006b) Chem Phys Lett 419:44–47
Yamada M, Akasaka T, Nagase S (2010) Acc Chem Res 43:92–102
Yamada M, Wakahara T, Tsuchiya T, Maeda Y, Akasaka T, Mizorogi N, Nagase S (2008a) J Phys Chem A 112:7627–7631
Yamada M, Wakahara T, Tsuchiya T, Maeda Y, Akasaka T, Mizorogi N, Nagase S (2008b) J Phys Chem A 112:7627–7631
Yamaguchi H, Oshikiri T, Harada A (2006) J Phys Condens Matter 18:S1809–S1816
Yang C-K (2007) Carbon 45:2451–2453
Yang C-K (2008) Carbon 46:705–705
Yang H, Lu CX, Liu ZY, Che YL, Olmstead M.M, Balch AL (2008) J Am Chem Soc 130:17296–17300
Yang S, Chen CB, Popov AA, Zhang WF, Liu FP, Dunsch L (2009) Chem Commun 6391–6393
Yang S, Popov AA, Dunsch L (2007a) J Phys Chem B 111:13659–13663
Yang SF, Dunsch L (2006) Angew Chem Int Ed 45:1299–1302
Yang SF, Popov AA, Dunsch L (2007b) Angew Chem Int Ed 46:1256–1259
Yannoni CS, Hoinkis M, de Vries MS, Bethune DS, Salem JR, Crowder MS, Johnson RD (1992) Science 256:1191–1192
Yasutake Y, Shi ZJ, Okazaki T, Shinohara H, Majima Y (2005) Nano Lett 5:1057–1060
Yoon M, Yang SY, Wang E, Zhang ZY (2007) Nano Lett 7:2578–2583
Yoon M, Yang SY, Zhang ZY (2009) J Chem Phys 131:art. 64707
Yoshida Z-I, Dogane I, Ikehira H, Endo T (1993) Chem Phys Lett 201:481–484
Zettergren HA, Martin FM (2008) Chem Phys Chem 9:861–866
Zhang J, Porfyrakis K, Morton JJL, Sambrook MR, Harmer J, Xiao L, Ardavan A, Briggs GAD (2008a) J Phys Chem 112:2802–2804
Zhang Y, Li M, Hao C, Shi Z, Gu Z (2006) Carbon 44:475–479
Zhang ZX, Han PD, Liu XG, Zhao JF, Jia HS, Zeng FG, Xu BS (2008b) J Phys Chem C 112:19158–19161
Zhao Y, Heben MJ, Dillon AC, Simpson LJ, Dorn HC, Zhang SB (2007) J Phys Chem C 111:13275–13279
Zhao Z, Truhlar DG (2004) J Phys Chem A 108:6908–6918
Zimmermann U, Malinowski N, Näher U, Frank S, Martin TP (1994) Phys Rev Lett 72:3542–3545
Zope RR (2008) J Phys B 41:art. no. 085101
Zuo T, Beavers CM, Duchamp JC, Campbell A, Dorn HC, Olmstead MM, Balch AL (2007) J Am Chem Soc 129:2035–2043
Zuo T, Walker K, Olmstead MM, Melin F, Holloway BC, Echegoyen L, Dorn HC, Chaur MN, Chancellor CJ, Beavers CM, Balch AL, Athans AJ (2008) Chem Commun 1067–1069

Chapter 8
Detailed Atlas of Kekulé Structures of the Buckminsterfullerene

Damir Vukičević and Milan Randić

Abstract Buckminsterfullerene has 12500 Kekulé structures grouped in 158 isomorphic classes. In this paper we reproduce the results of paper (Vukičević et al. Croatica Chemica Acta 78: 223, 2005) with some extensions. Namely, for each Kekulé structure we provide: number of structures isomorphic to it, the average number of π-electrons that belong to hexagon, the average number of π-electrons that belong to pentagon, the number of conjugated cycles of lengths 6, 10 and 14, the number of all conjugated cycles, degree of freedom, maximum number of independent conjugate cycles and maximum number of independent conjugated hexagons.

8.1 Introduction

Buckminsterfullerene, C_{60}, is the first fullerene that was theoretically conceived and experimentally obtained (Kroto et al. 1985). Theoretically, C_{60} may be viewed as being constructed by the leap-frog transformation of the fullerene C_{20}. The leap-frog transformation (Fowler 1986; Manolopulos et al. 1992; Fowler and Manolopoulos 1995) is the transformation in which the dual of the original fullerene is formed and then truncated on all vertices.

It is highly symmetric fullerene and it has 158 automorphism. Among its 12500 Kekulé structures, there are only 158 non-isomorphic ones (Klein et al. 1986). These Kekulé structures have been recently extensively studied in the series of papers (El-Basil 2000; Vukičević et al. 2005; Vukičević and Randic 2005; Randić et al. 2007). In this paper, we extend results of the paper (Vukičević et al. 2005) in which atlas of Kekulé structures has been presented. Here, we extend this atlas by providing some additional data and reorganize presentation in such a way that all data is given next to the figure that represents corresponding Kekulé structure.

D. Vukičević (✉)
Department of Mathematics, University of Split, Nikole Tesle 12, HR-21000 Split, Croatia
e-mail: vukicevi@pmfst.hr

F. Cataldo et al. (eds.), *The Mathematics and Topology of Fullerenes*,
Carbon Materials: Chemistry and Physics 4, DOI 10.1007/978-94-007-0221-9_8,

The following parameters are presented:

(1) Number of structures isomorphic to the presented one
We are interested in invariant values of Kekulé structures. Since they are equal for all isomorphic structures we do not need to present all 12500, but only 158 non-isomorphic ones.

(2) Degree of freedom (forcing number)
Let K be any Kekulé structure and let S be the set of its double bonds. We say that S is the forcing set of K if K is the only Kekulé structure that contains S. Cardinality of the smallest set forcing set S is called forcing number (or degree of freedom). Alternatively, we could say that degree of freedom is the smallest number of double bonds that determines the Kekulé structure completely (Klein and Randić 1987). It is expected that the significance of the Kekulé structure increases as the degree of freedom increases.

(3) The average number of π-electrons that belong to the hexagon and the average number of π-electrons that belong to the pentagon
Recently, the concept of algebraic Kekulé structures of bezenoids has been presented (Balaban and Randić 2004a, b). Namely, let K be any Kekulé structure. π-electron content of the hexagon H is the number $K(H)$ of π-electrons of K that belong to H. $K(H)$ is calculated in the following way: every double bond of K that belong exclusively to H contributes 2 to $K(H)$ and every double bond shared by H and another hexagon contributes one to $K(H)$. The function that assigns to each H value $K(H)$ is the algebraic Kekulé structure that corresponds to H. Analogous concept can be applied to fullerenes, too. Just note that here we observe both hexagons and pentagons; and that each double bond is shared between two faces. Now, we can calculate for each face F its average π-electron content $APEC(F) = \sum_{K \in KS} K(F) / card(KS)$, where KS is the set of all Kekulé structures and $card(KS)$ is their number. Further let us denote set of all pentagons by SP and set of all hexagons by SH. The average number of π-electrons that belong to the hexagon AH is calculated by $AH = \sum_{F \in SH} K(F) / card(SH)$ and the average number of π-electrons that belong to the hexagon AP is calculated by $AP = \sum_{F \in SP} K(F) / card(SP)$. It can be easily seen that it holds $20 \cdot AH + 12 \cdot AP = 60$, hence increment of AH implies decrement of AP and vice versa. It is expected that the significance of Kekulé structure increases with the increment of AH and decreases with the increment of AP.

(4) Numbers of conjugated cycles (of lengths 6, 10 and 14; and total number)
The cycle is called conjugated cycle is the cycle in which double and single bonds alternate. It is known that resonance energy of benzenoids can be estimated from the numbers of conjugated cycles of length 6, 10 and 14. Several

formulas for such calculations are given in Fries (1927). Hence, here we give for the each Kekulé structure numbers of conjugated cycles of lengths 6, 10 and 14. Also we provide a total number of conjugated cycles. Especially significant value is the value of the number of conjugated cycles of length 6. The structures with the largest number of conjugate hexagons are called Fries structures (Fries et al. 1935; Randić 2003). It is expected that the importance of the Kekulé structure increases with the increment of the number of conjugated hexagons, i.e. that Fries structures (usually) correspond to the most important Kekulé structures.

(5) Maximum number of independent conjugated hexagons and maximum number of independent conjugated cycles
Two cycles are independent if they do not share a common vertex. The set of the cycles is the set of independent cycles if they are pairwise independent. Structures with the largest number of independent conjugated hexagons are named Clar's structures and it is expected that (usually) these are the most important Kekulé structures (Randić 2003). Hence, it is expected that importance of Kekulé structure increases with the increment of the number of independent conjugated hexagons. Analogous claim holds for independent conjugated cycles also.

In this paper, similarly as in paper (Vukičević et al. 2005) we present 158 non-isomorphic Kekulé structures in the same ordering as there. The additional parameters given here are: number of conjugated cycles, number of independent conjugated hexagons and the number of independent conjugate cycles. The results are given in the table form and organized as follows:

Drawing of the Kekulé structure	Its number (coinciding with the number in paper (Vukičević et al. 2005)	The number of Kekulé structures isomorphic to the observed one	The average number of π-electrons that belong to the hexagon	The average number of π-electrons that belong to the pentagon
	Number of conjugated hexagons	Number of conjugated cycles of length 10	Number of conjugated cycles of length 14	Number of conjugated cycles
	Degree of freedom		Maximum number of independent conjugated hexagons	Maximum number of independent conjugated cycles

Atlas of Kekulé Structures

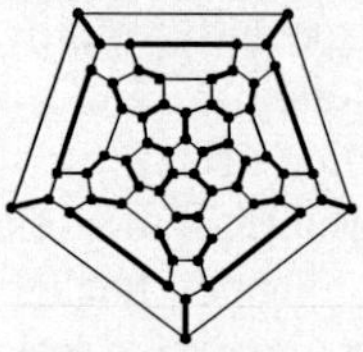	1	1	3.000	0.000	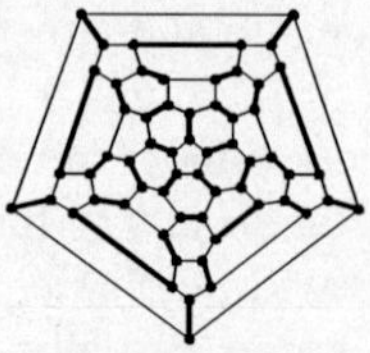	2	20	2.850	0.250
	20	0	0	542		17	3	0	596
		10	8	8			9	8	8
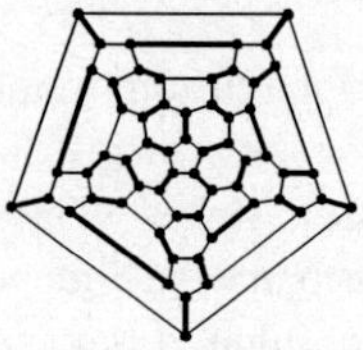	3	60	2.700	0.500	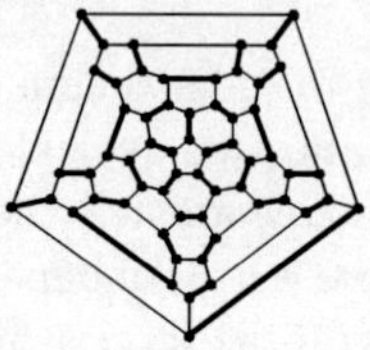	4	120	2.400	1.000
	15	4	1	692		11	6	3	913
		9	8	8			8	7	7
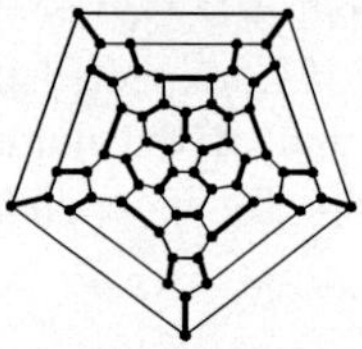	5	12	2.250	1.250	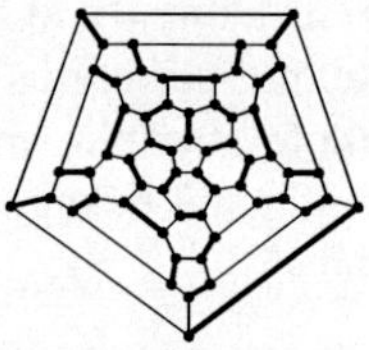	6	60	2.250	1.250
	10	5	5	1072		10	5	5	1066
		8	7	7			8	7	7
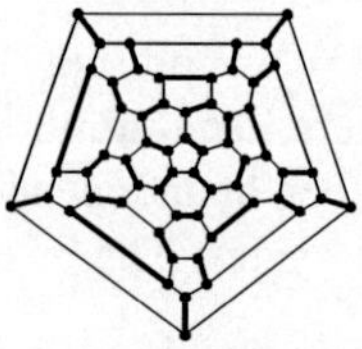	7	120	2.400	1.000	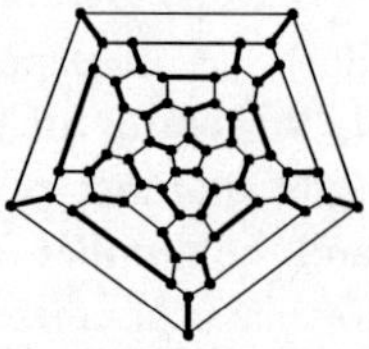	8	60	2.250	1.250
	11	7	1	951		10	6	3	1137
		8	7	7			8	7	7
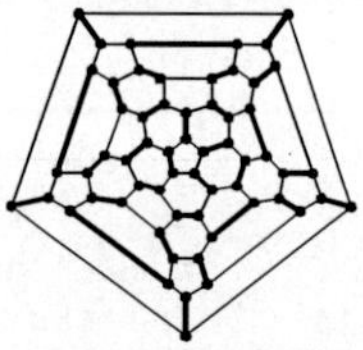	9	120	2.550	0.750	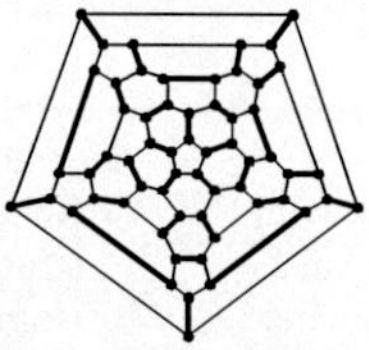	10	120	2.550	0.750
	12	7	1	803		12	7	1	833
		8	7	7			8	8	8
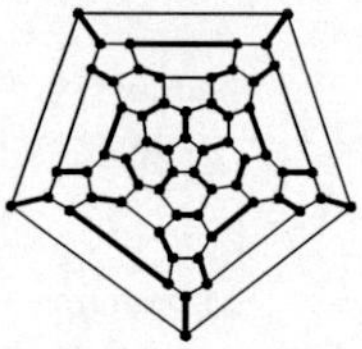	11	60	2.550	0.750	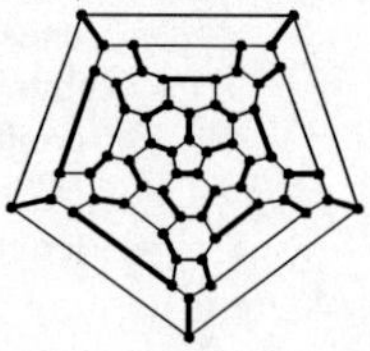	12	120	2.400	1.000
	12	7	1	851		11	6	3	957
		8	7	7			8	7	7

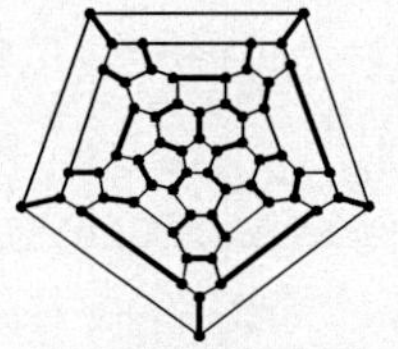	13	30	2.400	1.000	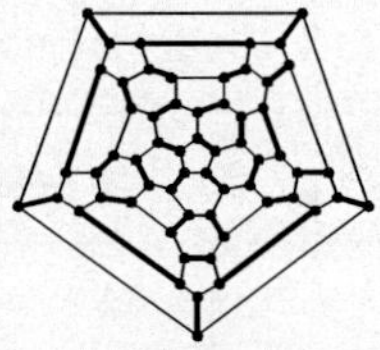	14	60	2.550	0.750
	10	8	2	1038		12	7	1	760
		8	8	8			8	7	7
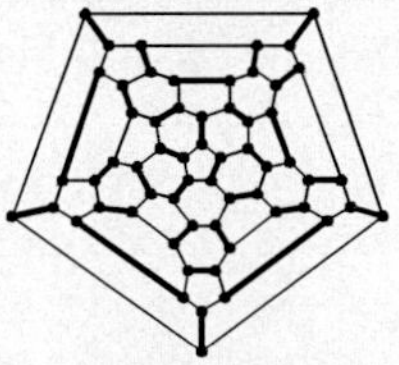	15	120	2.400	1.000	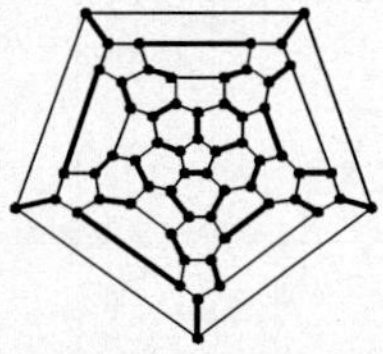	16	120	2.550	0.750
	11	6	3	991		13	5	2	799
		8	8	8			8	8	8
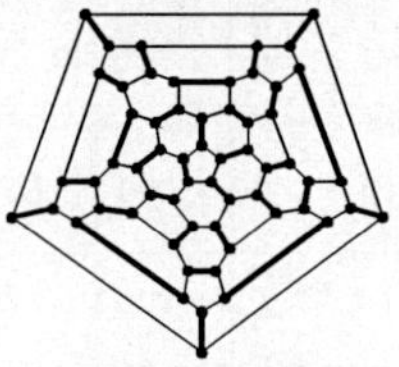	17	120	2.250	1.250	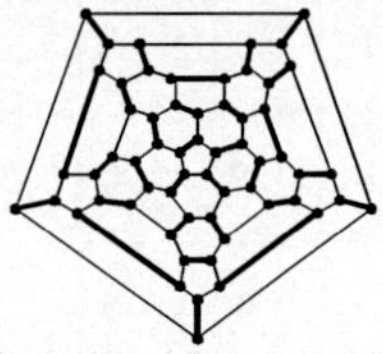	18	120	2.400	1.000
	9	7	4	1215		10	8	2	944
		8	8	8			8	8	8
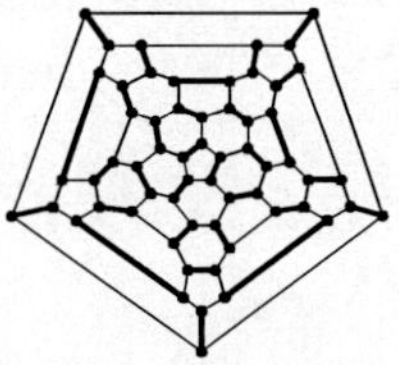	19	120	2.250	1.250	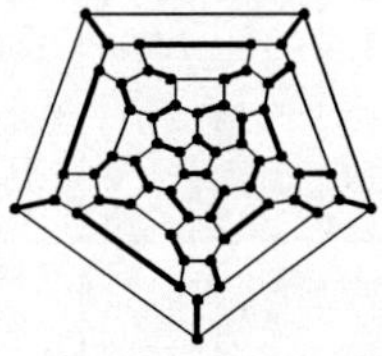	20	30	2.400	1.000
	10	5	5	1175		12	4	4	964
		8	8	8			8	8	8
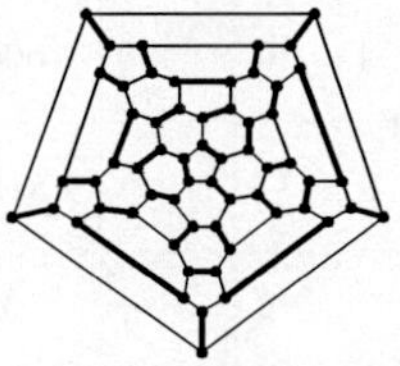	21	60	2.100	1.500	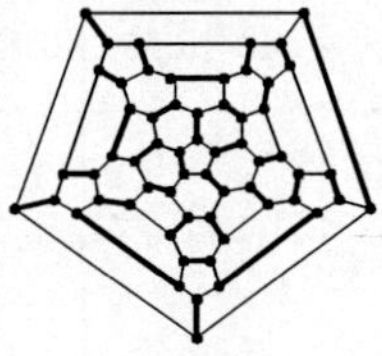	22	20	2.100	1.500
	9	4	7	1492		8	6	6	1541
		8	8	8			8	8	8
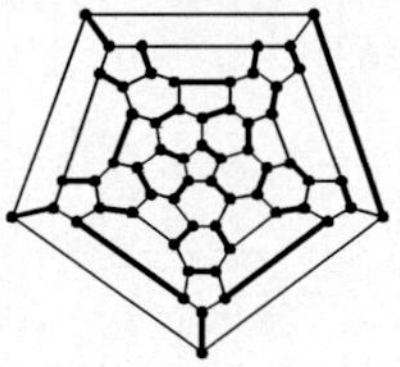	23	40	2.250	1.250	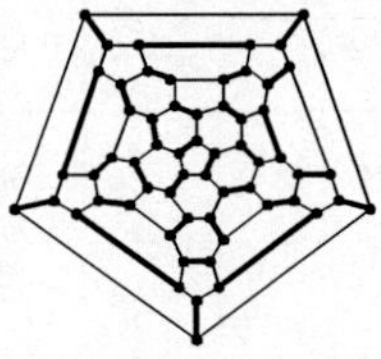	24	40	2.400	1.000
	8	9	3	1067		11	6	3	918
		8	8	8			8	8	8

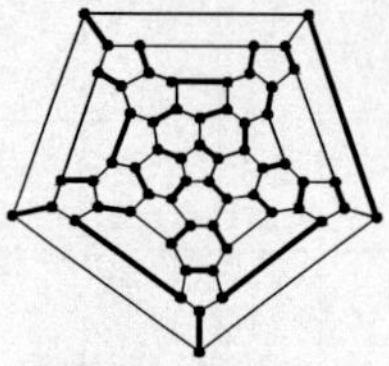	25	60	2.100	1.500	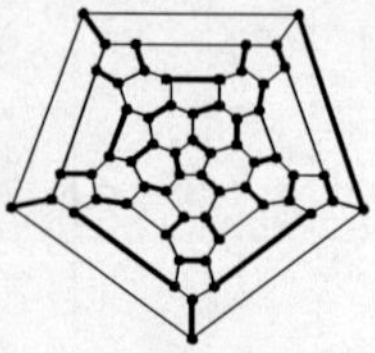	26	40	1.950	1.750
	8	6	6	1419		8	3	9	1866
		8	8	8			8	8	8
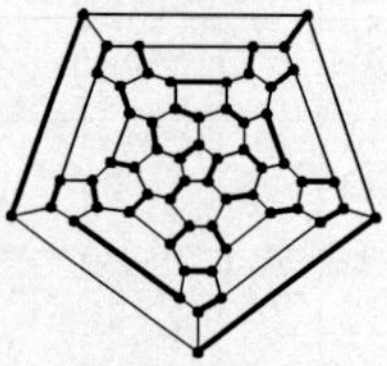	27	5	1.800	2.000	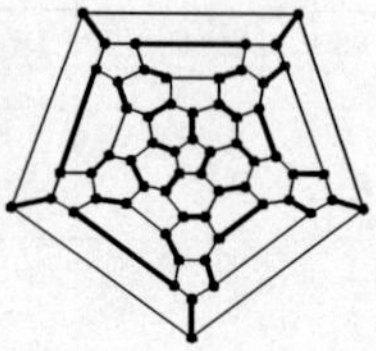	28	20	2.550	0.750
	8	0	12	2406		13	6	0	832
		8	8	8			8	7	7
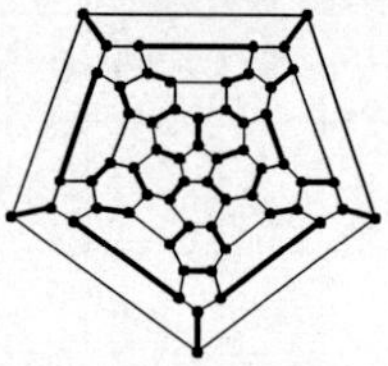	29	60	2.700	0.500	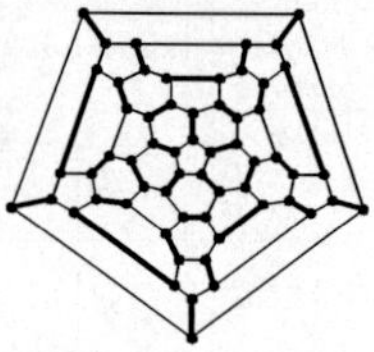	30	30	2.700	0.500
	14	6	0	659		14	6	0	697
		8	8	8			8	7	7
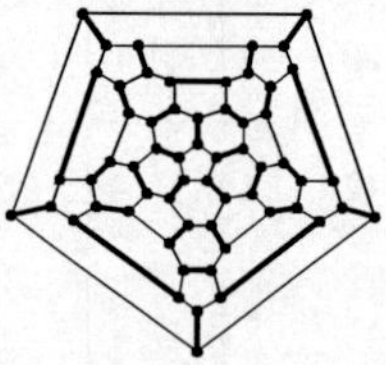	31	10	2.700	0.500	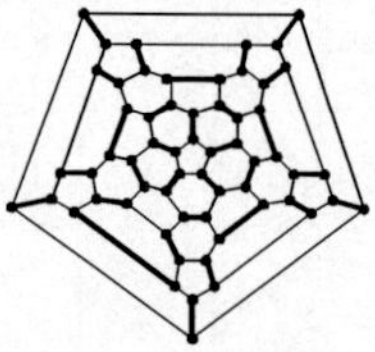	32	60	2.400	1.000
	14	6	0	733		11	6	3	961
		8	8	8			8	7	7
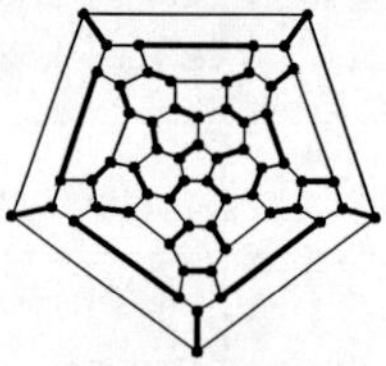	33	40	2.550	0.750	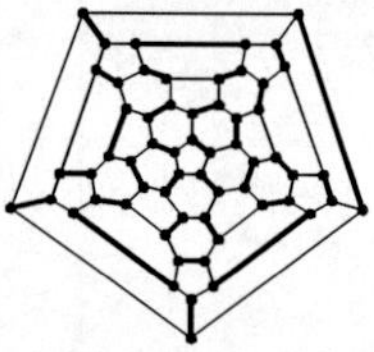	34	10	2.400	1.000
	11	9	0	731		8	12	0	812
		8	8	8			8	8	8
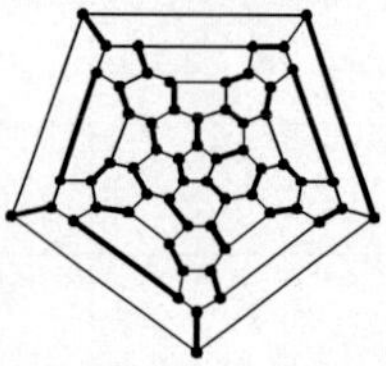	35	6	2.500	0.833	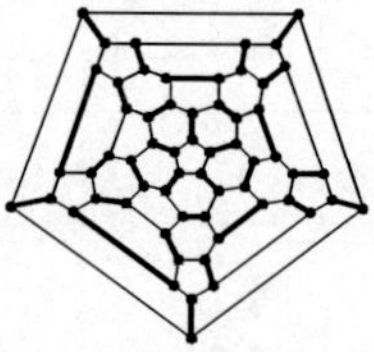	36	60	2.550	0.750
	10	0	0	1083		13	5	2	792
		8	5	4			8	7	7

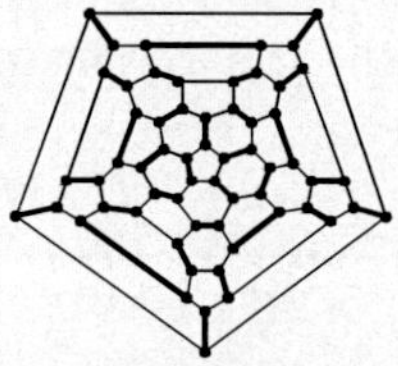	37	60	2.400	1.000	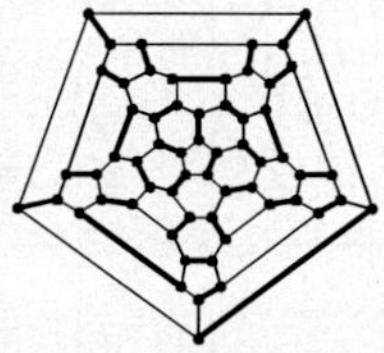	38	120	2.100	1.500
	10	8	2	1022		8	6	6	1378
		7	7	7			7	7	7
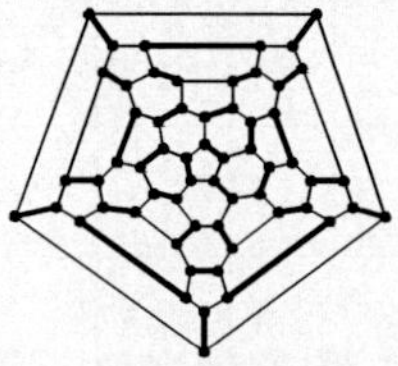	39	120	2.250	1.250	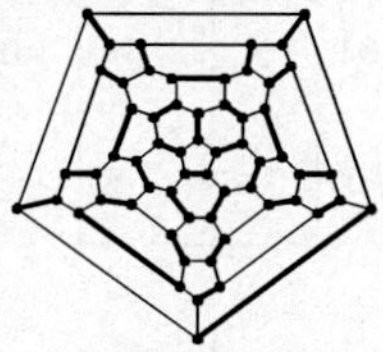	40	120	2.250	1.250
	9	7	4	1093		8	9	3	1043
		7	7	7			7	7	7
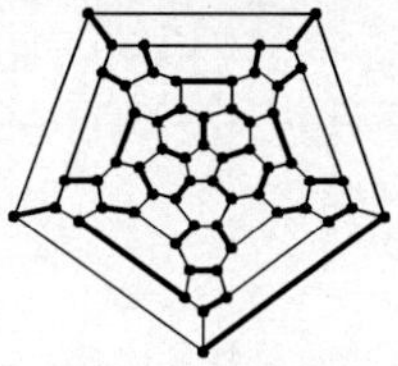	41	120	2.250	1.250	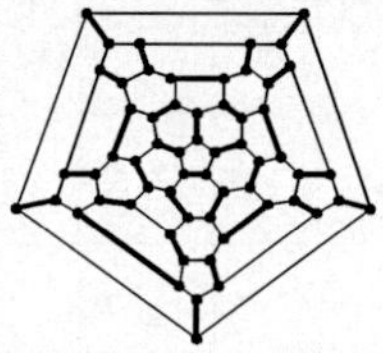	42	120	2.250	1.250
	9	7	4	1120		9	7	4	1145
		7	7	7			7	7	7
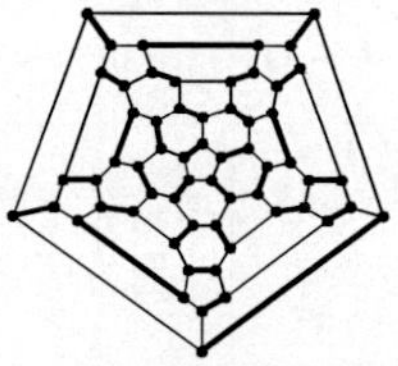	43	120	2.250	1.250	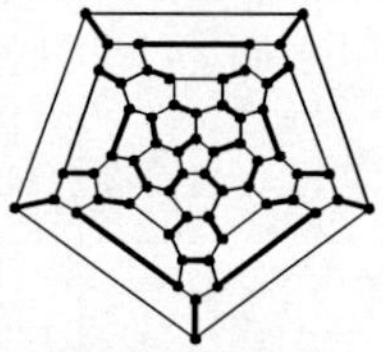	44	120	2.400	1.000
	8	9	3	1116		9	10	1	879
		7	7	7			7	7	7
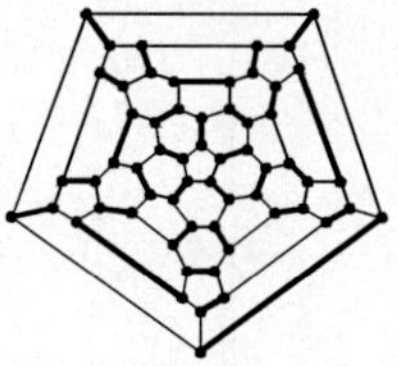	45	60	2.400	1.000	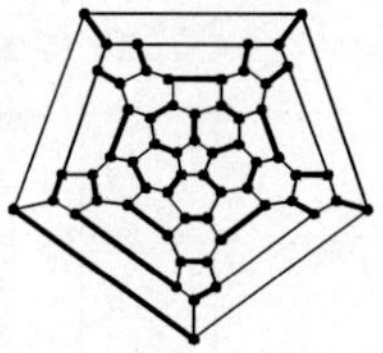	46	60	2.200	1.333
	10	8	2	947		7	2	1	1475
		7	7	7			7	5	4
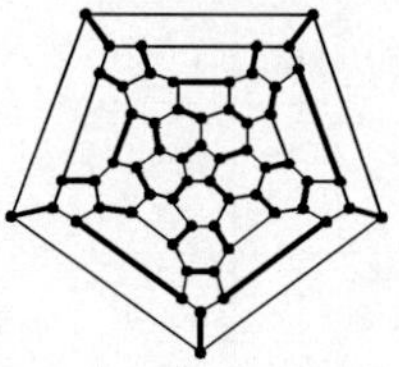	47	60	2.250	1.250	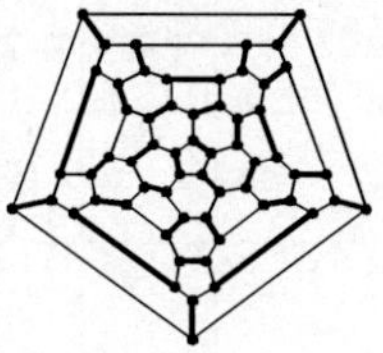	48	120	2.250	1.250
	8	10	1	1208		9	8	2	1164
		7	7	7			7	7	7

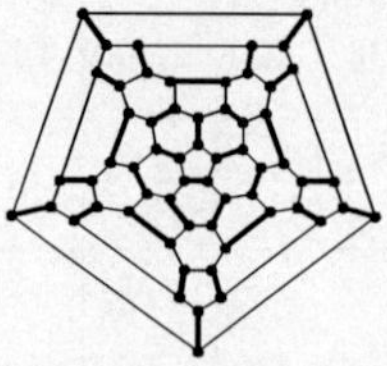	49	60	2.100	1.500	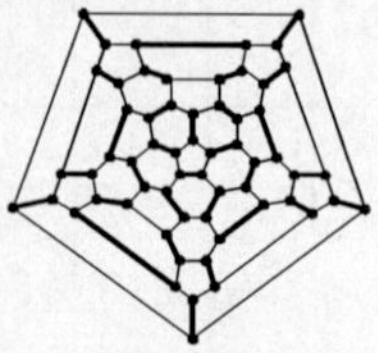	50	120	2.400	1.000
	8	7	4	1333		10	8	2	970
		7	7	7			7	7	7
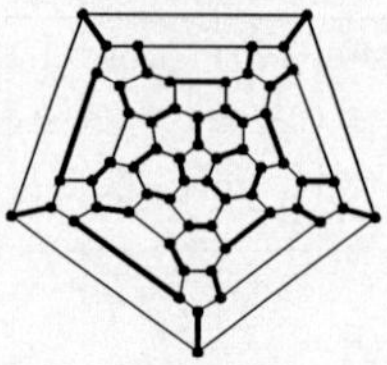	51	60	2.400	1.000	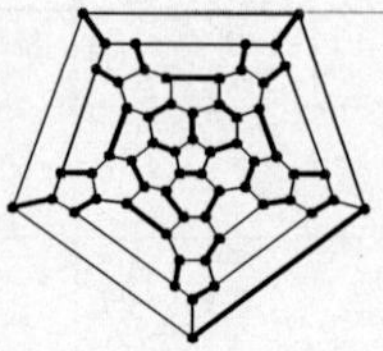	52	120	2.100	1.500
	10	8	2	915		8	6	6	1267
		7	7	7			7	7	7
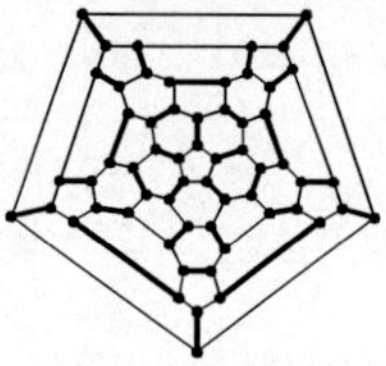	53	60	2.400	1.000	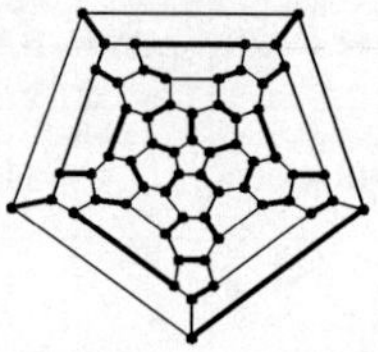	54	120	2.400	1.000
	10	8	2	938		10	8	2	917
		7	7	7			7	7	7
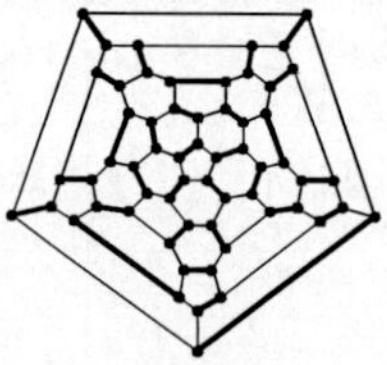	55	120	2.100	1.500	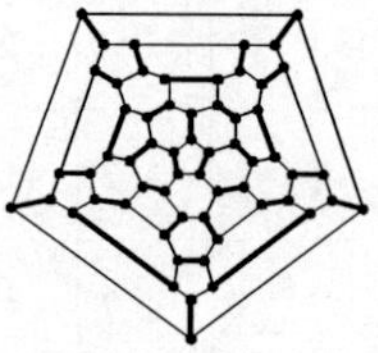	56	60	2.100	1.500
	7	9	3	1409		8	7	4	1348
		7	7	7			7	7	7
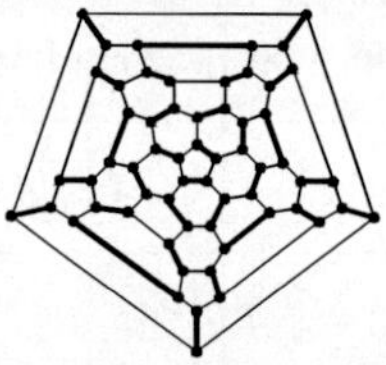	57	60	2.250	1.250	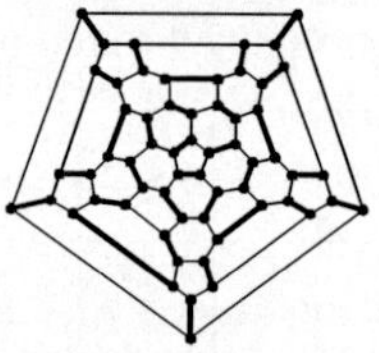	58	120	2.100	1.500
	9	7	4	1149		8	7	4	1329
		7	7	7			7	7	7
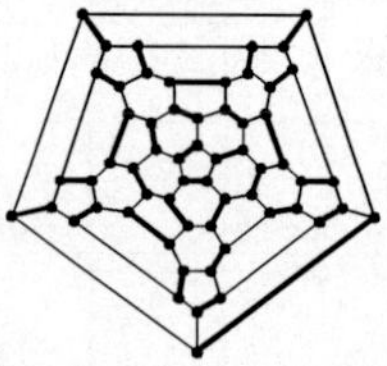	59	120	1.950	1.750	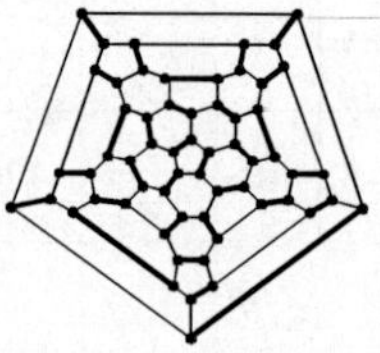	60	60	1.950	1.750
	7	6	6	1588		7	6	6	1725
		7	7	7			7	7	7

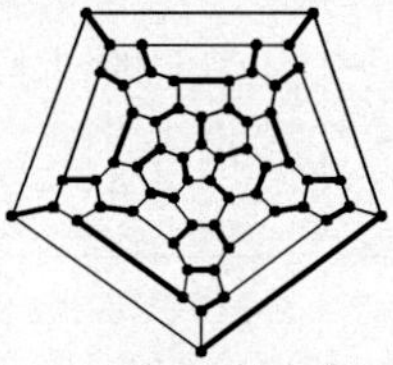	61	60	2.100	1.500	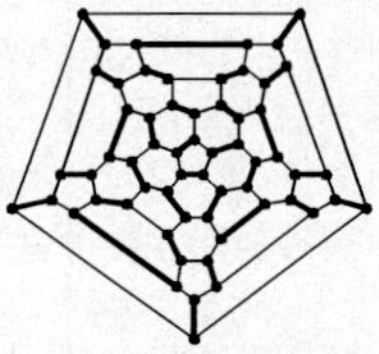	62	120	2.250	1.250
	8	6	6	1368		9	7	4	1146
		7	7	7			7	7	7
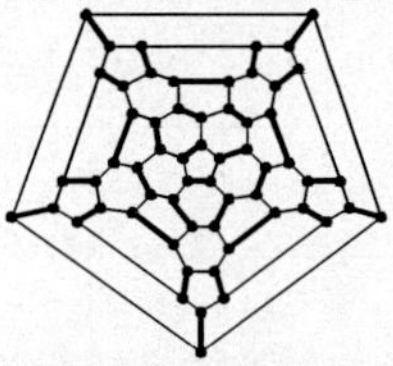	63	60	1.950	1.750	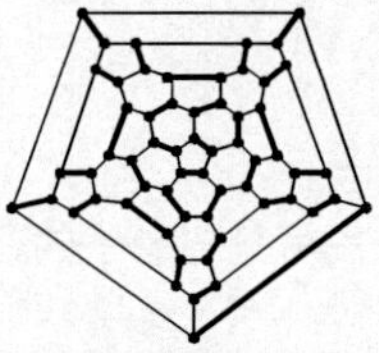	64	40	1.950	1.750
	7	7	4	1556		7	6	6	1464
		7	7	7			7	7	7
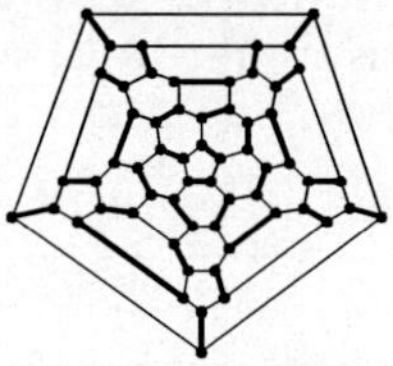	65	120	2.100	1.500	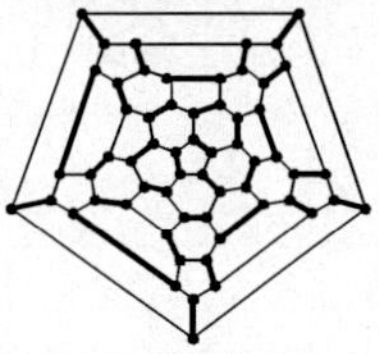	66	120	2.250	1.250
	8	7	4	1356		9	8	2	1088
		7	7	7			7	7	7
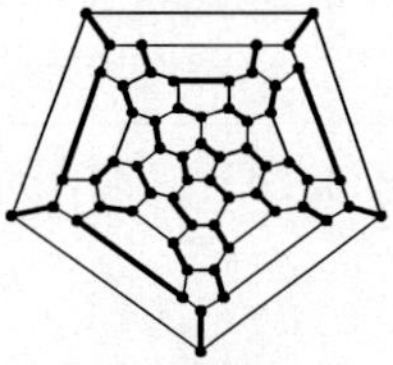	67	120	2.350	1.083	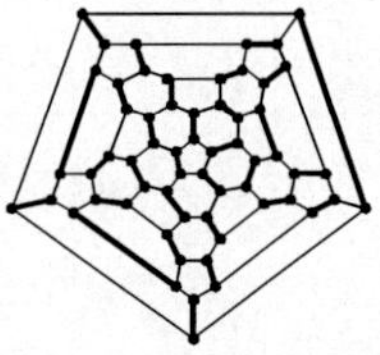	68	60	2.350	1.083
	8	2	0	1369		8	2	0	1219
		7	5	4			7	5	4
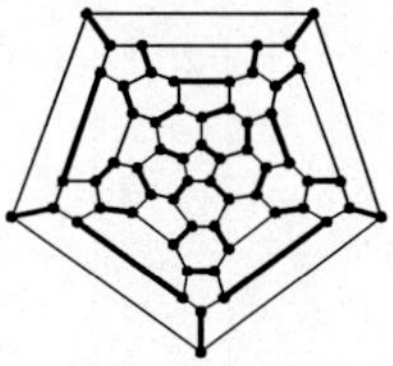	69	60	2.400	1.000	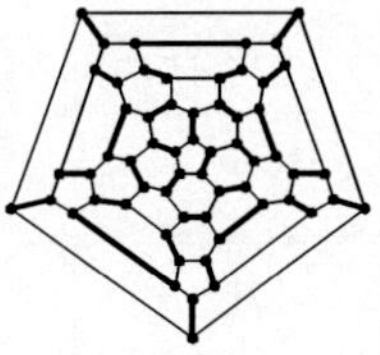	70	20	2.400	1.000
	10	9	0	985		10	9	0	1066
		7	7	7			7	7	7
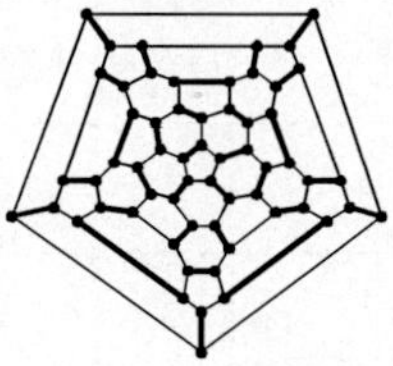	71	120	2.250	1.250	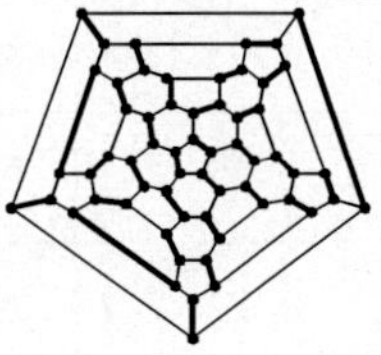	72	60	2.150	1.417
	8	10	1	1109		5	0	0	1870
		7	7	7			7	4	2

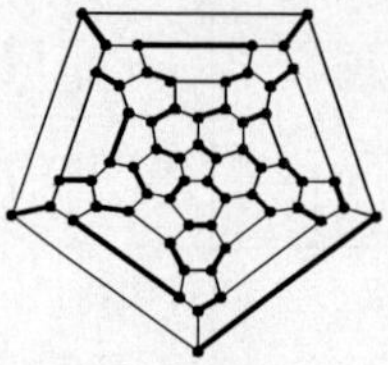	73	60	2.350	1.083	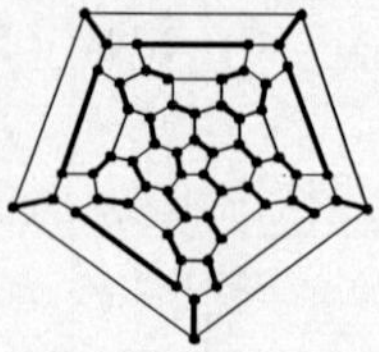	74	30	2.500	0.833
	7	3	0	1303		10	0	0	1105
		7	5	4			7	5	4
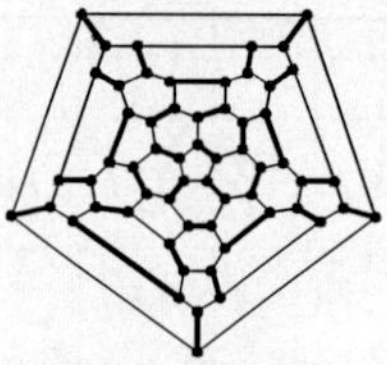	75	60	2.250	1.250	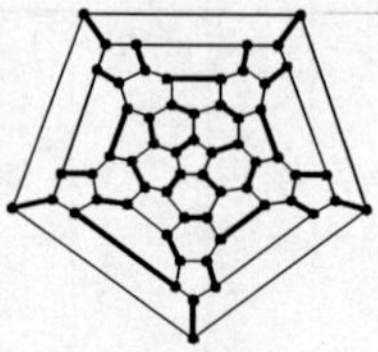	76	120	2.250	1.250
	9	8	2	1144		9	8	2	1198
		7	7	7			7	7	7
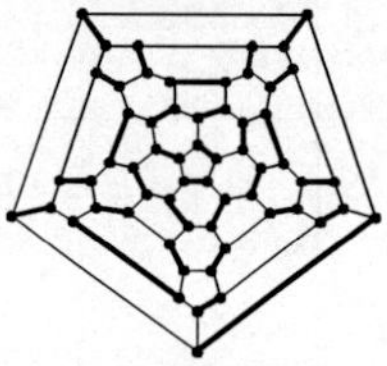	77	40	2.100	1.500	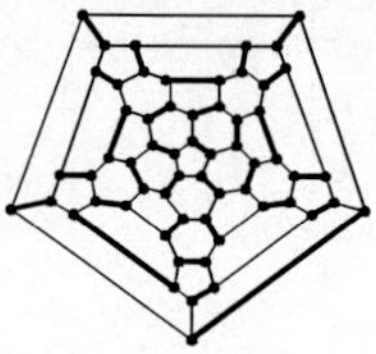	78	120	2.100	1.500
	7	9	3	1245		7	9	3	1317
		7	7	7			7	7	7
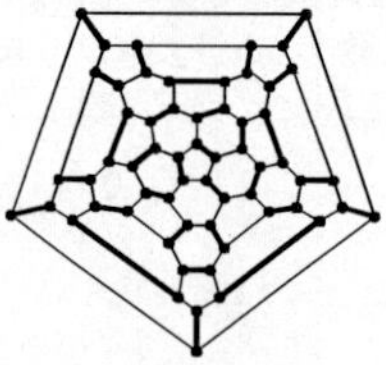	79	120	2.100	1.500	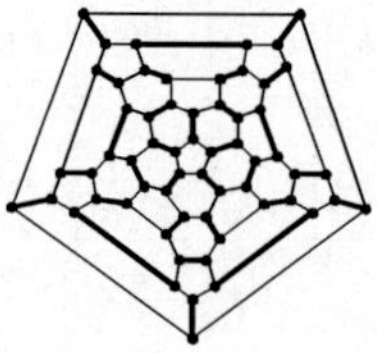	80	60	2.550	0.750
	8	7	4	1373		11	9	0	768
		7	7	7			7	7	7
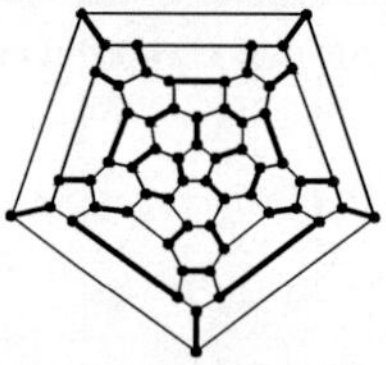	81	120	2.250	1.250	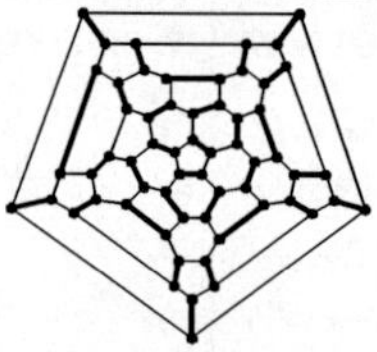	82	30	2.100	1.500
	9	7	4	1164		8	8	2	1413
		7	7	7			7	7	7
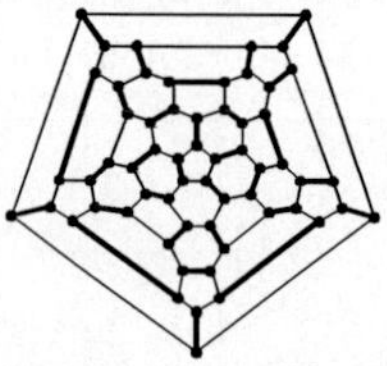	83	60	2.400	1.000	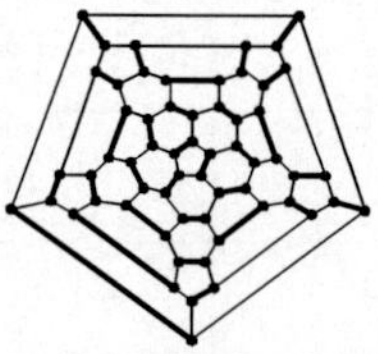	84	30	1.900	1.833
	10	8	2	989		4	4	2	2147
		7	7	7			6	5	4

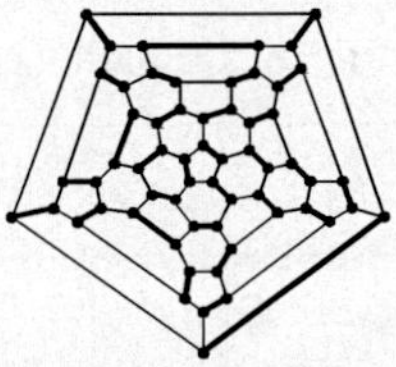	85	60	2.050	1.583	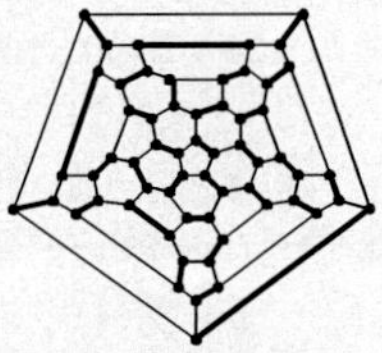	86	60	2.200	1.333
	5	4	0	1921		6	4	0	1720
		6	5	4			6	5	4
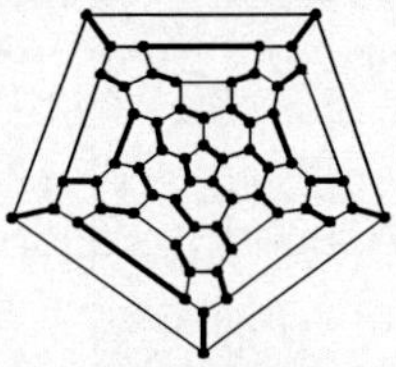	87	120	2.200	1.333	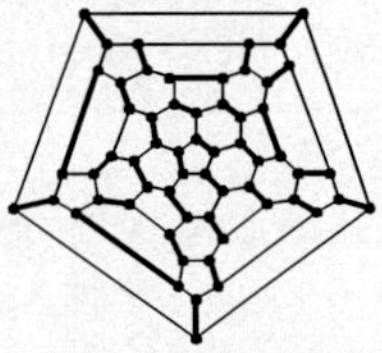	88	120	2.200	1.333
	6	4	0	1720		7	2	1	1526
		6	5	4			6	5	4
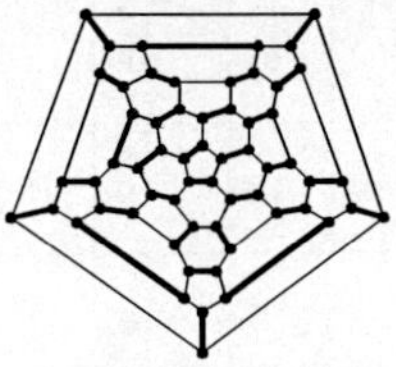	89	120	2.200	1.333	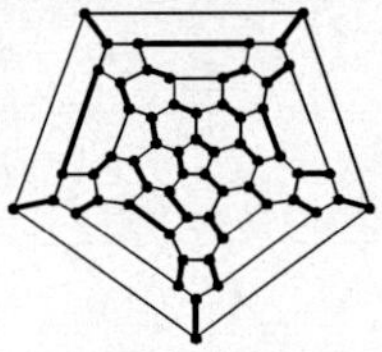	90	60	2.200	1.333
	6	4	0	1687		6	4	0	1693
		6	5	4			6	5	4
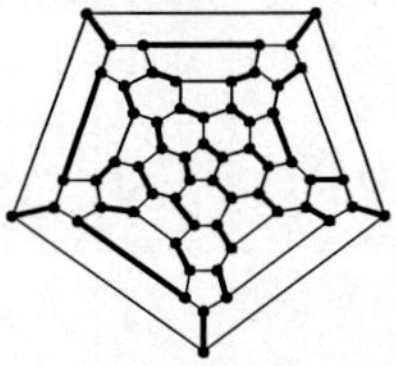	91	120	2.350	1.083	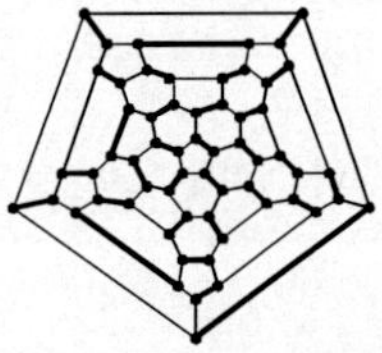	92	120	2.200	1.333
	8	2	0	1371		5	5	0	1595
		6	5	4			6	5	4
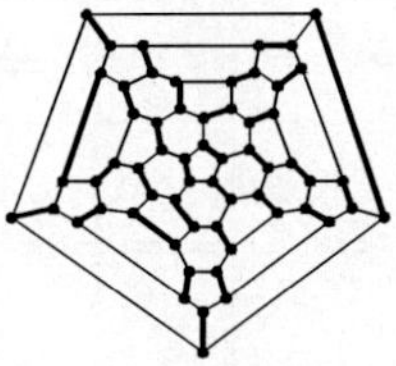	93	120	2.000	1.667	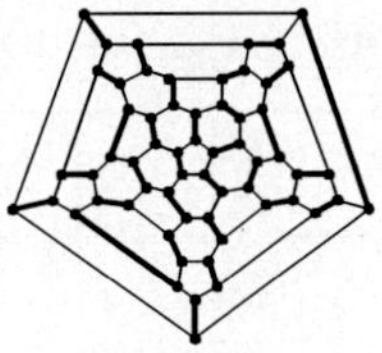	94	60	2.200	1.333
	3	2	0	2508		6	4	0	1468
		6	4	2			6	5	4
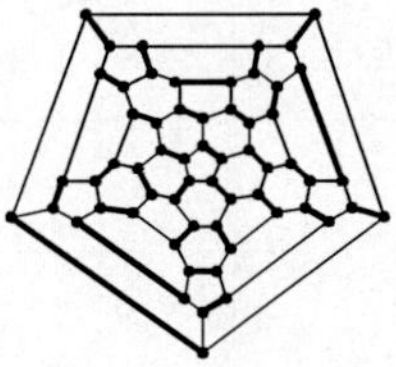	95	30	2.200	1.333	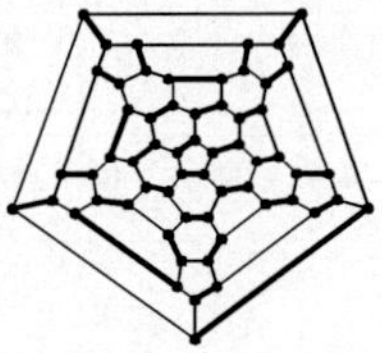	96	120	2.050	1.583
	6	4	0	1404		5	4	1	1892
		6	5	4			6	5	4

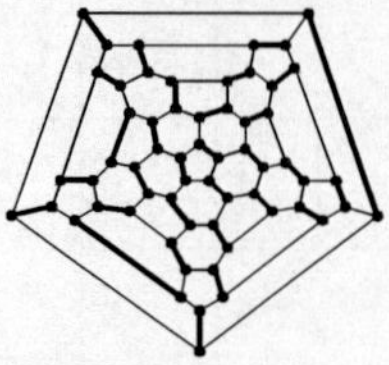	97	120	2.000	1.667	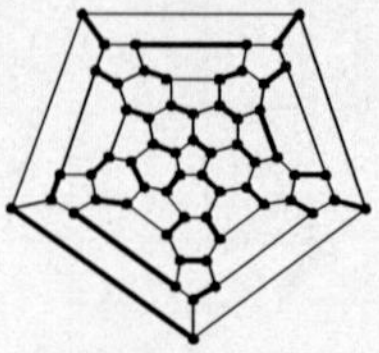	98	60	2.200	1.333
	3	2	0	2409		6	4	0	1359
		6	4	2			6	5	4
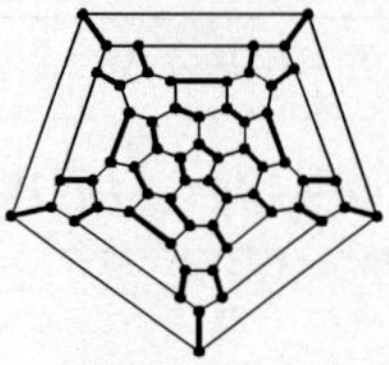	99	60	1.900	1.833	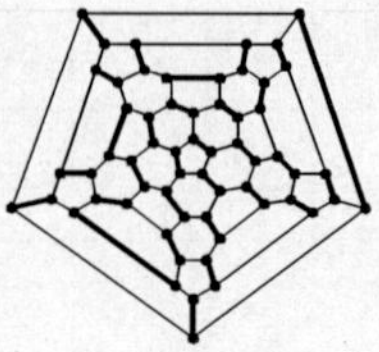	100	120	2.050	1.583
	4	4	2	2237		5	4	1	1922
		6	5	4			6	5	4
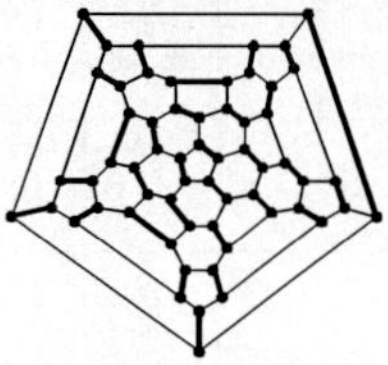	101	120	1.900	1.833	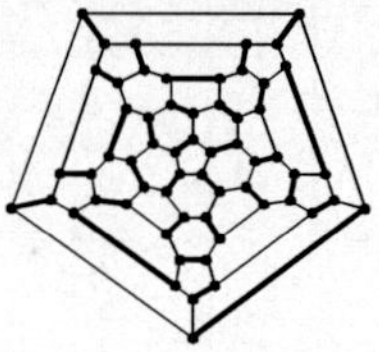	102	10	2.100	1.500
	4	4	2	2169		6	12	0	1523
		6	5	4			6	6	6
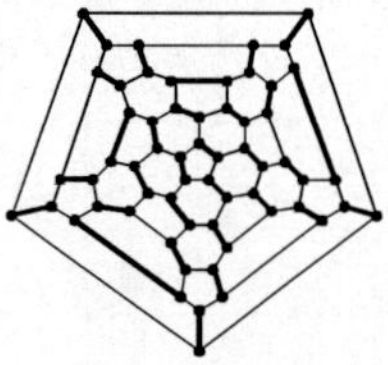	103	60	2.200	1.333	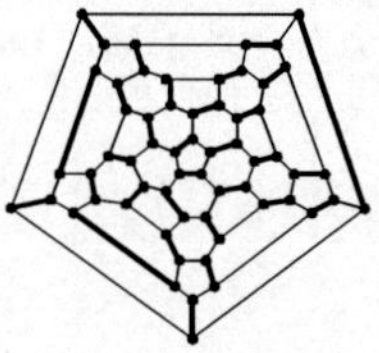	104	60	2.000	1.667
	6	4	0	1768		3	2	0	2159
		6	5	4			6	4	2
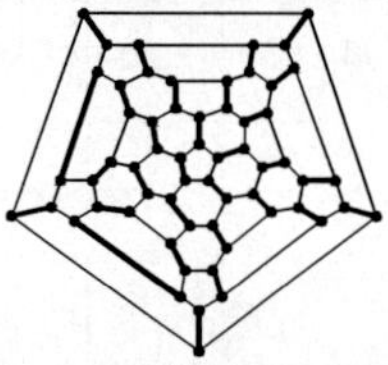	105	60	2.250	1.250	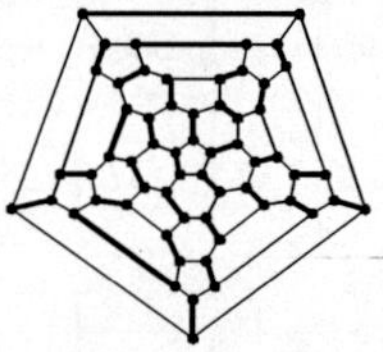	106	30	2.200	1.333
	5	0	0	2558		4	0	0	4143
		6	4	2			6	4	2
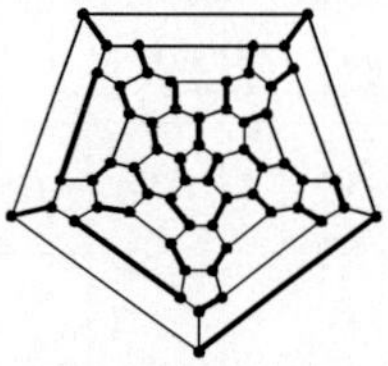	107	60	2.150	1.417	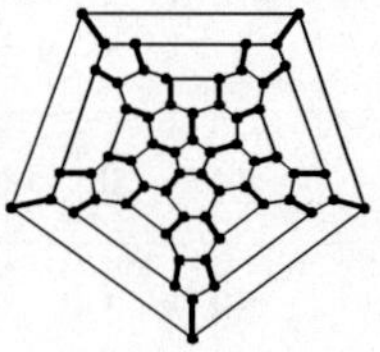	108	6	2.000	1.667
	4	0	0	3604		0	0	0	7918
		6	4	2			6	4	0

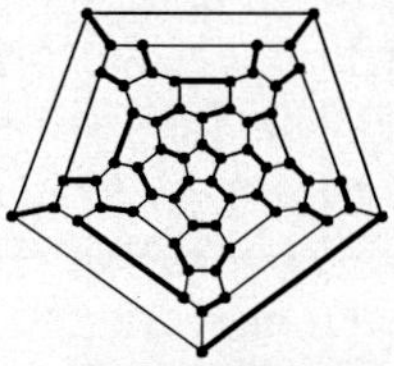	109	120	2.200	1.333	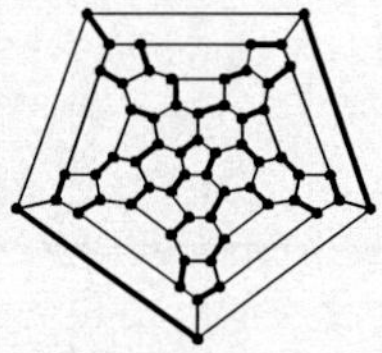	110	30	1.800	2.000
	6	3	1	1656		0	0	0	4659
		6	5	4			6	4	0
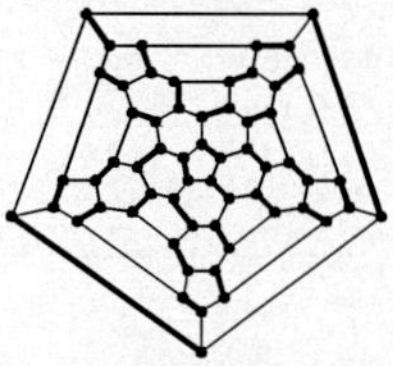	111	10	1.800	2.000	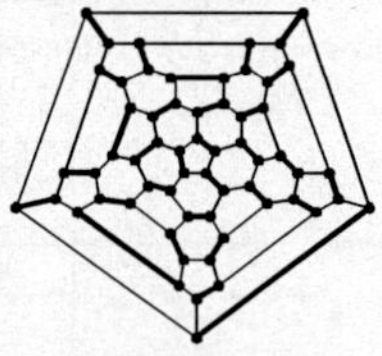	112	120	2.050	1.583
	0	0	0	5884		5	3	2	2107
		6	4	0			6	5	4
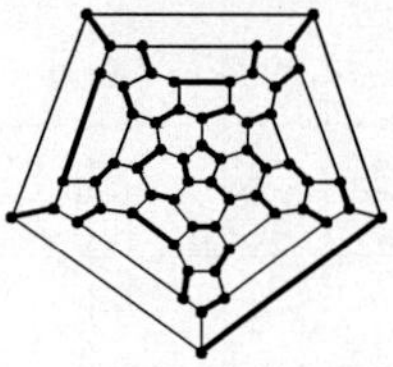	113	120	2.050	1.583	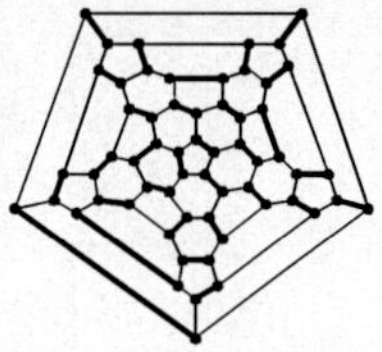	114	60	1.900	1.833
	5	4	1	1857		4	4	2	1728
		6	5	4			6	5	4
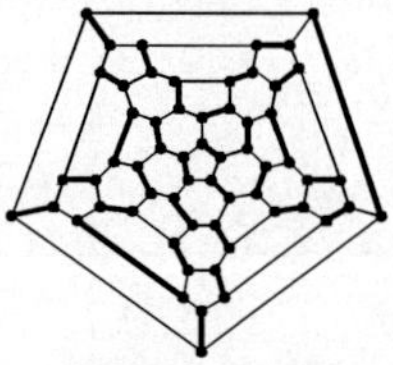	115	120	2.050	1.583	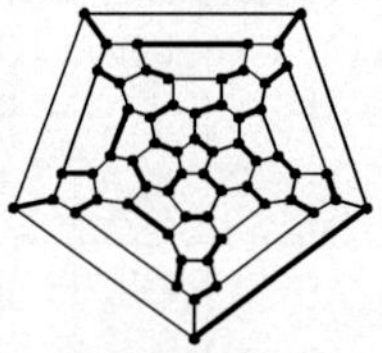	116	120	2.200	1.333
	5	4	1	1557		6	3	1	1671
		6	5	4			6	5	4
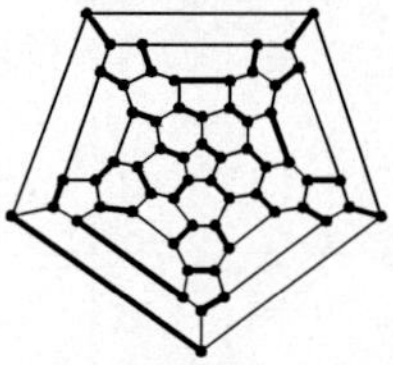	117	120	2.050	1.583	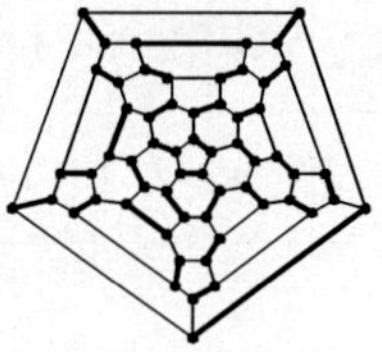	118	120	2.050	1.583
	5	4	1	1501		5	3	2	1868
		6	5	4			6	5	4
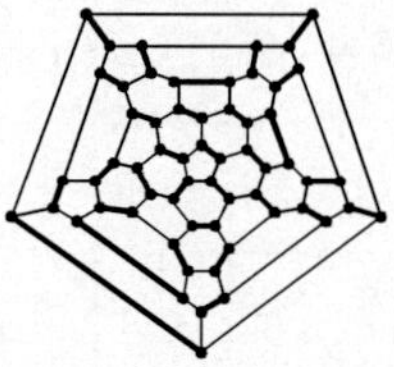	119	60	2.200	1.333	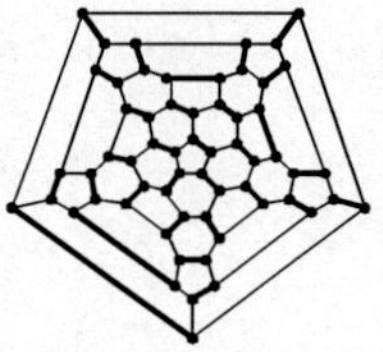	120	30	1.900	1.833
	7	2	1	1297		4	4	2	1685
		6	5	4			6	5	4

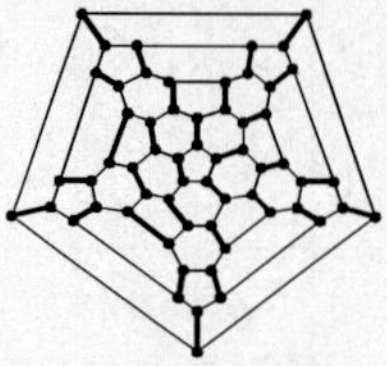	121	60	1.950	1.750	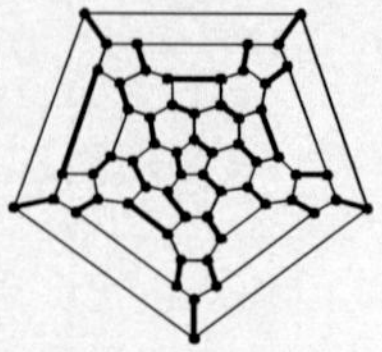	122	120	2.050	1.583
	2	2	1	3738		5	4	1	1914
		6	4	2			6	5	4
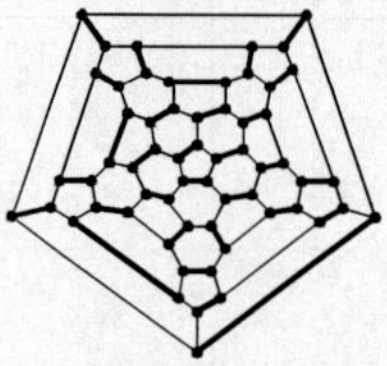	123	60	1.900	1.833	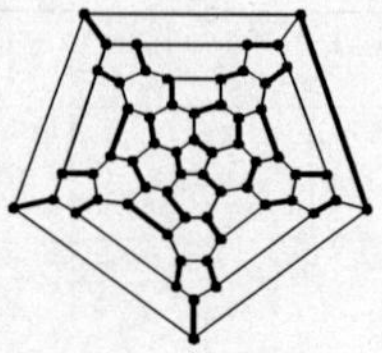	124	60	1.900	1.833
	4	4	2	2212		4	4	2	1831
		5	5	4			5	5	4
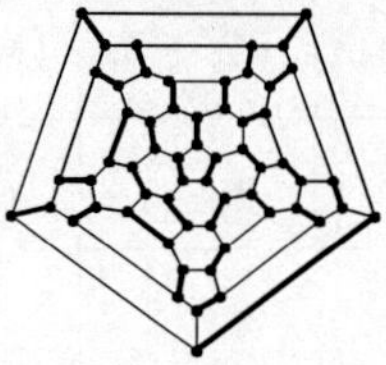	125	120	1.850	1.917	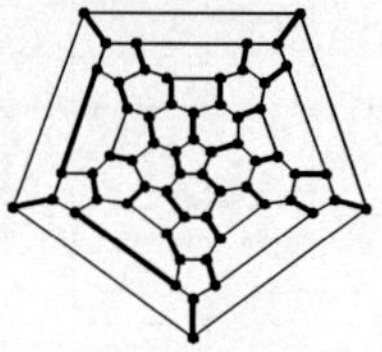	126	60	2.100	1.500
	2	1	1	4195		3	2	0	3492
		5	4	2			5	4	2
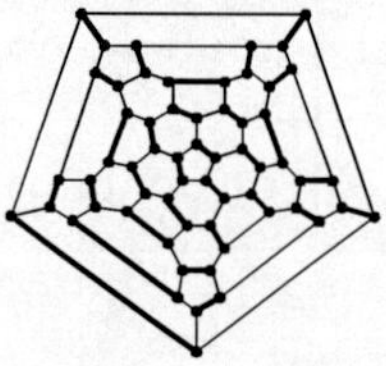	127	120	1.850	1.917	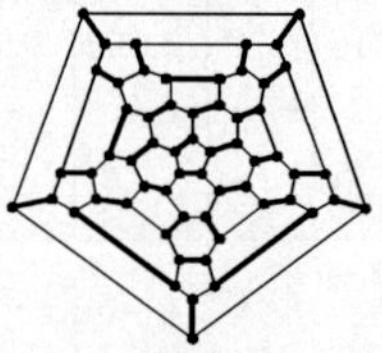	128	120	2.050	1.583
	2	2	1	2733		5	4	1	1966
		5	4	2			5	5	4
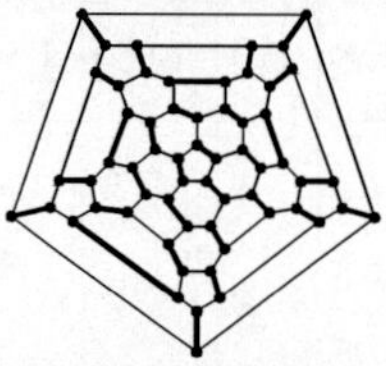	129	120	2.050	1.583	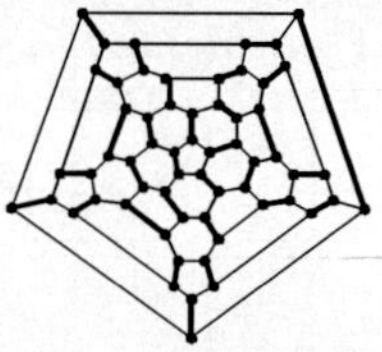	130	60	2.050	1.583
	5	4	1	1973		5	4	1	1639
		5	5	4			5	5	4
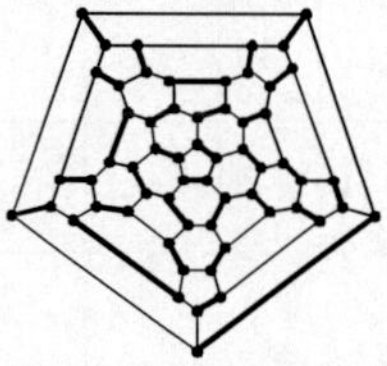	131	120	2.050	1.583	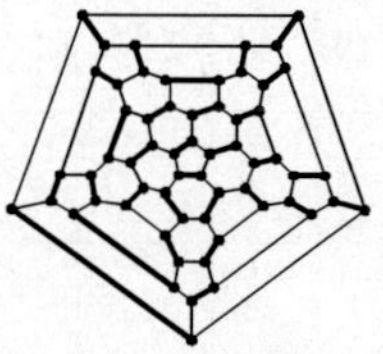	132	60	1.900	1.833
	4	5	1	2019		2	2	0	4387
		5	5	4			5	4	2

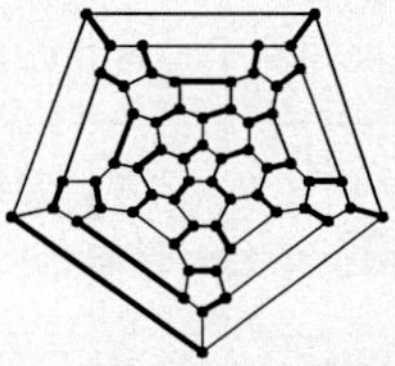	133	60	1.900	1.833	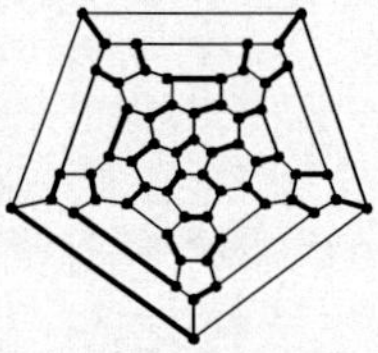	134	120	2.050	1.583
	2	2	0	4229		3	1	0	4151
		5	4	2			5	4	2
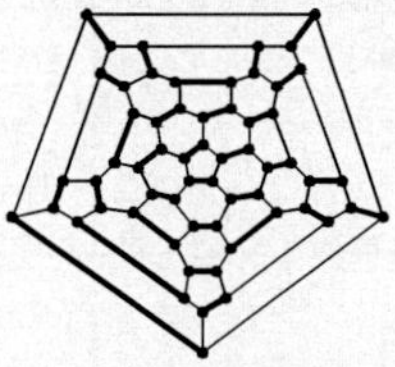	135	120	1.850	1.917	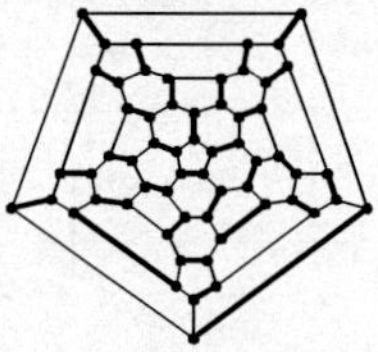	136	30	2.000	1.667
	1	1	0	3534		0	0	0	9894
		5	4	1			5	4	0
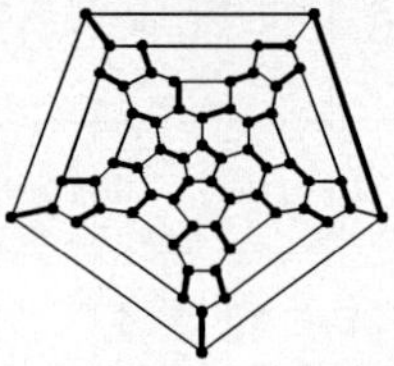	137	60	1.900	1.833	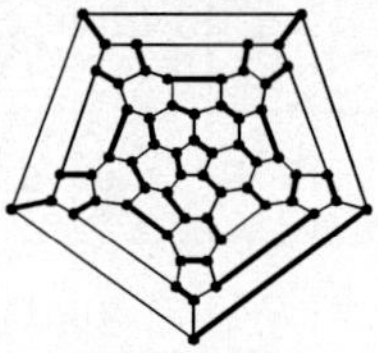	138	60	1.850	1.917
	0	0	0	6296		2	2	0	3864
		5	4	0			5	4	2
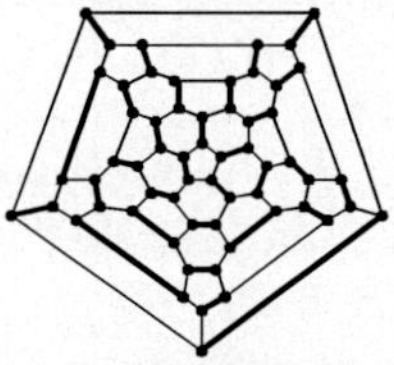	139	20	2.100	1.500	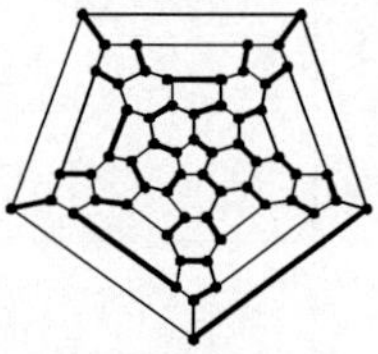	140	120	2.050	1.583
	2	0	0	8173		4	5	1	2099
		5	4	2			5	5	4
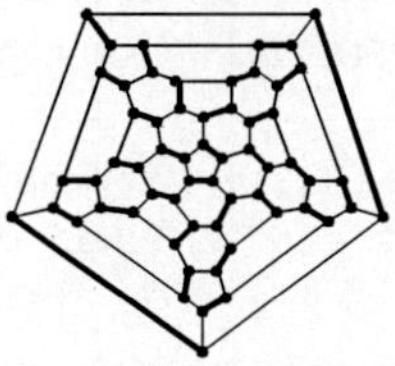	141	60	1.800	2.000	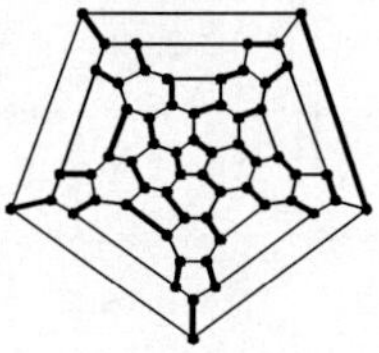	142	120	1.850	1.917
	0	0	0	5129		2	2	1	2941
		5	4	0			5	4	2
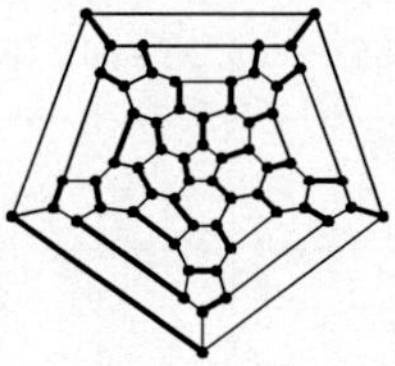	143	60	1.900	1.833	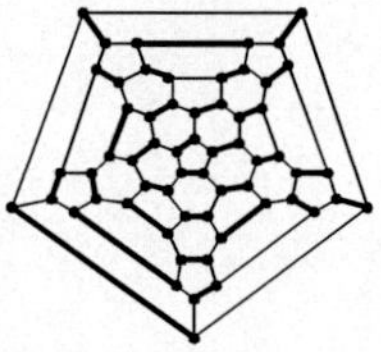	144	60	2.000	1.667
	0	0	0	6310		2	0	0	3133
		5	4	0			5	4	1

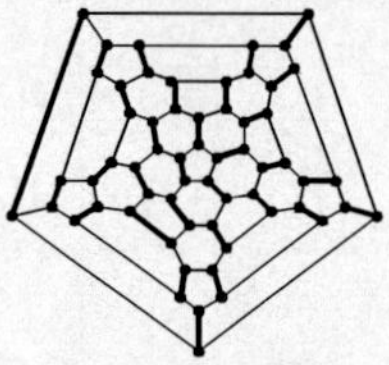	145	120	1.950	1.750	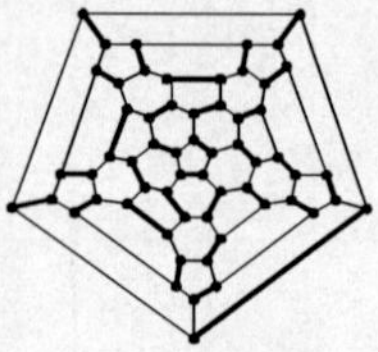	146	120	1.900	1.833
	2	2	1	3858		4	3	3	2212
		5	4	2			5	5	4
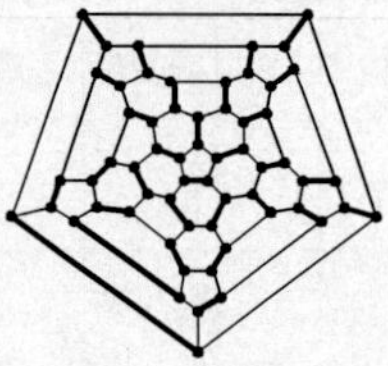	147	120	2.100	1.500	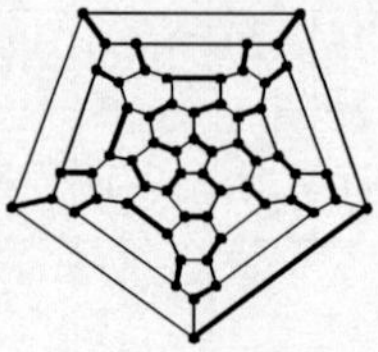	148	120	2.050	1.583
	3	2	0	3402		5	3	2	1990
		5	4	2			5	5	4
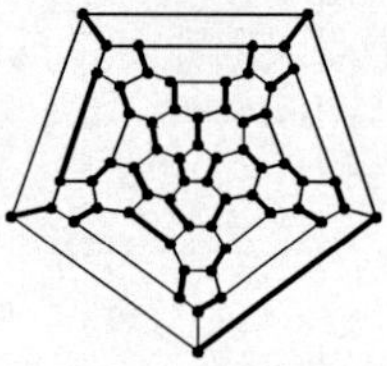	149	120	2.000	1.667	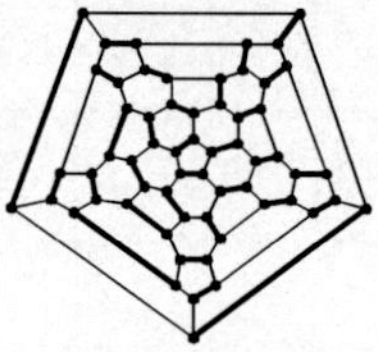	150	20	1.800	2.000
	3	1	0	3758		2	0	0	5509
		5	4	2			5	4	2
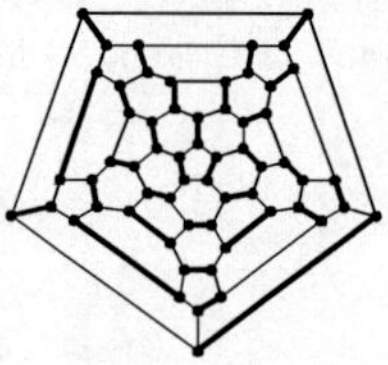	151	40	1.950	1.750	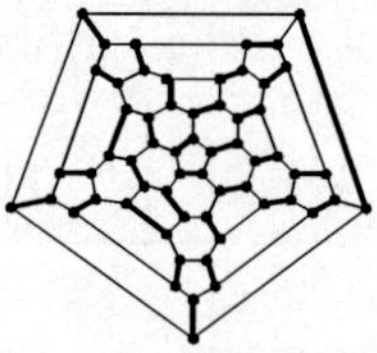	152	60	1.850	1.917
	2	0	0	6796		2	2	1	3020
		5	4	2			5	4	2
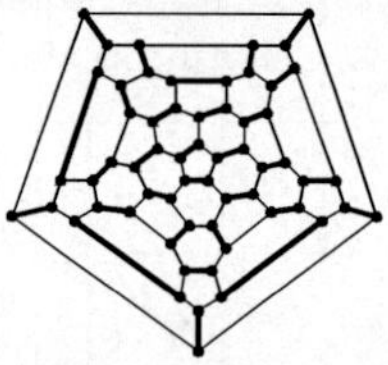	153	60	2.200	1.333	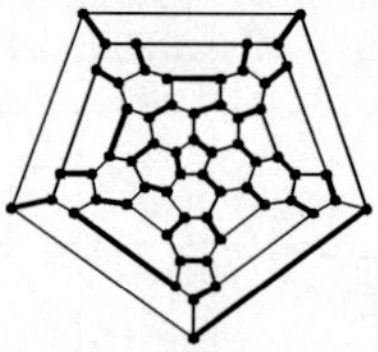	154	120	1.900	1.833
	6	4	0	1764		4	3	3	2436
		5	5	4			5	5	4
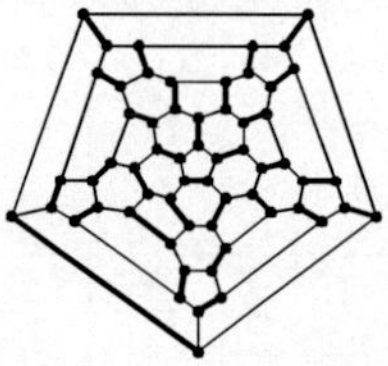	155	120	1.950	1.750	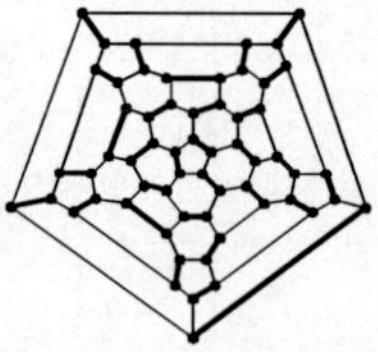	156	120	1.900	1.833
	2	2	1	4240		4	4	1	2287
		5	4	2			5	5	4

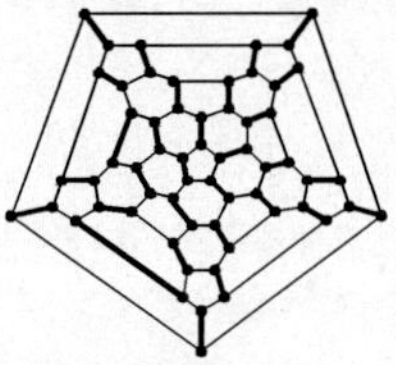	157	120	2.100	1.500	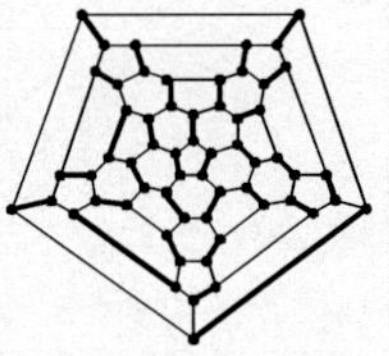	158	120	2.000	1.667
	3	2	0	3198		2	2	0	4407
		5	4	2			5	4	2

Acknowledgment The partial support of Croatian Ministry of Science, Education and Sport (grants no. 177-0000000-0884 and 037-0000000-2779) is gratefully acknowledged.

References

Balaban AT, Randić M (2004a) J Chem Comput Sci 44:50
Balaban AT, Randić M (2004b) Polycyclic Arom Comp 24:173
El-Basil S (2000) J Mol Struc – Theochem 531:9
Fowler PW (1986) Chem Phys Lett 131:444
Fowler PW, Manolopoulos DE (1995) An atlas of fullerenes. Clarendon Press, Oxford
Fries K (1927) J Liebigs Ann Chem 454:121
Fries K, Walter R, Schilling K (1935) J Liebigs Ann Chem 516:248
Klein DJ, Randić M (1987) J Comput Chem 8:516
Klein DJ, Schmalz TG, Hite GE, Steitz WA (1986) J Am Chem Soc 108:1301
Kroto HW, Heath JR, O'Brian SC, Curl RF, Smalley R (1985) Nature 318:162
Manolopulos DE, Woodal DR, Fowler PW (1992) J Chem Soc Faraday Trans 88:2427
Randić M (2003) Chem Rev 103:3449
Randić M, Kroto H, Vukičević D (2007) J Chem Inf Model 47:897
Vukičević D, Kroto HW, Randić M (2005) Croatica Chemica Acta 78:223
Vukičević D, Randić M (2005) Chem Phys Lett 401:446

Chapter 9
A Graph Theoretic Approach to Atomic Displacements in Fullerenes

Ernesto Estrada, Naomichi Hatano, and Adelio R. Matamala

Abstract The recently developed idea of analyzing complex networks in terms of node displacement due to vibration (Estrada and Hatano, Chem Phys Lett 486:166–170, 2010a) is applied to fullerenes. The fact that the ramafullerenes (fullerenes of Ramanujan graphs) are limited to fullerenes with relatively small number of C atoms is explained from the point of view of the node displacement. The node displacement is also shown to indicate the stability of isomers of C_{40} fullerenes. It is suggested from the analysis of local node displacement that instability of fullerenes mainly comes from pentagon-rich areas of the molecules.

9.1 Introduction

Most of us have been aware of graphite since we were children. We can remember how useful our pencils were in learning to write and the advantage of deleting our errors simply by using rubber erasers. Fewer, however, have had the opportunity of admiring the bright and perfection of diamond. For people involved in the study of molecular structures, nothing has been more wonderful than contemplating the structure of fullerenes (see Fig. 9.1). The simplicity, elegance and beauty of this molecular structure have captivated many natural scientists and mathematicians in the last decades (Aldersey-Williams 1995). These three allotropes of carbon: graphite, diamond and fullerenes, are examples of how the combinatorial organization of atoms can produce very different structures with remarkable distinct properties (Pierson 1993).

More formally, fullerenes are 3-regular polyhedral graphs. A graph is an object formed by a set of nodes, which are joined together by links or edges. Regular

E. Estrada (✉)
Department of Mathematics and Statistics, Department of Physics, Institute of Complex Systems, University of Strathclyde, Glasgow G1 1XQ, UK
e-mail: ernesto.estrada@strath.ac.uk

F. Cataldo et al. (eds.), *The Mathematics and Topology of Fullerenes*,
Carbon Materials: Chemistry and Physics 4, DOI 10.1007/978-94-007-0221-9_9,

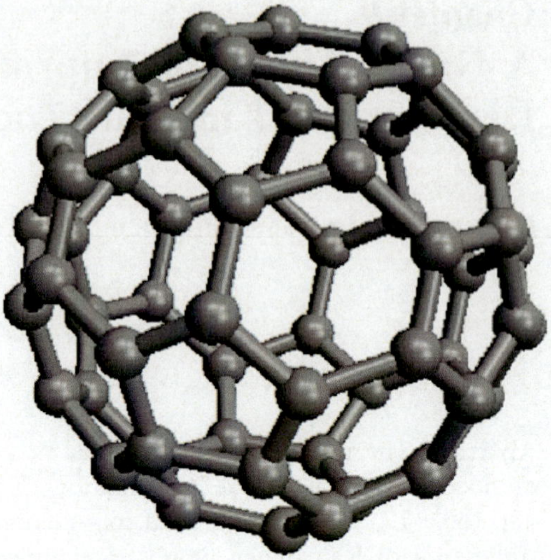

Fig. 9.1 Illustration of the molecular structure of buckminsterfullerene, C_{60}

graphs, in particular, are those having the same number of links per node. In "classical" fullerenes only pentagons and hexagons form the structure, while in the non-traditional ones cycles of other sizes are also allowed. Due to the many interesting mathematical results existing for regular graphs, it is not surprising that many researchers have paid attention to the graph-theoretic properties of fullerenes (Fajtlowitz and Larson 2003; Doslić 2005, 2008; Fowler, 2002, 2003; Manolopoulos et al. 1991; Zhang and Balasubramanian 1993). Many invariants, old and new, have been studied for this family of molecules, and many important conclusions about their structure, stability, function and reactivity have been obtained on the basis of such topological ideas.

Here we propose the study of atomic displacements in fullerenes due to small vibrations in the molecule as a whole. We use a graph-theoretic approach based on physically sounded ideas taken from mechanics. For the first time we show here a connection between some isoperimetric properties of graphs and vibrational properties and we extract important conclusions about the stability of these molecules. We also give a theoretical justification for the empirical evidence that the most stable fullerenes are those displaying the smallest number of adjacent pentagons. What we show here is that such pentagonal isolation confers more vibrational rigidity to the molecule, which is translated in larger stability.

9.2 Preliminary Definitions

Let G be a connected graph without loops or multiple links having n nodes. Then the adjacency matrix of G, $\mathbf{A}(G) = \mathbf{A}$, is a square, symmetric matrix of order n, whose elements A_{ij} are ones or zeroes if the corresponding nodes are adjacent or not, respectively. The sum of a row or column of this matrix is known as the degree of corresponding node i and designated here by δ_i. This matrix has n (not necessarily distinct) real-valued eigenvalues, which are denoted here by $\lambda_1, \lambda_2, \ldots, \lambda_N$. The i th component of the j th eigenvector of the adjacency matrix are designated here by $\varphi_j(i)$. Here the eigenvalues are usually labelled in a non-increasing manner:

$$\lambda_1 > \lambda_2 \geq \cdots \geq \lambda_n. \tag{9.1}$$

The Laplacian matrix of a graph is defined as $\mathbf{L} = \mathbf{D} - \mathbf{A}$, where $\mathbf{D}$ is the diagonal matrix of degrees δ_i and $\mathbf{A}$ the adjacency matrix of the graph. The eigenvalues of the Laplacian matrix are ordered here as follows:

$$0 = \mu_1 < \mu_2 \leq \cdots \leq \mu_{n-1} \leq \mu_n. \tag{9.2}$$

The i th component of the j th eigenvector of the Laplacian matrix are designated here by $U_j(i)$.

The Moore-Penrose generalised inverse (or the pseudo-inverse) $\mathbf{L}^+$ of the graph Laplacian $\mathbf{L}$ has been proved to exist for any connected graphs. Using $\mathbf{L}^+$ a graph metric known as the resistance distance can be computed. The resistance distance (Klein and Randić 1993) between a pair of nodes can be obtained by using the following formula (Xiao and Gutman 2003):

$$\Omega_{ij} = \left(\mathbf{L}^+\right)_{ii} + \left(\mathbf{L}^+\right)_{jj} - \left(\mathbf{L}^+\right)_{ij} - \left(\mathbf{L}^+\right)_{ji} \tag{9.3}$$

for $i \neq j$.

9.3 Atomic Displacements in Molecules

Let us consider a molecular graph in which atoms represent unit mass balls and bonds are identified with springs of a common spring constant k (Estrada and Hatano 2010a, b). The vibrational potential energy from the static position of the molecule can be expressed as

$$V(\vec{x}) = \frac{k}{2}\vec{x}^T \mathbf{L}\vec{x}, \tag{9.4}$$

where $\vec{x}$ is the vector whose i th entry x_i is the displacement of the atom i from its equilibrium position.

Under these assumptions two of the present authors (EE and NH) have derived a topological formula for the mean displacement of a node i when the molecule is immersed into a thermal bath of inverse temperature $\beta = 1/k_B T$, where k_B is the Boltzmann constant. The procedure followed by Estrada and Hatano (Estrada and Hatano 2010a, b) is sketched below. First we can express the atomic displacements as

$$\Delta x_i \equiv \sqrt{\langle x_i^2 \rangle} = \sqrt{\int x_i^2 P(\vec{x})\, d\vec{x}}, \tag{9.5}$$

where $\langle \cdots \rangle$ denotes the thermal average and $P(\vec{x})$ is the probability distribution of the displacement of the nodes given by the Boltzmann distribution. The normalization factor that appear in the expression of $P(\vec{x})$ represents the partition function of the molecule and can be expressed as

$$\begin{aligned} Z &= \int d\vec{y} \exp\left(-\frac{\beta k}{2} \vec{y}^T \Lambda \vec{y}\right) \\ &= \prod_{\mu=1}^{n} \int_{-\infty}^{+\infty} dy_\mu \exp\left(-\frac{\beta k}{2} \lambda_\mu y_\mu^2\right). \end{aligned} \tag{9.6}$$

It is well known that the smallest eigenvalue of the discrete Laplacian matrix is equal to zero, $\mu_1 = 0$. This is interpreted in this context as the translational movement of the molecule as a whole, the coherent motion in one direction. Here we remove this motion of the centre of mass and focus on the relative atomic motion only. In this case we obtain the following modified partition function

$$\begin{aligned} \tilde{Z} &= \prod_{\mu=2}^{n} \int_{-\infty}^{+\infty} dy_\mu \exp\left(-\frac{\beta k}{2} \mu_\mu y_\mu^2\right) \\ &= \prod_{\mu=2}^{n} \sqrt{\frac{2\pi}{\beta k \mu_\mu}}. \end{aligned} \tag{9.7}$$

Then, after some algebraic manipulation we finally arrive at the expression for the topological atomic displacement:

$$\Delta x_i \equiv \sqrt{\langle x_i^2 \rangle} = \sqrt{\sum_{\nu=2}^{n} \frac{[U_\nu(i)]^2}{\beta k \mu_\nu}}, \tag{9.8}$$

We have also shown that the topological atomic displacements can be obtained directly from the Moore-Penrose generalized inverse of the Laplacian matrix (Estrada and Hatano 2010a, b).

In our previous works we also showed that the Kirchhoff index of a molecule can be expressed as the sum of the squared atomic displacements produced by small molecular vibrations multiplied by the number of atoms in the molecule:

$$Kf = n \sum_{i=i}^{n} (\Delta x_i)^2 = n^2 \overline{(\Delta x)^2}. \tag{9.9}$$

Furthermore, the sum of resistance distances for a given atom and any other atom in the molecule $R_i = \sum_{j=1}^{n} \Omega_{ij}$, can be expressed in terms of the atomic displacements as

$$R_i = n\,(\Delta x_i)^2 + \sum_{i=1}^{n} (\Delta x_i)^2 = n\left[(\Delta x_i)^2 + \overline{(\Delta x)^2}\right].$$

On the other hand, the mean square displacement of a node i is given by

$$(\Delta x_i)^2 \equiv \left\langle x_i^2 \right\rangle = \int x_i^2 P\,(\vec{x})\, d\vec{x} \tag{9.10}$$

and the correlation between the displacements of nodes i and j is given by

$$\left\langle x_i x_j \right\rangle = \int x_i x_j P(\vec{x}) d\vec{x}, \tag{9.11}$$

where $\langle \cdots \rangle$ denotes the average with respect to $P(\vec{x})$. The function (11) can be represented using the Moore-Penrose generalized Laplacian as follows:

$$\left\langle x_i x_j \right\rangle = \frac{1}{\beta k} \left(\mathbf{L}^+\right)_{ij}. \tag{9.12}$$

Finally, Eq. (9.11) is followed by the thermal average of the vibrational potential energy Eq. (9.6) in the form

$$\langle V\,(\vec{x})\rangle = \frac{1}{2}\sum_{i=1}^{n} k_i \left\langle x_i^2 \right\rangle - \sum_{i,j\in E} \left\langle x_i x_j \right\rangle = \frac{1}{\beta k}\sum_{i=1}^{n} k_i \left(\mathbf{L}^+\right)_{ii} - \sum_{i,j\in E} \left(\mathbf{L}^+\right)_{ij}. \tag{9.13}$$

9.4 Atomic Displacements and Expansion in Regular Graphs

In a regular graph it is known that the following relationship exists between the eigenvalues of the Laplacian and the eigenvalues of the adjacency matrix of a graph (ordered as in Section 9.2):

$$\mu_j = \lambda_1 - \lambda_j.$$

It is also known that for these graphs the eigenvectors of the adjacency and Laplacian matrix coincide. Then it is straightforward to realise that the atomic displacements in molecules whose graphs are regular can be written in terms of the spectra of their adjacency matrix as follow:

$$(\Delta x_i)^2 = \frac{1}{\beta k} \sum_{j=2}^{n} \frac{[\varphi_j(i)]^2}{\lambda_1 - \lambda_j}. \tag{9.14}$$

Let us consider for the sake of simplicity, the case where $\beta k \equiv 1$ and let $\Delta = \lambda_1 - \lambda_2$ be the *spectral gap* of the graph. Then, we can write (10) as follows

$$(\Delta x_i)^2 = \frac{[\phi_2(i)]^2}{\Delta} + \sum_{j=3}^{n} \frac{\phi_j(i)^2}{\lambda_1 - \lambda_j}. \tag{9.15}$$

Then, the average atomic displacement in a molecule can be expressed as

$$\overline{(\Delta x_i)^2} = \frac{1}{n} \sum_{i=1}^{n} \left(\frac{[\phi_2(i)]^2}{\Delta} + \sum_{j=3}^{n} \frac{\phi_j(i)^2}{\lambda_1 - \lambda_j} \right) = \frac{1}{n} \left(\frac{1}{\Delta} + \sum_{j=3}^{n} \frac{1}{\lambda_1 - \lambda_j} \right). \tag{9.16}$$

Obviously, the first term of the RHS of Eq. (9.15) has the largest contribution to the atomic displacements of a given molecule. Then, for a given regular molecule the magnitude of the atomic displacements due to molecular vibration/oscillations depends very much on the magnitude of the spectral gap. Those molecules having large spectral gaps are expected to display the smallest atomic displacements.

There is a family of graph named good expansion (GE) graphs. A graph is considered to have GE if every subset S of nodes ($|S| \leq 1/2\,|V|$) has a neighborhood that is larger than some "edge expansion ratio" $h(G)$ multiplied by the number of nodes in S. A neighborhood of S is the set of nodes which are linked to the nodes in S. Formally, for each vertex $v \in V$ (where V is the set of nodes in the network), the neighborhood of v, denoted as $\Gamma(v)$ is defined as: $\Gamma(v) = \{u \in V\,|(u, v) \in E\}$ (where E is the set of edges in the graph). Then, the neighborhood of a subset $S \subseteq V$ is defined as the union of the neighborhoods of the nodes in S: $\Gamma(S) = \bigcup_{v \in S} \Gamma(v)$ and the network has GE if $\Gamma(v) \geq h(G)|S|\ \forall S \subseteq V$.

The edge expansion ratio $h(G)$ of a graph is defined as (Hoory et al. 2006)

$$h(G) \stackrel{def}{=} \min_{S \subseteq V,\, |E(S)| \leq |E|/2} \frac{|E(S, \bar{S})|}{|S|}, \tag{9.17}$$

where $|E(S, \bar{S})|$ denotes the number of links that have one endpoint in S and another endpoint in $\bar{S}$. The connection between good expansion and algebraic graph theory comes from the celebrated Alon-Milman theorem (Alon and Milman 1985), which states that for a finite, connected δ-regular graph G, with spectral gap Δ, the expansion constant is bounded as follows:

$$\frac{\Delta}{2} \leq h(G) \leq \sqrt{2\delta\Delta}. \tag{9.18}$$

Accordingly, high expansion necessarily means large spectral gap Δ. Consequently, we can resume our results concerning atomic displacements and good expansion as follows:

> Among all graphs with *n* nodes, those having good expansion properties display the smallest topological displacements for their nodes.

In order to illustrate our results for some artificial graphs we selected the set of all cubic graphs with 10 nodes. These graphs are illustrated in Fig. 9.2

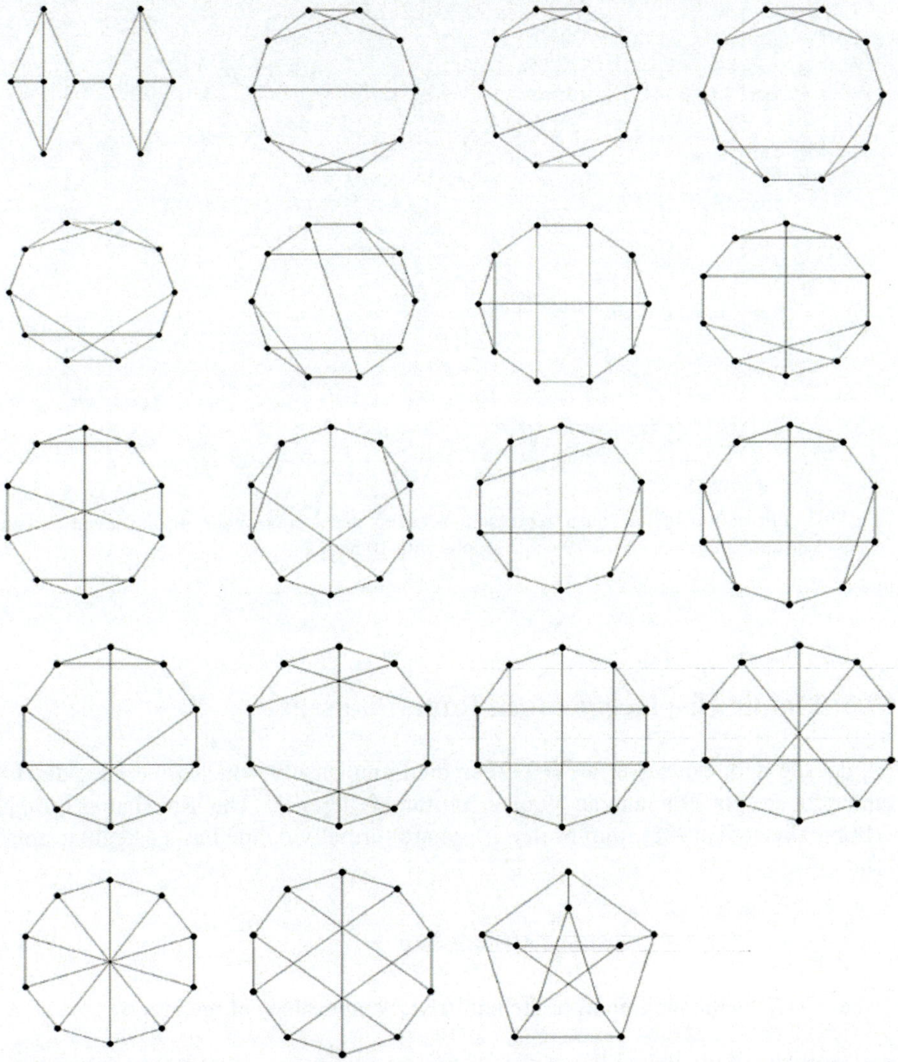

Fig. 9.2 Illustration of all cubic graphs ($\delta = 3$) with 10 nodes. The last graph depicted is known as the Petersen graph

When we plot the values of the average node displacement $\overline{(\Delta x_i)^2}$ against the inverse spectral gap $1/\Delta$ for these 3-regular graphs with 10 nodes we obtain a straight line of slope 0.093 and intercept 0.271 as illustrated in Fig. 9.3. As can be seen the graph displaying the smallest average displacement of nodes in the Petersen graph (last graph in Fig. 9.2), which has the largest spectral gap Δ among all cubic graphs of 10 nodes.

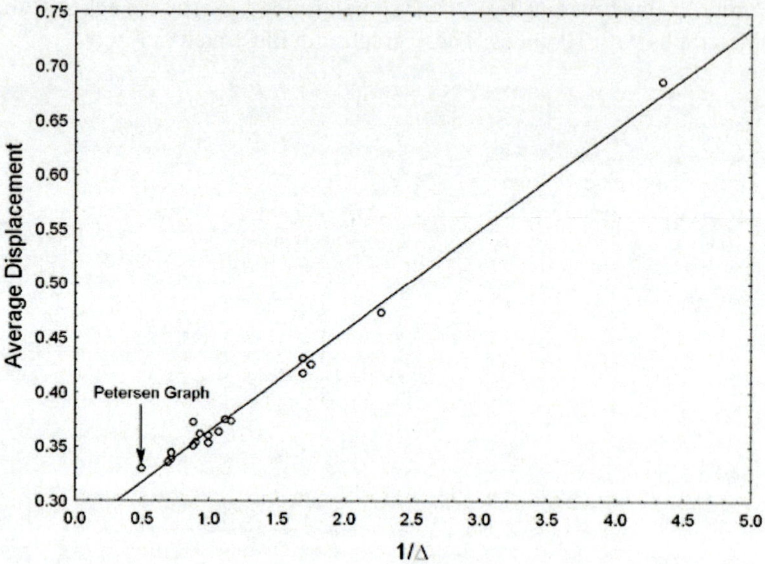

Fig. 9.3 Illustration of the linear regression between the average node displacement and the inverse spectral gap $1/\Delta$ for the 19 cubic graphs with 10 nodes

9.5 Atomic Displacements in Ramafullerenes

A decade ago, Fowler et al. (1999) studied empirically which fullerenes display the property of Ramanujan graphs, or ramafullerenes. The Ramanujan graphs (Lubotzky et al. 1988; Ram Murty 2003) are formally defined as a δ-regular graph for which

$$\lambda(G) \leq 2\sqrt{\delta - 1}, \tag{9.19}$$

where $\lambda(G)$ is the maximum of the non-trivial eigenvalues of the graph

$$\lambda(G) = \max_{|\lambda_i| < \delta} |\lambda_i| \,. \tag{9.20}$$

In the case of fullerenes, $\lambda(G) \leq 2\sqrt{2}$. It has been proved that Ramanujan graphs are good expanders. Using the Alon-Boppana theorem (Alon 1986) it can be shown that for a δ-regular graph with n nodes,

$$\lambda_2 \leq \lambda(G) \leq 2\sqrt{\delta - 1}, \tag{9.21}$$

which shows that Ramanujan graphs are good expanders. In the mentioned paper of Fowler et al. (1999) it was found that a relatively large number of ramafullerenes exists among fullerenes having between 50 and 76 atoms. The distribution of ramafullerenes is displayed in Fig. 9.4 for fullerenes having between 20 and 100 atoms. Based on these empirical finding it was conjectured that there is no ramafullerene for $n > 84$.

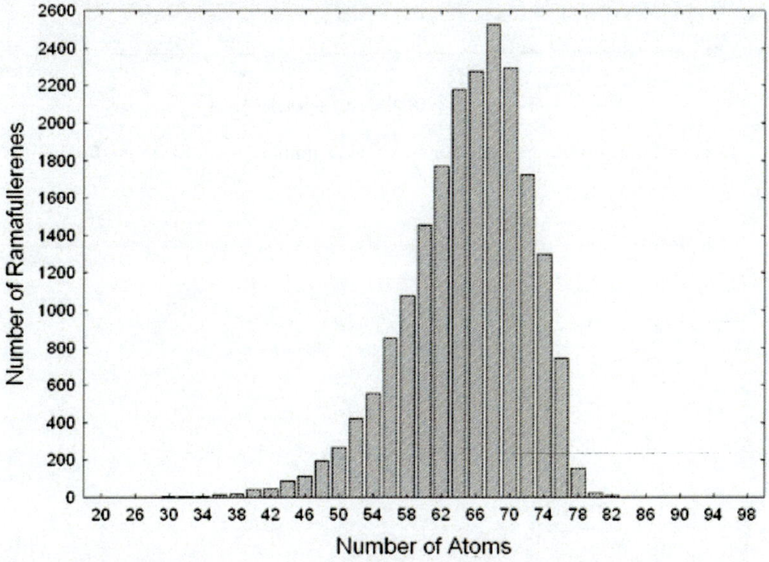

Fig. 9.4 Distribution of the number of ramafullerenes as a function of the number of atoms (Fowler et al. 1999)

A plausible explanation for this finding is that the spectral gap decays very fast with the number of atoms in the fullerenes. For instance, in Fig. 9.5 we plot the spectral gap of some fullerenes having between 20 and 540 atoms, where we also show the line below which no ramafullerene exists, i.e., $\Delta > 3 - 2\sqrt{2}$.

The immediate implication of this decay of the spectral gap with the number of nodes is that the average atomic displacement $\overline{(\Delta x_i)^2}$ in fullerenes increases as a power law with the number of atoms. This situation is illustrated in Fig. 9.6, where the best fit obtained indicates that $\overline{(\Delta x_i)^2} \sim n^{0.042}$.

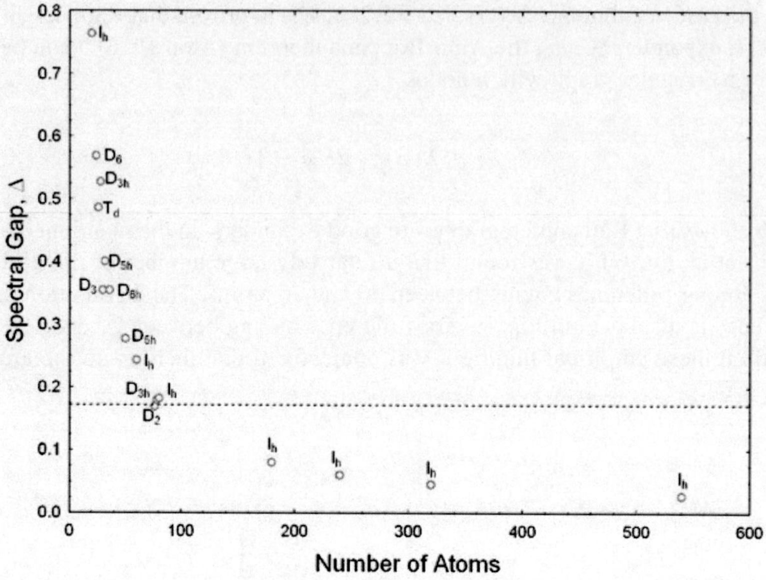

Fig. 9.5 Decay of the spectral gap as a function of the number of atoms in the fullerenes

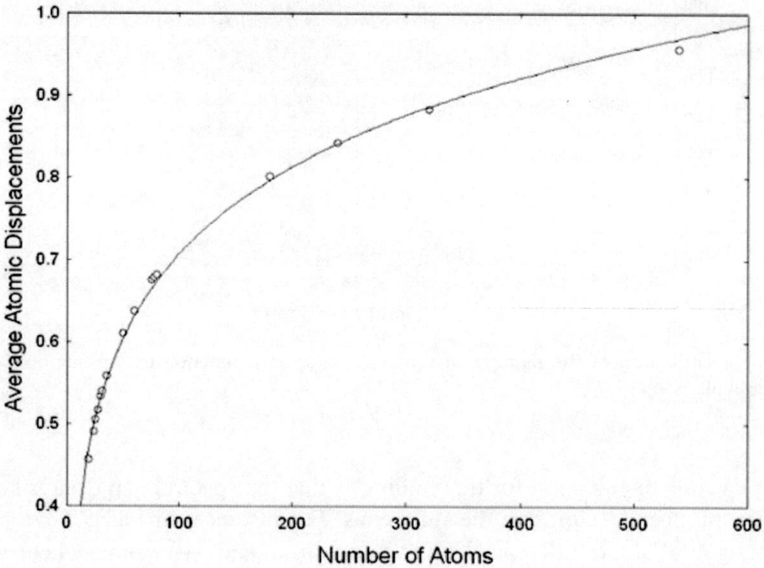

Fig. 9.6 Power law increase of the average atomic displacement as a function of the number of atoms in the fullerenes

9.6 Atomic Displacements in Isomers of Fullerene C_{40}

In order to understand in a better way the relation among the spectral gap, the atomic displacements and the energetics of fullerenes we are going to study 40 isomers of C_{40}. When plotting the inverse of the spectral gap for these fullerenes versus the vibrational potential (see Fig. 9.7) or the average atomic displacements (graphic not displayed) we observe that the smallest value of $1/\Delta$, i.e., the largest spectral gap, corresponds to the fullerene C40:40. Here we denote fullerenes by C40:X, where X corresponds to the labeling given by Fowler and Manolopoulos in their Atlas of Fullerenes (Fowler and Manolopoulos 1995). The smallest vibrational potential, however, corresponds to C40:38 followed by C40:39. The isomer C40:38 has been identified by 11 out of 12 computational methods as the most stable one among C_{40} fullerenes (Albertazzi et al. 1999), while C40:39 has been identified as the second most stable by 9 of these methods.

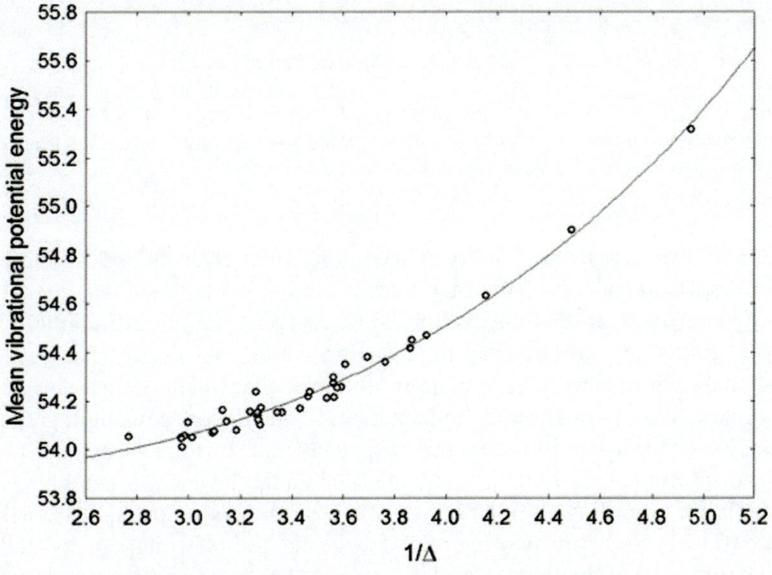

Fig. 9.7 Increase of the thermal average of the vibrational potential energy as a function of the inverse of the spectral gap in C_{40} fullerenes

In Fig. 9.8 we plot the thermal average of the vibrational potential $\langle V(\vec{x})\rangle$ of all C_{40} isomers versus the relative energy calculated by a hybrid density functional method with a minimal STO-3G basis as reported by Albertazzi et al. (1999). A good correlation exists between both magnitudes with a correlation coefficient $r = 0.961$ and equation: $E = 5851.99\,(-0.4933 + 0.00916\,\langle V(\vec{x})\rangle)^{0.2} - 1532$. The importance of this relationship goes beyond the possibility of predicting stability of fullerenes. For instance, this relationship indicates a possible cause for the difference in stability of fullerene isomers. That is, the largest the rigidity of

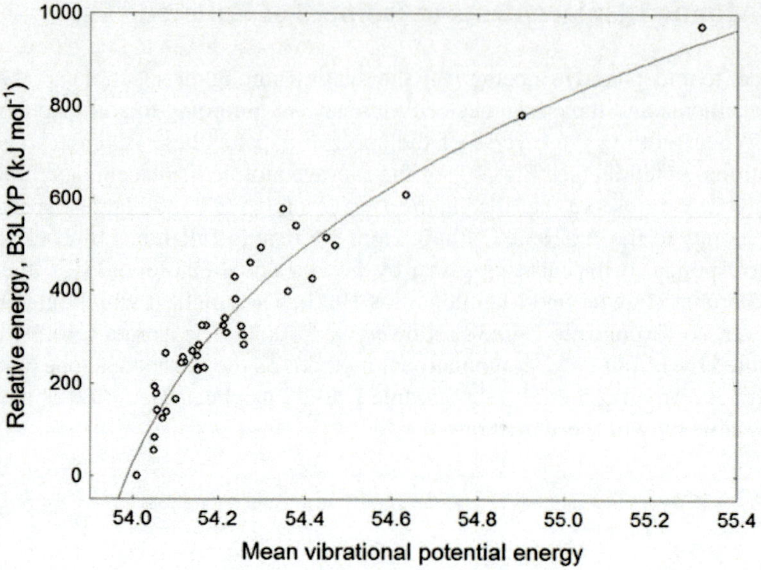

Fig. 9.8 Relationship between the mean vibrational potential energy $\langle V(\vec{x})\rangle$ (Eq. (9.13)) and the relative energy calculated by density functional theory for C_{40} fullerenes. B3LYP energies are relative to fullerene C40:38 as taken from Albertazzi et al. (1999)

a fullerene the largest its stability. The rigidity here is measured by the average atomic displacements or the thermal average vibrational potential energy.

In agreement with this observation is the fact that the largest atomic displacements in fullerenes are observed for atoms in pentagonal rings. That is, atoms in pentagonal rings display in general more atomic displacements than atoms in hexagonal cycles. Among those atoms in pentagonal rings the ones fusing together show the largest flexibility, i.e., the largest displacements. In Fig. 9.9 we illustrate the atomic displacements for two isomers of C_{40} with the lowest (*top*) and largest (*bottom*) stability according to B3LYP energies. As can be seen in the least stable C_{40} isomer (C40:1) there are two regions of large flexibility which are located at the left and right part of the figure (top-left image). These two regions are formed completely by fused pentagons in which a central pentagon is surrounded by other six. This central pentagon has the largest flexibility among all cycles in this molecule (see Fig. 9.10a). In the case of the most stable C_{40} fullerene, C40:38 the largest atomic displacements are observed for the atoms in the very centre of six fused pentagonal rings as can be seen in Fig. 9.10b. Such flexibility decreases as soon as the atoms are far from the centre of this system, which implies that they are in contact with hexagonal rings.

An interesting conclusion that we can extract from these findings is that there is not a plausible geometric explanation for why pentagonal rings are more flexible than hexagonal ones. That is, from geometric intuitive reasoning we could expect that hexagons are more flexible than pentagons. As we have not considered

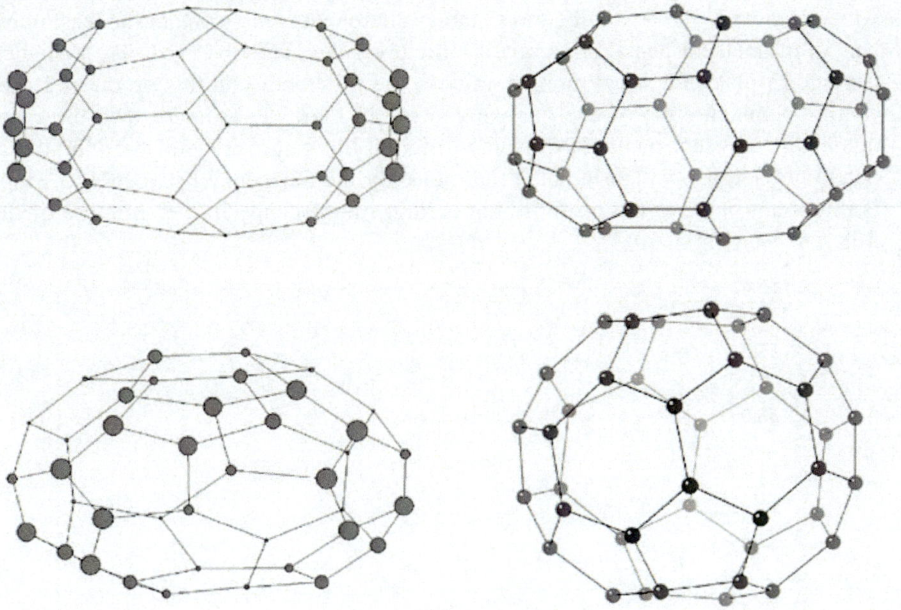

Fig. 9.9 Illustration of the atomic displacements (*left graphics*) for the isomers C40:1 (*top*) and C40:38 (*bottom*) with the lowest and largest stability, respectively, according to B3LYP energies. The graphics at the right hand side are three-dimensional embeddings of these fullerenes. The radii of the nodes in the graphics on the left-hand side are proportional to the atomic displacements

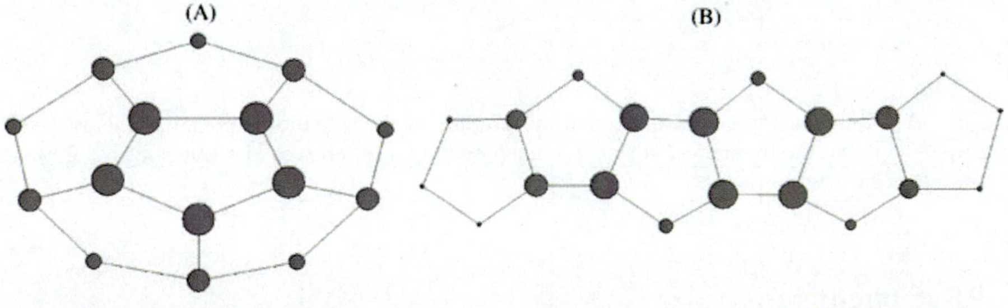

Fig. 9.10 Illustration of the atomic displacements of fused pentagonal rings in C_{40} fullerenes. (**a**) System of seven fused pentagonal rings in C40:1, the least stable C_{40} isomer. (**b**) System of six fused pentagonal rings in C40:38, the most stable C_{40} isomer. The radii of the nodes in both graphics are proportional to the atomic displacements

any geometric or electronic characteristic of fullerenes in deriving our atomic displacement measures, we can conclude that the cause of the observed differences in flexibility/rigidity between pentagonal and hexagonal rings is a purely topological one.

Finally, we would like to remark that our current findings support the hypothesis that the pentagon adjacency number is a good predictor of the stability of fullerenes. It has been shown in several studies (Balaban et al. 1995; Campbell et al. 1996;

Albertazzi et al. 1999) that the most stable fullerene isomers contain the least number N_p of adjacent pentagons in their structures. We have observed here that such pentagon isolation confers more rigidity to the fullerenes and this produces larger energy stabilization. However, as we can see in Fig. 9.11 isomers with the same number of adjacent pentagons display different thermal average of the vibrational potential energies, which indicate that not only the adjacency between such rings is important but also the position that certain rings occupy in the structure of the fullerene (see Fig. 9.10).

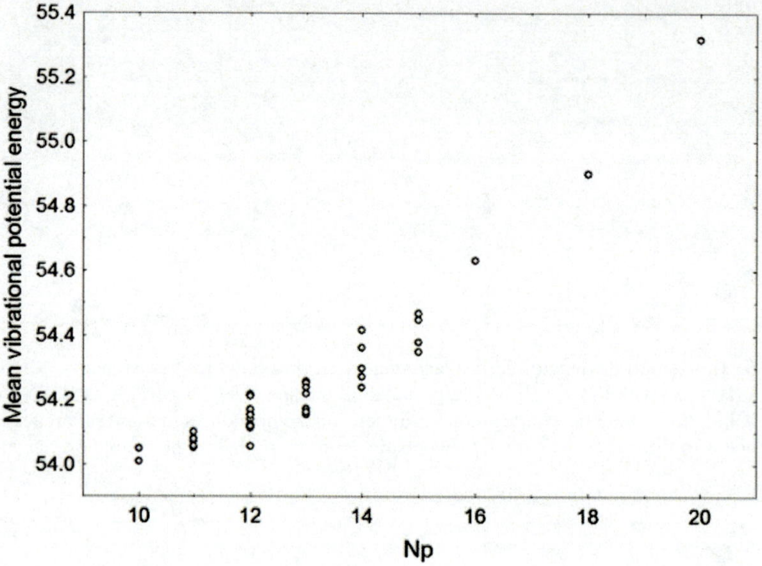

Fig. 9.11 Relationship between the thermal average of the vibrational potential energy $\langle V(\vec{x})\rangle$ (Eq. (9.13)) and the number of adjacent pentagons in C_{40} fullerenes. The values of N_p are taken from Réti and László (2009)

9.7 Conclusions

In the present article, we applied to fullerenes, a recently developed idea of analyzing complex networks in terms of the vibrational potential energy of the atomic displacements. After defining the mean atomic (or node) displacement, we argued that a small atomic displacement means a large spectral gap of the graph (Fig. 9.3), which in turns means that the graph has good expansion.

We demonstrated these relations in fullerenes. Fullerenes with the property of Ramanujan graphs, or ramafullerenes, are good expanders. The ramafullerenes are limited to fullerenes with relatively small numbers of C atoms. We explained this fact from the above two points of view, namely, the spectral gap and the atomic displacement. We demonstrated that as the number of atoms increases, the spectral gap decreases (Fig. 9.5) and the atomic displacement increases (Fig. 9.6).

As another application of the atomic displacement, we examined isomers of fullerenes C_{40}. We showed that the thermal average of the vibrational potential energy of our simple definition has strong correlation with the energy obtained from elaborate calculation of density functional theory. Since the latter tells us the stability of each isomer, we claim that our vibrational energy also indicates the isomers' stability. We went further and showed that the atomic displacement is generally larger in the area of pentagons than in the area of hexagons. This suggests that the instability of a fullerene isomer is originated in pentagon-rich areas confirming the pentagon isolation rule.

Acknowledgements EE thanks P. Fowler for providing the dataset of C_{40} isomers used in this study. EE thanks New Professor's Fund of the University of Strathclyde for partial financial support. ARM thanks FONDECYT (Chile) under grant No. 1080561

References

Albertazzi A, Domene C, Fowler PW, Heine T, Seifert G, Van Alsenoy C, Zerbetto F (1999) Phys Chem Chem Phys 1:2913
Aldersey-Williams H (1995) The most beautiful molecule: the discovery of the buckyball. Wiley, London
Alon N (1986) Combinatorica 6:83
Alon N, Milman VD (1985) J Combin Theory Ser B 38:73
Balaban AT, Liu X, Klein DJ, Babić D, Schmalz TG, Seitz WA, Randić M (1995) J Chem Inf Comput Sci 35:396
Campbell EEB, Fowler PW, Mitchell D, Zerbetto F (1996) Chem Phys Lett 250:544
Doslić T (2005) Chem Phys Lett 412:336
Doslić T (2008) J Math Chem 43:647
Estrada E, Hatano N (2010a) Chem Phys Lett 486:166–170
Estrada E, Hatano N (2010b) In: Gutman I, Furtula B (eds) Novel molecular descriptors. Theory and applications I. Mathematical chemistry monographs No. 8. University of Kragujevac, Serbia, p 3
Fajtlowitz S, Larson SE (2003) Chem Phys Lett 377:485
Fowler PW (2002) Croat Chem Acta 75:401
Fowler PW (2003) MATCH Commun Math Comput Chem 48:87–96
Fowler PW, Manolopoulos DE (1995) An atlas of fullerenes. Clarendon Press, Oxford. Reprinted: Dover, New York, NY (2006)
Fowler PW, Rogers KM, Fajtlowick S, Hansen P, Capporosi G (1999) Facts and conjectures about fullerene graphs: leapfrog, cylinder and ramanujan fullerenes. Les Cahiers du GERAD, G-99-66
Hoory S, Linial N, Wigderson A (2006) Bull Am Math Soc 43:439
Klein DJ, Randić M (1993) J Math Chem 12:81
Lubotzky A, Phillips R, Sarnak P (1988) Combinatorica 9:261
Manolopoulos DE, May JC, Down SE (1991) Chem Phys Lett 181:105
Pierson HO (1993) Handbook of carbon, graphite, diamond and fullerenes. properties, processing and applications. Noyes Pub, New Jersey
Ram Murty M (2003) J Ramanujan Math Soc 18:1
Réti T, László I (2009) Acta Polytech Hung 6:85
Xiao W, Gutman I (2003) Theor Chem Acc 110:284–289
Zhang H, Balasubramanian K (1993) J Phys Chem 97:10341

Chapter 10
Counting Spanning Trees in Toroidal Fullerenes

E.C. Kirby, R.B. Mallion, and P. Pollak

Abstract The Cycle Theorem was introduced under that name by Kirby, Klein, Mallion, Pollak and Sachs in 2004. It provides a formula for calculating how many spanning trees a graph has (its complexity) but, for some graphs, even quite small ones, the calculation is a laborious and error-prone process to do by hand. A simple algorithm for computer application is developed, which uses the Cycle Theorem to calculate the complexities of some non-planar graphs that can be embedded on a torus, forming symmetrical trivalent tessellations on that surface.

10.1 Introduction

Kirby et al. (2004) introduced what they called the Cycle Theorem. Applicable to a graph of any genus, this theorem provides an expression for calculating how many spanning trees that graph has (a number that will subsequently be referred to as the graph's *complexity*). However, actually applying the Cycle Theorem to any graph that is of a size likely to be of interest within the field of fullerene chemistry would be a daunting and rather error-prone task using "pencil & paper" methods. In this chapter, we develop and present an algorithm suitable for implementation as a computer program that uses the Cycle Theorem to calculate the complexities of some non-planar graphs that can be embedded in a torus to form symmetrical trivalent tessellations on that surface.

R.B. Mallion (✉)
School of Physical Sciences, University of Kent, Canterbury, CT2 7NH, England, UK
e-mail: R.B.Mallion@kent.ac.uk

F. Cataldo et al. (eds.), *The Mathematics and Topology of Fullerenes*,
Carbon Materials: Chemistry and Physics 4, DOI 10.1007/978-94-007-0221-9_10,

10.2 The Cycle Theorem

The Cycle Theorem states that the number of spanning trees, $t(G)$, in a graph G is given by the expression

$$t(G) = \frac{(\det \mathbf{Z})(\det \mathbf{Z}^{\mathrm{T}})}{(\det \mathbf{U})(\det \mathbf{U}^{\mathrm{T}})} \qquad (10.1),$$

in which (Kirby et al. 2004) **Z** is a ($\mu \times e$) cycles $\rightarrow$ edges incidence-matrix and **U** is a non-singular ($\mu \times \mu$) sub-matrix of **Z**, where μ is the circuit rank of G and e is the number of edges in G. **U** is selected from **Z** in such a way that the edges that do not correspond to the columns of **U** form a spanning tree of G: those edges that do correspond to the columns of **U** are said to form a set of *chords*. There is a (1 – 1) correspondence between spanning trees and the non-singular ($\mu \times \mu$) matrices **U**. The determinants of all matrices **U** of this kind have the same absolute value (Kirby et al. 2004). We may define

$$\mathbf{M} = \mathbf{Z}.\mathbf{Z}^{\mathrm{T}} \qquad (10.2),$$

and since, obviously,

$$(\det \mathbf{U})(\det \mathbf{U}^{\mathrm{T}}) = (\det \mathbf{U})^2, \text{ and}$$

$$(\det \mathbf{Z})(\det \mathbf{Z}^{\mathrm{T}}) = \det \mathbf{Z}.\mathbf{Z}^{\mathrm{T}} \qquad (10.3),$$

Equation (10.1) may also conveniently be written

$$t(G) = \frac{\det \mathbf{M}}{(\det \mathbf{U})^2} \qquad (10.4).$$

As Kirby et al. (2004) observed, one of the attractive features of the theorem of Gutman et al. (1983) – which has as a drawback the limitation that it is applicable only to planar graphs – is that the data needed for its implementation can be "seen" in the drawn (embedded) graph. In particular, the matrix $\mathbf{M} = \mathbf{Z}.\mathbf{Z}^{\mathrm{T}}$ can be compiled, *without* first devising **Z**, by inspecting the cycles of the graph that form a basis (Kirby et al. 2004) and the edges that are common to pairs of such cycles.

Likewise, we feel that the more elegant form of the theorem of Kirby et al. (2004) is the one in which it can be expressed in terms of interpreting **M** as a "cycle-overlap" matrix. When two cycles of a graph G have an edge in common and at that edge their orientations agree (see Kirby et al. 2004), we say that there is a "match"; if they disagree, there is a "mis-match". Having identified μ cycles of G that form a basis (see Kirby et al. 2004), we can compile the (symmetrical) matrix **M** directly from the following definitions of its elements, m_{ij} $(1 \leq i,j \leq \mu)$:

$$m_{ii} = \text{number of edges in cycle } i,$$

and, for $i \neq j$,

$m_{ij} = \{(\text{number of matches}) - (\text{number of mis-matches})\}$, in cycles i and j. From this point of view, **M** is indeed aptly called a "cycle-overlap matrix".

Finally, Pollak (2010) has re-examined the Cycle Theorem from the point of view of Duality and, following the "cycle-overlap" approach, has replaced Eq. (10.4) with

$$t(G) = \frac{\det \mathbf{M}}{\det \mathbf{M}^*} \quad (10.5),$$

where $\mathbf{M}^*$ is a new cycle-overlap matrix of G in which only those edges that are the chords of some spanning-tree of G are taken into account. It is immaterial which spanning tree is used. This is, again, a statement of the general Cycle Theorem (Kirby et al. 2004): however, just as Kirby et al. (2004) showed that it was possible to arrange matters so that $|\det \mathbf{U}| = 1$, so Pollak (2010) has demonstrated that it is always possible, in a systematic way, to ensure that $\det \mathbf{M}^* = 1$. When this occurs, Eqs. (10.4) and (10.5) reduce to the elegantly simple expression embodied in Eq. (10.6):

$$t(G) = \det \mathbf{M} \quad (10.6).$$

10.3 Generic Circuits

Graphs of genus 1, which can be embedded without self-crossings in a torus, can also be represented in the plane with no self-crossings if some suitable convention is adopted, such as repeating certain labelled vertices and edges, or, equivalently, by drawing an appropriate edge outwards to the edge of a rectangle on one side. The edge is then treated as having circled (invisibly) "behind" the diagram and so reappears on the opposite side to connect with its other vertex. Such conventions are amongst those described in detail in, for example, the paper of Kirby et al. (2004). For our purpose here, we adopt the former method, where it is sufficient to say that, in a plane – *e.g.*, on a page – the torus is shown as a parallelogram whose opposite sides are *identified*, point by point. More physically, we may think of the torus as having been cut to form a tube and this tube then being cut open: the resulting object is flattened out. The parallelogram so formed may then be repeated indefinitely often, in both directions, to form a grid in the plane. Within this grid, it is relatively easy to follow the connections "by eye".

When a planar graph G is presented as a diagram on a sheet of paper, it is, *ipso facto*, embedded in a plane (or, what is topologically equivalent, on a sphere (O'Leary and Mallion 1987; Brown et al. 1991)). Circuits bounding empty areas of the plane are among the circuits that we need in order to count the number of spanning trees. We call such circuits *rings*, in accordance with the standard terminology of organic chemistry. These rings are then equipped with a sense of circulation. Such cycles do not, in general, supply elements of a suitable set of cycles, called a fundamental cycle-base (Kirby et al. 2004). Note that this definition intentionally includes the circuit that separates the entire embedded graph from the (empty)

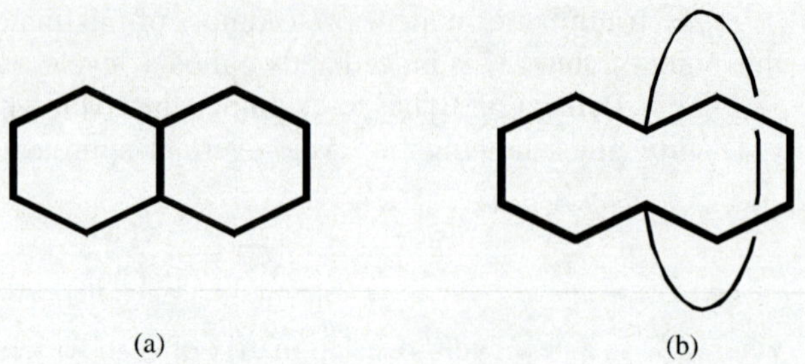

Fig. 10.1 The naphthalene graph (a) embedded in the plane, and (b) with its central edge (bond) distorted out of the plane for a possible near-spherical embedding

rest of the plane or sphere, *i.e.*, its "periphery". In strictly chemical usage, when the embedding is in the plane, rather than in the surface of a sphere, this periphery would usually not be considered to be a ring. For example, Fig. 10.1a shows a representation of the hydrocarbon naphthalene, and there is no difficulty in viewing the two hexagons as rings from either standpoint (mathematical or chemical), but the "peripheral case" is less obvious to a chemist. If one distorts the central edge/bond out of the plane (Fig. 10.1b) the periphery does indeed become an obvious "ring", as described above. In terms of molecular geometry, however, this is a preposterous notion. The periphery of a larger structure – say a planar graphite sheet – may, perhaps, more easily be seen as a ring, especially if it is distorted towards becoming hemispherical in shape. There will then be many hexagons, and one large ring.

However, with some molecules that are, by their geometrical nature, three-dimensional, there is no problem at all – for example, Buckminsterfullerene (Fig. 10.2) – because, here, it is in the *planar* embedding that edges appear distorted and unreal in a chemical sense.

The Euler-Poincaré Theorem (Wilson 1972) states that $f+v-e=2(1-g)$, where f is the number of faces, e is the number of edges, v is the number of vertices, and g is the genus, of a graph G. Euler's Theorem is a special case of the Euler-Poincaré

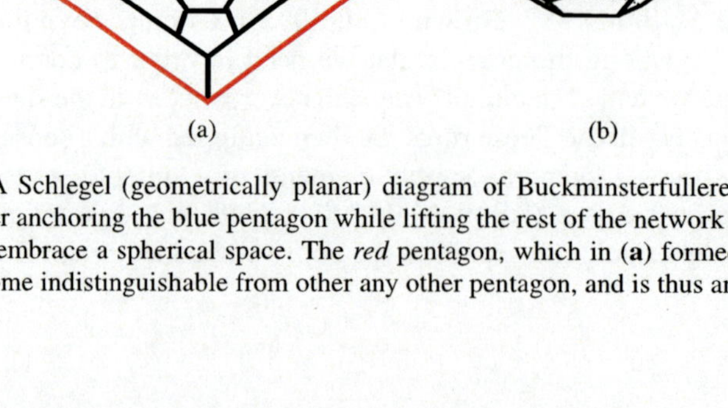

Fig. 10.2 (a) A Schlegel (geometrically planar) diagram of Buckminsterfullerene, and (b) the same object after anchoring the blue pentagon while lifting the rest of the network out of the plane of the paper to embrace a spherical space. The *red* pentagon, which in (a) formed the periphery, has, in (b), become indistinguishable from other any other pentagon, and is thus an obvious ring

Theorem and it can be applied to the torus (see, for example, László 2009), for which $g = 1$: it then becomes $f = e - v$. For a graph of genus 1, embedded in a torus, we then have that $\mu = e - v + 1 = f + 1$. It is clear that only (f–1) rings of G are independent, so that a cycle-base for G (Kirby et al. 2004) must include two cycles in addition to the (f–1) that are sited in the set of rings and are easy to select "by eye". These two cycles must, of course, be independent of those in the rings and of each other. They can be chosen from amongst the so-called "generic" cycles (Kirby et al. 2004) of the embedded graph, namely, those that cannot be shrunk to a point without leaving the surface of the torus. More generally, a generic circuit, on any surface, does not divide it into two parts. To be independent of each other, one of them must "go through the hole" of the torus and the other must "go around the hole". It is further convenient (with future regard to values of det $\mathbf{M}^*$ and/or det $\mathbf{U}$) that each be *simple*. This means that they do not wind needlessly round the torus: when shown on the parallelogram diagram, each contains just one pair of identified points. One generic cycle will have this pair of identified points on one pair of opposite sides, and the other on the other pair. In passing, we mention that similar considerations can be extended to more complicated surfaces.

When such a cycle-base has been selected for G, it can be shown (Kirby et al. 2004; Pollak 2010) that $\det \mathbf{M}^* = 1$ and that, once again, as *per* Eq. (10.6),

$$t(G) = \det \mathbf{M} \qquad (10.6).$$

It is immaterial how senses are allocated to the cycles in this procedure (Kirby et al. 2004), although for the purpose of computation it may be more *convenient* to specify a convention.

10.4 An Algorithm

We now develop a computer-applicable approach for the calculation of the complexities of certain graphs embedded, without self-crossings, in a torus. In general, these are graphs of genus 1, but, as will be seen later, in Section 10.5, planar graphs – which are, of course, of genus 0 – such as the cube, may also be embedded in a torus in a "generic" way, if at least one of each kind of generic circuit (Section 10.3) is present.

The graphs that we have selected are, at least potentially, of chemical interest. All are regular, of degree 3; the advantage that the Cycle Theorem has in respect of the order of the determinants that need to be evaluated can be exploited, especially if the graph (molecule) contains many vertices (atoms) – see Pollak (2010).

We adopt Eq. (10.5)

$$t(G) = \frac{\det \mathbf{M}}{\det \mathbf{M}^*} \qquad (10.5),$$

but it reduces, in fact, to

$$t(G) = \det \mathbf{M} \qquad (10.6),$$

as shown in Section 10.2, above. The challenge is, therefore, how to compile the matrix **M** in a "computer-friendly" manner.

10.5 Toroidal Polyhexes

10.5.1 General Considerations

In this section we consider the class of toroidal polyhexes (TPHs); (Kirby 1993; Kirby et al. 1993; Kirby and Pollak 1998; Ceulemans et al. 2000, 2002; Fowler et al. 2000; John and Sachs 2009). These are symmetrical tessellations of the torus by means of hexagons. In the way described in Section 10.3, they may be represented as an unlimited bi-periodic pattern in the plane. As was shown by Kirby et al. (1993), three integers, a, b and d ($a \geq 1, 0 \leq b < a, d \geq 1$), suffice to specify the structure's connectivity. Note, however, that, while an ordered trio of (a, b, d) values specifies a unique TPH, there may be up to six trios that are equivalent (Kirby et al. 1993; Kirby and Pollak 1998), and reflect different ways of making an excision from the imagined lattice. Even then, given any specific ordered trio, (a, b, d), there are a number of ways in which an excision may be made to illustrate the information pictorially. Here, as elsewhere, we adopt the convention shown in Fig. 10.3.

The form of Fig. 10.3 is standard and, across the whole class of TPHs, the integers a, b and d are the only variables, and can account for all possible connectional isomers (but note that, in a few cases, more than one mapping can be distinguished for a given TPH; see Kirby et al. 1993, 2009, "Unpublished work"). This immediately suggests how we may choose a cycle base in a way that is both convenient for matrix compilation by hand and is also suitable for implementation as computer code.

Consider, again, Fig. 10.3. This shows this particular TPH as (a-b-d) = (5-3-2), but since, as emphasised above, there may be up to six equivalent trios, we need a rule that defines a unique trio, for reference. To arrive at this we choose (a-b-d)-trios with the smallest d-value, and from these pick out the smallest value for b. No attention to a is necessary because, as can be seen from Fig. 10.3, ad represents the number of hexagons, and is constant for a given TPH.

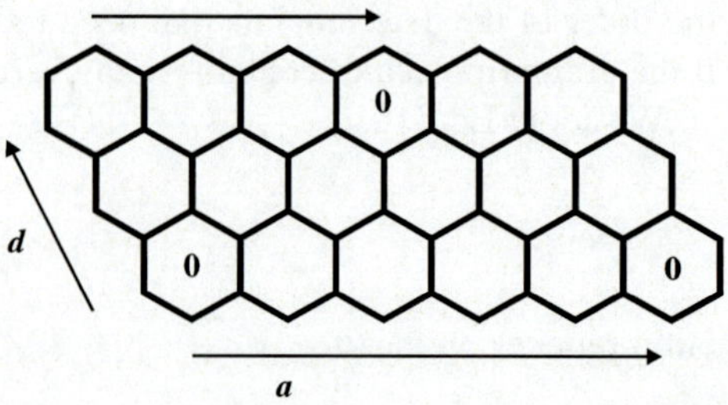

Fig. 10.3 The notation used in this paper and elsewhere, e.g., Kirby and Pisanski (2007). The hexagons labelled "0" define a TPH, and this particular example represents TPH (a-b-d) = (5-3-2)

As a first step we note that Fig. 10.3 can be read backwards, *i.e*., right-to-left, to give an equivalent code (5-4-2). Furthermore, by extending the pattern of Fig. 10.3 in the way that has been described above, similar patterns emerge that perhaps yield different values of *a*, *b* and *d*. These can be seen by considering the other two possible axes for *a*; there are, of course, three ways in which a line can be drawn to bisect opposite edges of a hexagon, and, in this case, we note that Fig. 10.3, which represents code TPH(5-3-2), is exactly equivalent to five other trios; namely (10-3-1), (10-8-1), (5-4-2), (10-4-1) and (10-7-1). Thus, the reference-code for cataloguing is TPH (10-3-1), as shown in Kirby and Pollak (1998).

10.5.2 Generating an Extended TPH Bi-Periodic Pattern

By hand, but with computer assistance, it is intuitive and straightforward to extract a suitable *tile*. This is a drawing of a single graph that contains every hexagon just once (one of which is labelled zero, as explained above) and is of a suitable shape, usually quadrilateral. By "suitable", we mean one where, if multiple copies of such a tile are made, they can be joined seamlessly in such a way as to reproduce a section of the potentially infinite bi-periodic pattern representing the TPH in question. This bi-periodic pattern can be built up on-screen, to the extent required, by simple "copy, drag & drop" methods. However, devising instructions for an automatic generator of TPH patterns on-screen by use of (*a*-*b*-*d*) keyboard-input is more convenient, and is not very complicated to write.

Figure 10.3 contains *ad* hexagons, together with some periodic repeats. Of these, (*ad* – 1) will be chosen to form a *full set* of independent ring-cycles; by this, we mean *any* set of rings that contains all but one of the hexagons. (Recall that when we use the method of Gutman et al. (1983) for calculating the complexity of a finite planar polyhex, all the rings that might be regarded as obvious are used, but not the periphery, because that is defined by other rings and so is not independent of them – see, for example, Fig. 10.2, above.) Any one of the rings (hexagons in this case) on the surface of the torus (Fig. 10.3) may be chosen as being the excluded, dependent one, and the one selected is labelled "0". A recipe for hand-generating a more-extensive sample of the bi-periodic pattern is as follows. Given a two-dimensional grid of fused regular-hexagons (on-screen or printed out), any hexagon may be marked as the first hexagon to be given the label "0", and it is then a simple matter to identify, and mark, all the places where such a "0"-marked hexagon of a specific toroidal-polyhex, encoded as TPH(*a*-*b*-*d*), repeats itself within a planar bi-periodic projection of its pattern. To do this –

(1) Label the first hexagon of the first row as "0". We shall subsequently refer to such a hexagon as a "zero-hexagon". Then label every *a*th -hexagon to the right as "0" until the limit of the grid is reached.
(2) From one or more zero-hexagons, count *d* hexagons down the 60° diagonal. At this point, both the *b*th hexagon to the left, and the ($a - b$)th one to the right, will be zero hexagons. These new zero-hexagons define where every

other zero-hexagon within this row should be marked because, within any horizontal row, all zero-hexagons must be a hexagons apart. This follows from the definition (see Fig. 10.3) recalling that, on the bi-periodic diagram, every zero-hexagon is not only identical but represents the *same* hexagon within the toroidal embedding. (This is also true of every other distinct hexagon, of course, whether or not it is so-labelled.)

Repetition of this second step quickly brings a large enough section of the bi-periodic pattern into view. However, if any straight-line zero-hexagon-to-zero-hexagon distances are too large to be accommodated within the boundaries of the pattern on display, then non-pictorial methods must be used to generate all the possible TPH equivalents and isomers (such as the program described in Kirby and Pollak 1998). In the course of generating these equivalents it should be borne in mind that there will always be six to be determined, but that some of the six may be repeats, thus giving a set containing fewer distinct-but-equivalent trios that all represent the same TPH (in terms of connectivity).

The algorithm just described is a digression as far as the counting of spanning trees is concerned. The point of our drawing attention to it is that a sufficiently large sample of the repetitive pattern can provide an alternative, pictorial, method for picking out the different members of this small set (up to six) of TPH (a-b-d)-trios.

Each set of equivalent trios represents a single graph, but, for a given torus, there are a number of what we refer to as connectional isomers, which may be envisaged as being produced by cutting the torus tube, twisting it, and then reconnecting it. The pairing of vertices/atoms connected across the plane of the cut will then be different. It will be a different isomer, in the original chemical sense that bonds will connect different pairs of atoms. However, if, before re-joining the tube ends, we imagine twisting more than one complete revolution, we shall have duplicated one of the previous isomers in a connectional sense, but still have a different structural isomer because of its twist. There would, in principle, be an infinite number of these – although, obviously, not in practice. A third possibility is of conformational isomerism without any cutting, should there be any energy minima when one simply twists the torus, but whether any of this has significance in practice is unknown at present, as far as we are aware.

10.5.3 Counting the Spanning Trees

To return to our main task, we require only the parallelogram described earlier (Fig. 10.3), which can be seen as a planar representation – what might be thought of as a "cut & skinned-torus". Note, however, that, for computational convenience, the diagram that we actually use is Fig. 10.4 (on the next page). Like Fig. 10.3, this has the first column repeated as the last column, and the top row is the same as the bottom row, but displaced b columns to the right in a cyclic permutation, in order to show information on its twist. If $b = 0$ there is no twist in a connectional

Fig. 10.4 A ring- and generic-cycle labelled TPH (in this case (5-3-2)) exemplifying the standard conventions adopted (Cf. Fig. 10.3)

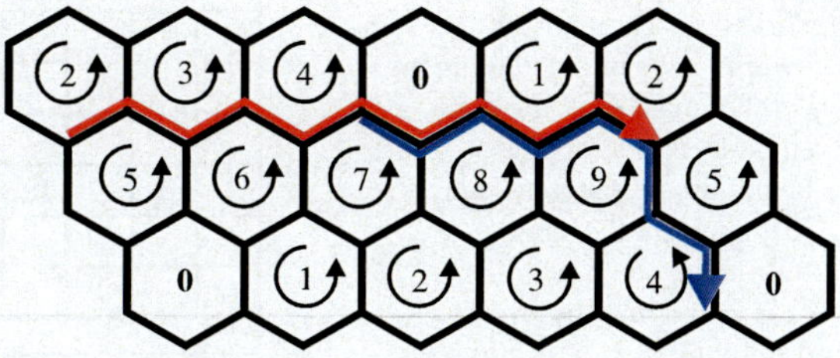

sense although, from that, we cannot be certain that there is no simple physical twist (in a real molecule). What is new about Fig. 10.4, compared with Fig. 10.3, is that we have labelled all the rings and added all the circuits that we need in order to compile an overlap matrix, and we have chosen a specific convention that is simple to apply to any toroidal polyhex.

The two generic cycles will be called "red" and "blue" ("R" and "B" for brevity). They must, of course, satisfy the criteria set out in Section 10.3. R must traverse the representative parallelogram from left to right, just once, and B from top to bottom, just once, as in Fig. 10.4.

Having arrived at a fully ring-labelled TPH diagram, we may map this onto a square matrix for computer manipulation. A labelled diagram such as Fig. 10.4 is convenient for computation of complexity by hand, and it also illustrates the conventions that we adopt while writing an algorithm for implementation as computer code.

When compiling **M**, there are six kinds of (symmetrical) overlap to consider: (a) ring-ring pairs, (b) red-circuit and rings, (c) blue-circuit and rings, (d) red-circuit and blue-circuit, (e) red-circuit and red-circuit, (f) blue-circuit and blue-circuit. Note that each leading-diagonal element may be thought of as representing the overlap of a circuit with itself. Consequently, such overlaps are always positive. The first, (a), forms a sub-matrix of size $(ad - 1) \times (ad - 1)$ – *i.e.*, (9×9) in the example of Fig. 10.4 – that can easily be compiled, since the leading-diagonal elements will all be 6 and, for each edge that a given ring-pair has in common, – 1 will be contributed to the corresponding matrix-element. This follows from our convention that every ring-circuit is traversed in the same (anti-clockwise) sense. Note that if $a \leq 2$ and/or $d \leq 2$, two hexagons may have more than one common edge, as in Fig. 10.5. Overlaps (b) to (d) (as mentioned and labelled above) involving the generic cycles, may be compiled by inspection. These compilations need particular care: they are surprisingly error-prone to write by hand, since every overlap-entry for the matrix may be the sum of several specific overlap-components (Section 10.2), each of which may be positive or negative. (This is the case, for example, for the element in Row 2, Column B in the matrix featuring in Fig. 10.5a and for those in Row 1, Column B and Row 2, Column B in the matrix depicted in Fig. 10.5c.) However, such a compilation, once written in generalised terms as computer code, generates any cycle-overlap matrix within the computer's capacity, virtually instantaneously.

We emphasise again that the red and blue paths represent the two "generic" ones that we use, and there is no particular merit in choosing any specific route, so long

(A)	1	2	3	R	B
1	6	−2	−2	2	4
2	−2	6	−2	−2	0
3	−2	−2	6	−2	−4
R	2	−2	−2	4	3
B	4	0	−4	3	6

(B)	1	2	3	R	B
1	6	−2	−2	2	2
2	−2	6	−2	−2	2
3	−2	−2	6	−2	−2
R	2	−2	−2	4	−1
B	2	2	−2	−1	4

(C)	1	2	3	R	B
1	6	−2	−2	2	−1
2	−2	6	−2	−2	1
3	−2	−2	6	−2	−3
R	2	−2	−2	4	3
B	−1	1	−3	3	6

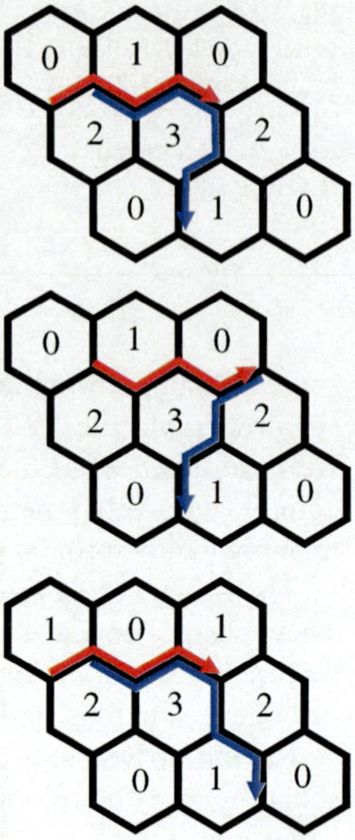

Fig. 10.5 Three of the several ways of calculating an overlap matrix for the cube embedded in a torus as a boundless polyhex. (**a**) and (**b**) represent the same four-hexagon embedding (2-0-2) but use slightly different generic-circuits. (**c**) shows the TPH(2-1-2) embedding that arises by rotation of one side of a cube about one of its diagonal axes (Kirby et al. 1993). The determinant of all three overlap-matrices is 384, as expected, for example, from Kirby et al. (2004)

as each of them starts and finishes at the same vertex (and so is a circuit), and that they are independent and simple in the sense that one goes through, and one goes around, the "hole" of the torus just once.

The toroidal-polyhex networks are, of course, particularly easy to deal with – they are regular tessellations of hexagons, and all connectional isomers can be described by the three integer parameters a, b, and d. The same general principles can also be applied to other networks of interest.

10.6 Application to Pentaheptite (Azulenoid) Networks

Pentaheptite (or azulenoid) networks have been the subject of some attention; (Crespi et al. 1996; Deza et al. 2000; Diudea et al. 2003). These consist solely of pentagons and heptagons in equal numbers, and can often be generated by the well-known Stone-Wales rearrangement of pyrene units (Stone and Wales 1986; Kirby and Pisanski 2007).

As we have seen above, in Fig. 10.5, the cube, which, as a polyhedron, has six square faces, can be represented as a toroidal embedding of four hexagons. On the boundless surface of a torus, this can equally well be seen as two naphthalene-units or one pyrene unit – the latter being amenable to a Stone-Wales (1986) transformation into two fused azulenes. As chemical entities, the azulenes were first named – after the colour of azulene itself – in 1864, by Piesse, but more than 70 years then elapsed before the essential structure was established (St. Pfau and Plattner 1936).

	1	2	3	R	B
1	7	−3	−2	3	3
2	−3	7	−2	−3	−3
3	−2	−2	5	−1	1
R	3	−3	−1	4	1
B	3	−3	1	1	4

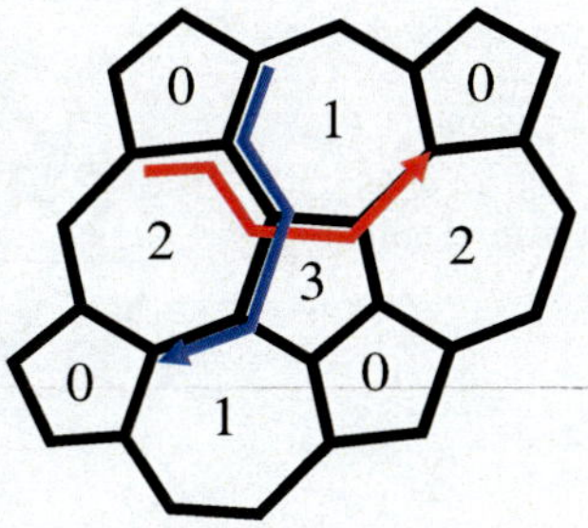

Fig. 10.6 A pentaheptite analogue of the cube (cf. Fig. 10.5), but, unlike the corresponding toroidal-polyhex embedding, this is not topologically planar. Rather, it can be seen as a "cube" where one of its faces has a Möbius twist. Note that this should not be confused with the isomerism possible in certain toroidal polyhexes, including the cube, where a four-ring can be rotated around a diagonal axis, giving rise to a different embedding, that leaves connectivity unchanged (Kirby et al. 1993, 2009, "Unpublished work"). The complexity of this structure, calculated as the determinant of the overlap matrix shown (and confirmed by application of the Matrix Tree Theorem – see, for example, Mallion 1974/1975; Mallion 1975), is 392

Ever since then, however, chemists have been intrigued by the many ways in which azulenes and benzenoids are, as the pre-Socratic philosopher Parmenides (*ca.* 475 BC) might have put it, "the same and yet not the same". Here we show another (graph-theoretical) instance where they differ. The pyrene graph and its azulenic analogue may both be embedded without crossings in the surface of the torus; (see Fig. 10.6).

However, while the embedded pyrene-graph is equivalent to the (six-faced) cube, a topologically planar object that can equally be considered to be a band of four squares (Fig. 10.7a), this is not the case for the corresponding azulenic object. This is *not* planar: rather, it can seen as being embedded on a twisted – i.e., Möbius – band. (See Fig. 10.7b, and the Appendix). A Klein Bottle embedding is also possible in concept.

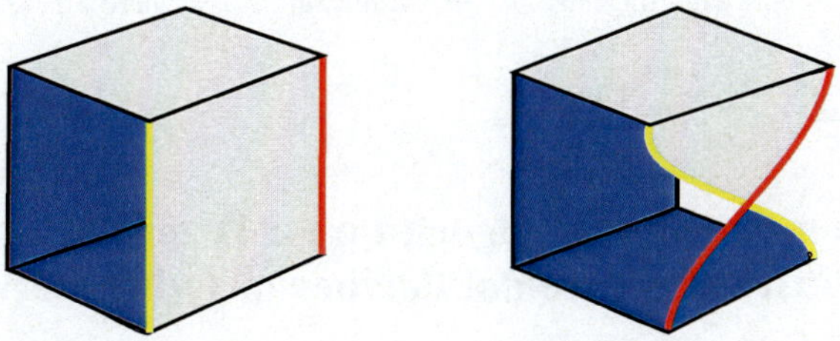

(a) (b)

Fig. 10.7 (**a**) The cube, which, as noted in Fig. 10.5, above, can also be embedded as a network of four hexagons in the surface of the torus. The cube is a polyhedron with six square faces, but here we treat it as a band of four squares, coloured *grey* on the outside, while inner "surfaces" are blue. The cube is topologically planar. In (**b**) we have cut the bottom of the nearest face, twisted it by 180°, and re-joined it, thus forming a Möbius band**.** This (now non-planar) object is isomorphic with the torus tiled with two azulenes shown in Fig. 10.6 (See also the Appendix.)

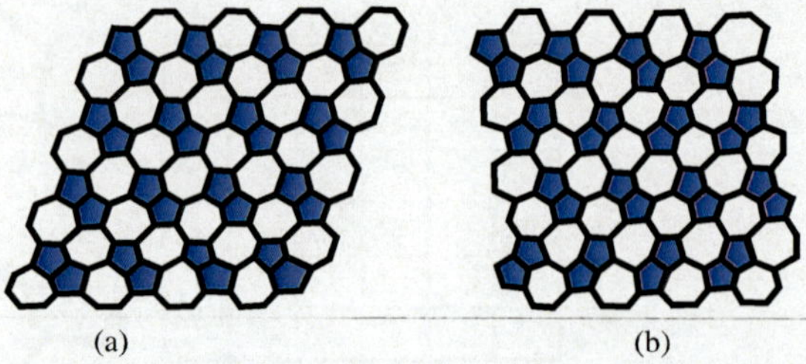

Fig. 10.8 The two types of pentaheptite (azulenic) network that are possible above a certain size. In order to show the distinction, only the five-membered rings are coloured

Note also that, for larger pentaheptite networks, there are two possible arrangements (see Fig. 10.8), and, in addition, the workings will clearly be a little more intricate since, for example, the ring sizes alternate between 5 and 7 along any horizontal sequence.

Another group of networks, of especial interest to one of us (Kirby 2006), consists of those that are *fully-azulenoid* (Fig. 10.9, whose defining characteristic is that all vertices can be accounted for by a set of disjoint pentagon-heptagon pairs that have a common edge, and are thus analogous to Clar structures).

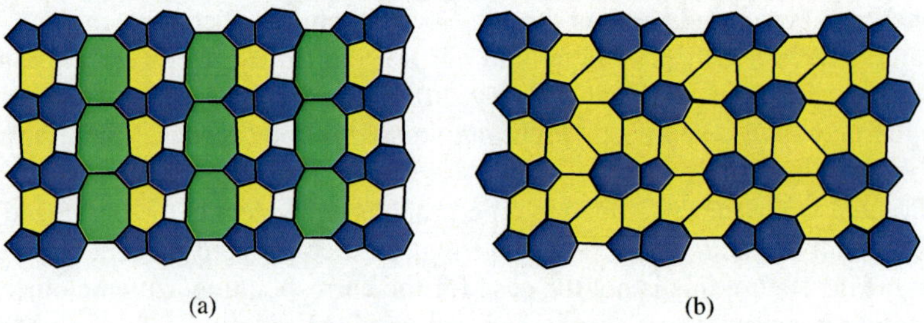

Fig. 10.9 Two types of fully resonant azulenoid-network. There are many possible arrangements (Kirby 2005; Brinkmann et al. 2009) – 1274 if the largest "empty" ring (that is, a ring that is not part of an azulene moiety) is 16-membered. In structure **(a)**, the azulene moieties are depicted in a parallel configuration whilst, in structure **(b)**, the azulenes are arranged in a boustrophedon manner. The latter represents the only pattern within this class where all non-azulenic (empty) rings are hexagons

10.7 Notes on Writing an Algorithm for Determining the Complexity of a Toroidal Polyhex or Other Network

10.7.1 Terminology

Here, we use the traditional terminology for navigation using "points of the compass" as abbreviations for spatial relationships on the page: N = north = up; S = south = down; E = east = right; W = west = left. Similarly, NE = up & right; SW = down & left, and so on.

These eight possible directions conveniently describe the positions of the eight neighbours of any matrix element within the interior of a rectangular matrix. The polyhex section of a TPH, such as that shown in Fig. 10.4, can be regarded merely as a distorted version of a rectangular matrix, where all the columns have been "pushed over", from the right, like a row of dominoes – although not to the ground. Note again that, when Fig. 10.4 is treated in this way, the first and last columns are identical, and the top row is the same as the bottom one, but displaced towards the right. This means that every labelled ring has six neighbours, although the identity of some of these for peripheral rings must be sought towards the opposite side of the matrix of rings.

10.7.2 An Outline Algorithm for Counting the Number of Spanning Trees (The Complexity) of a Toroidal Polyhex

For manual investigation of very small systems it is easier to work with a standard, ring-labelled, polyhex-section, but, if the process is to be automated, it is more convenient to transcribe the overlap information to a simple square-matrix, which we term a *ring-adjacency matrix*. This holds all the ring labels, *except* "0", of the defining structure-diagram, such as is shown in Figs. 10.4 and 10.5, *as well as* path labels "R" (red) and "B" (blue). Horizontal spatial-relationships within rows remain exactly as before, but strips of rings in the polyhex with a NW-SE alignment become the columns of the ring-adjacency matrix.

The complete algorithm has three stages. First, the ring-adjacency matrix is derived from the chemical graph. This ring-adjacency matrix is then used to construct a cycle-overlap matrix, and, finally, the complexity is evaluated as the determinant of that cycle-overlap matrix.

Consider, for example, a ring-adjacency matrix for Fig. 10.4. For each element, up to six out of the eight surrounding elements may be adjacent. A rule is provided for deciding *which* six of an element's neighbours are potentially adjacent. There are fewer than six if one or more of them is labelled "0", and therefore treated as the zero-hexagon. In practice, for most of the networks that, currently, are of the greatest chemical interest, this apparent limitation is no great privation. A systematic search of this ring-adjacency matrix gives the information needed to complete all entries of the cycle-overlap matrix. This is followed by a search for all overlaps involving the two generic cycles (red and blue).

At this stage, all off-diagonal elements are in place. It remains to fill the elements of the leading diagonal of the overlap matrix. For a TPH, each has a value of 6, except the last two, which vary according to the size of the generic cycles, but which are easily computed from a, b and d (see Section 10.5). This procedure, which sounds rather complicated when described, is in practice quite simple to encode.

For this first stage we have a comprehensive, compact and efficient method for characterising and encoding the toroidal polyhexes (Kirby et al. 1993; Kirby and Pollak 1998). Other networks with reasonably simple patterns could obviously be encoded in a similar way, albeit with less elegance. The nearer that the set of rings approaches an aperiodic spatial-distribution, the more complicated, and the less useful, will any such encoding method be. By an "aperiodic distribution" is meant an array of rings for which no sub-group of them can be defined as a tile that can be repeated to reproduce the whole entity. Thus, the polyhexes are not random or aperiodic (they are tiled with hexagons), nor are the pentaheptites – Figs. 10.6 and 10.8a – which may be tiled with pentagon-heptagon pairs. The Penrose tilings (Klee and Wagon 1991) represent the infinite case but, for finite ones, the complete ring-set may be the only tile. Either such a case could not be encoded in the way in which a TPH is encoded, or an attempt to do so would be no less compact than just describing the whole entity as a matrix. The second stage, generation of a cycle-overlap matrix, described earlier (Section 10.5), is one of the subjects of this chapter and, provided that we can obtain a ring-adjacency matrix that fully characterises the toroidal graph, we can always use it to generate the corresponding cycle-overlap matrix, adopting a "mechanised" version of the technique described earlier. To evaluate the required determinant, many well-known routines are available. We prefer to use one that involves manipulation only of integers but, with increasing size, the number of bits allocated to the storage of an integer on a modestly sized computer is soon exceeded, and the memory allocation for that integer overwhelmed, so that *complete* accuracy is then lost. The use of products of powers of prime numbers (Brown et al. 1991; John and Mallion 1994) is being investigated as a way of postponing the occurrence of this problem.

As noted, this approach can be adapted to include non-hexagonal rings, although passing all necessary structural information in concise form, and constructing the cycle-overlap matrix, will both be more complicated.

10.8 Concluding Remarks

As described in the body of this chapter, this work arose from the wish to be able to apply the quite-complicated procedure described in our earlier paper (Kirby et al. 2004) by, where possible, specifying conventions and then outlining an algorithm that can be implemented as a computer program. We have successfully done this, and have tested it. A number of results (while running under Windows XP) were checked against those obtained by other standard methods. However, for extensive use of the program, further attention to such matters as Windows-version compatibility, precise size-limit checking, general ease of use, and so on, would be desirable.

In the approach presented here, we have specifically and deliberately emphasised what we consider to be the intuitively appealing "cycle-overlap" aspects of the theorem that we proposed (on p. 267) in Kirby et al. 2004, rather than the formalism based on the **Z**-matrix, also described (on p. 266) in that paper. The "cycle-overlap" concept, which we and others (*e.g.*, Klein 1994, Personal discussions with R.B.M., Department of Theoretical Chemistry, University of Oxford; Haigh 2004, Personal correspondence with R.B.M) consider to be much the most elegant part of the whole method, arises in the version of our theorem embodied in Eq. (10.6) of the present chapter.

Finally, it might be appropriate to conclude with some remarks about the proposed algorithm's range of validity. As is well known, the numerical value of a graph's complexity – which is, of course, an exact integer – increases very rapidly with the number of vertices in the graph (see, for example Haigh 1996; Brown et al. 1996; Mallion and Trinajstić 2003) while, with structures like fullerenes *etc.*, the sizes of molecule that are of potential interest are becoming larger and larger. Most Personal Computers (PCs) now deal in 32 bits to the byte, giving an accessible positive-integer range of 0 to $(2^{32} - 1)$. PCs with 64 bit-bytes are steadily spreading, but, even so, given the rapidity, referred to above, with which complexity escalates as the size of molecule increases, there will continue to be a relatively modest limit on the maximum size for which an exact value can be derived. If difficulties like this are encountered in the future it is just possible that larger numbers than might normally be expected might be teased out of a given PC by the adoption of devices such as operating in integers expressed to a given (prime) modulus and by the judicious exploitation of theorems in Number Theory such as the Chinese Remainder Theorem, as was demonstrated in, for example, Brown et al. 1991. At the other end of the size range, it is noteworthy that, even with small molecules, a well-tested machine-algorithm can be of great help. The authors of this chapter have learned by bitter experience how surprisingly error-prone a pencil & paper computation can prove to be – even, for example, for a system as small as the cube.

Acknowledgement One of us (ECK) is grateful to Professors T. Pisanski and P.W. Fowler for help in attending the 6th Slovenian International Conference on Graph Theory – held at Bled, Slovenia in June 2007 – where a preliminary account of the work described in this chapter was presented.

Appendix: The Relationship of a Toroidal Embedding of a Double-Azulene to the Cube

For a section of the pentaheptite pattern (shown in Fig. 10.10) we label all rings with numbers in the range 0–3, in such a way that the infinitely repeating tile represents a fused double-azulene (*i.e.*, a pyrene after having undergone a Stone-Wales

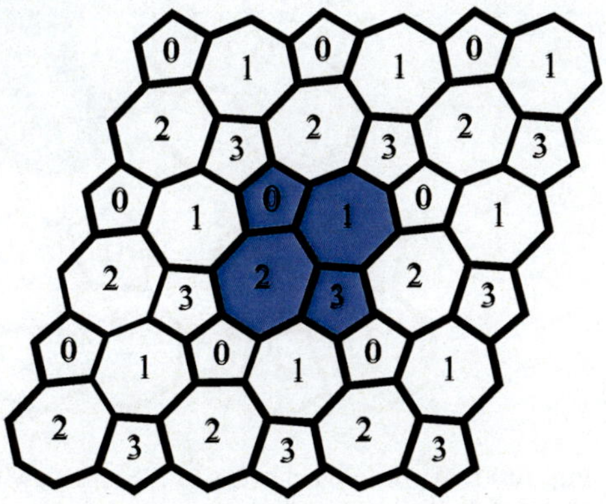

Fig. 10.10 A ring-labelled (3×3) tile-excerpt from the bi-periodic pattern of a double-azulene embedded in the surface of the torus. This allows the central, *blue*, tile to be fully viewed within the context of surrounding repeats of itself

transformation). One such tile is coloured blue in order to distinguish it from all its reproductions.

We must now label the vertices of this tile. A unique vertex is identified as such, by being common to a trio of rings with a specific orientation. As expected by comparison with the cube, there are eight distinct vertices. Also as expected – since this is a toroidal embedding – most vertices (those labelled 1, 2, 3, and 5) on the periphery of a tile appear twice, but two of them (those labelled 4 and 8) appear thrice, while internal vertices (the ones labelled 6 and 7) appear just once.

In Fig. 10.11, we give an "exploded" view of Fig. 10.10, for clarity – with vertex labelling extended throughout the diagram

We now compile the Connection Table of the azulene tile, and conduct obvious arithmetical checks.

Vertex	Connections			
1	3	7	8	
2	4	5	7	
3	1	4	6	
4	2	3	8	
5	2	6	8	
6	3	5	7	
7	1	2	6	
8	1	4	5	
	17	36	55	108

Check: Sum of vertex labels as integers 1 . . . 8 = 36.
Sum of the three connection-columns = 108 (= 3 × 36).

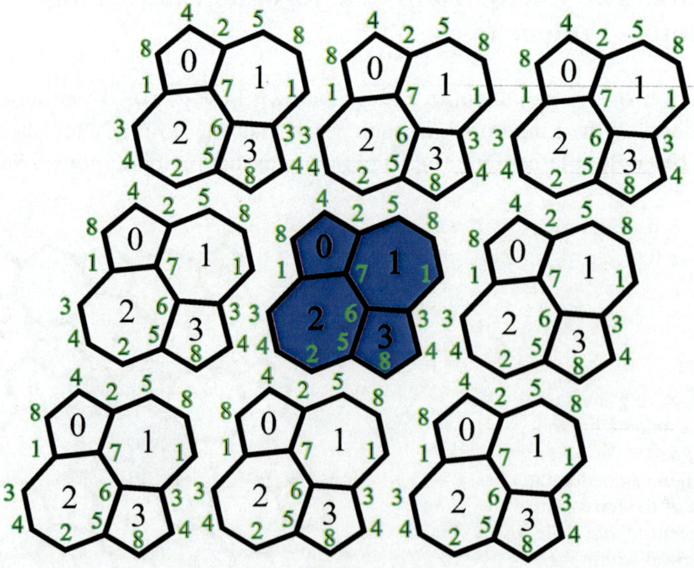

Fig. 10.11 An "exploded" view of Fig. 10.10, with full labelling of both rings and vertices

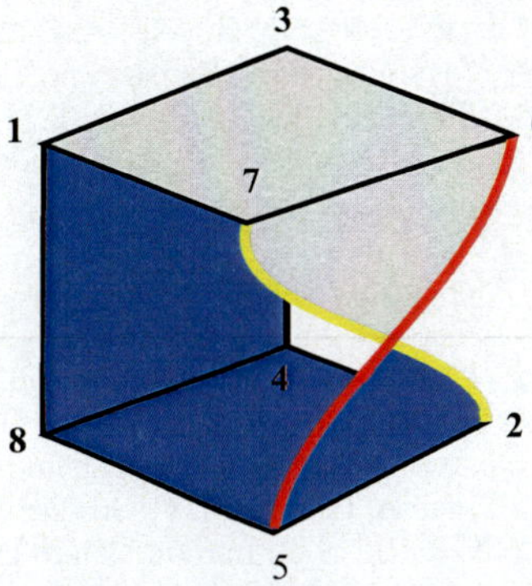

Fig. 10.12 A Möbius Band related to the cube. This is the same as Fig. 10.7b, but with its vertices labelled, and it is isomorphic to the toroidal embedding of a double-azulene shown in Fig. 10.6

Finally, we first use pencil & paper to draw a (literally) planar graph with these connections, but pay minimal attention to the occurrence of crossings. If we then examine this rough drawing and experiment with trying to redraw it with a three-dimensional perspective, it soon becomes apparent that it will not fit a cube; (again, this is only to be expected). On the other hand, a little more manipulation by trial & error reveals that if we (a) treat the cube as a *band* with four square-faces, (b) cut an internal edge of one of these faces and then (c) twist it by 180° before re-joining, the connections do fit, and the structure is isomorphic with a Möbius Band (Fig. 10.12) that is closely related to the cube.

References

Brinkmann G, Delgado-Friedrichs O, Kirby EC, Van Cleemput N (2009) Croat Chem Acta 82:781
Brown TJN, Mallion RB, Pollak P, de Castro BRM, Gomes JANF (1991) J Comput Chem 12:1118
Brown TJN, Mallion RB, Pollak P, Roth A (1996) Discrete Appl Math 67:51
Ceulemans A, Chibotaru LF, Bovin SA, Fowler PW (2000) J Chem Phys 112:4271
Ceulemans A, Compernolle S, Delabie A, Somers K, Chibotaru LF, Fowler PW, Margańska MJ, Szopa M (2002). Phys Rev B 65:115412
Crespi VH, Benedict LX, Cohen ML, Louie SG (1996) Phys Rev B 53:R13303
Deza M, Fowler PW, Shtogrin M, Vietze K (2000) J Chem Inf Comput Sci 40:1325
Diudea MV, Pârv B, Kirby EC (2003) MATCH Commun Math Comput Chem 47:53
Fowler PW, John PE, Sachs H (2000) DIMACS Ser Discrete Math Theor Comput Sci 51:139
Gutman I, Mallion RB, Essam JW (1983) Mol Phys 50:859
Haigh CW (1996) MATCH Commun Math Comput Chem 33:139
John PE, Mallion RB (1994) J Math Chem (a) 15:261;(b) J Math Chem 16:389 (Erratum)
John P, Sachs H (2009) Discrete Math 309:2663
Kirby EC (1993) Croat Chem Acta 66:13
Kirby EC (2005) In: Diudea MV (ed) Nanostructures: novel architecture. Nova Science, New York, NY, p 175
Kirby EC (2006) MATCH Commun Math Comput Chem 55:363
Kirby EC, Klein DJ, Mallion RB, Pollak P, Sachs H (2004) Croat Chem Acta 77:263
Kirby EC, Mallion RB, Pollak P (1993) J Chem Soc, Faraday Trans 89:1945
Kirby EC, Pisanski T (2007) MATCH Commun Math Comput Chem 57:411
Kirby EC, Pollak P (1998) J Chem Inf Comput Sci (a) 38:66; (b) 38:1256 (Erratum)
Klee V, Wagon S (1991) Old and new unsolved problems in plane geometry and number theory. Mathematical Association of America, Washington, DC

László I (2009) In: Graovac A, Gutman I, Vukičević D (eds) Mathematical methods and modelling for students of chemistry and biology. Hum naklada d.o.o, Zagreb, p 157
Mallion RB (1974/1975) Proc R Soc Lond 341:429
Mallion RB (1975) Chem Phys Lett 36:170
Mallion RB, Trinajstić N (2003) MATCH Comm Math Comput Chem 48:97
O'Leary B, Mallion RB (1987) In: King RB, Rouvray DH (eds) Graph theory and topology in chemistry; a collection of papers presented at an international conference at the University of Georgia, Athens, Georgia, 16–20 March 1987. Stud Phys Theor Chem 51:544
Piesse S (1864) Compt Rend Acad Sci (Paris) 57:1016
Pollak P (2010) Work in progress on aspects of Duality in the Cycle Theorem
Stone AJ, Wales DJ (1986) Chem Phys Lett 128:501
St. Pfau A, Plattner PL (1936) Helv Chim Acta 19:858
Wilson RJ (1972) Introduction to graph theory. Oliver & Boyd, Edinburgh, p 70

Chapter 11
Topological Determination of ^{13}C–NMR Spectra of C_{66} Fullerenes

Ottorino Ori, Franco Cataldo, Damir Vukičević, and Ante Graovac

Abstract This article presents a general topological computational method for the exact determination of the number of ^{13}C–NMR resonance peaks and their relative intensities and applies it to the case of C_{66} isomers. Heuristically, the joint usage of Wiener-based topological invariants of dual and direct molecular graphs provides quick simulations of its ^{13}C–NMR spectrum of a given fullerene. Topological tools confirm their power by studying the automorphisms of the fullerene molecular graphs to determine molecular symmetry and ^{13}C–NMR resonance pattern of C_{66} cages. Computations produce eight molecules with the proper C_{2v} symmetry among 4478 distinct C_{66} isomers, including the C_{66}–C_{2v} isomer with two pairs of fused pentagons experimentally detected in metallofullerenes synthesis.

11.1 Introduction

Fullerenic C_n cages are hollow carbon molecules formed by 12 pentagons and $n/2 - 10$ hexagons and usually respect the well known isolated pentagon rule [IPR] stating that, for stable fullerenes, each pentagon has to be surrounded by one or more rings of hexagons (Kroto 1987). Many observed fullerenes like C_{60}, C_{70}, C_{76}, C_{78}, etc. strictly obey to this IPR topological constraint in such a way that stable fullerenic surfaces may be assimilated to closed networks of nanocones somehow interconnected by distorted portions of graphenic lattice.

This IPR rule applies to all the C_n molecules with $n = 60$ or $n = 70 + 2k$ for $k > 0$; for the intermediate cases C_{62}–C_{68} it is impossible that all the pentagons remain isolated, originating a group of fullerenes with non-IPR isomers only (Fowler and Manolopoulos 1995). IPR prevents excessive strain of chemical carbon bonds shared by touching pentagons that, in turn, would require very high deformations to form the fullerenic closed carbon cage. Generally, fullerenes that violate IPR are then highly reactive and difficult to synthesize as standalone

O. Ori (✉)
Actinium Chemical Research, Via Casilina 1626/A, 00133 Rome, Italy
e-mail: ottorino.ori@alice.it

F. Cataldo et al. (eds.), *The Mathematics and Topology of Fullerenes*,
Carbon Materials: Chemistry and Physics 4, DOI 10.1007/978-94-007-0221-9_11,

molecules. Theoretical principles leading to the chemical stabilization of fused-pentagon fullerene molecules are extensively treated in several recent relevant reviews (Tan et al. 2009; Cui et al. 2010). Schematically, non-IPR fullerenes may be favored by two basic chemical mechanisms: (i) forming fullerenic endohedral structures (endoclusters) encapsulating metal clusters in the fullerene cages with strong coordination of the metal ions to fused-pentagons regions or, alternatively (ii) forming exohedral hydrogenated or chlorinated fullerenic derivatives. The stability of non-IPR exoderivatives fullerenes involves both principles, the *strain-relief* generated by the sp^3 partial hybridization of the carbons of the fused pentagons and the *local-aromaticity* of the alternationg single/double bonds C–C/C=C persisting also after exoderivatization (Tan et al. 2009).

Experimentally, non-IPR C_{66} fullerenes have been synthesized in quantities sufficient for characterization as stable metallofullerene endoclusters $Sc_2@C_{66}$ (Wang et al. 2000) and, more recently (Tan et al. 2009), in the exohedral chlorinated forms $C_{66}Cl_6$ and $C_{66}Cl_{10}$, evidencing outstanding violations of IPR. Significantly, above products are moreover based on two symmetry-distinct C_{66} isomers, having different touching pentagons topology able to favor their formation and chemical stability. The isomeric space of the C_{66} fullerene accounts in fact for 4478 possible (non-IPR) symmetry-distinct isomers $2 \times D_3$, $1 \times C_{3v}$, $18 \times C_{2v}$, $112 \times C_s$, $211 \times C_2$ and $4134 \times C_1$ (Fowler and Manolopoulos 1995) with different configurations of adjacent pentagons. Considering the observed 19-lines (5×2; 14×4) in the high resolution ^{13}C–NMR pattern of $Sc_2@C_{66}$ endocluster (Wang et al. 2000), only 8 structural C_{66}–C_{2v} symmetry cages are compatible with the resonance spectra and with synchrotron radiation powder data (Takata et al. 2001). This observed C_{66}–C_{2v} cage has two pairs of fused pentagons and conventionally is called here C_{66}–$C_{2v}{}^{\#0011}$ fullerene according to the exhaustive studies in Cui et al. (2010); this specific molecule (Fig. 11.1) is similarly represented as (C_{66}–C_{2v}) (5,6) fullerene in the IUPAC recommendations about fullerenes numbering (Cozzi et al. 2005). It coincides with the isomer #C_{66}:4348 of other scientific reports (Réti and László

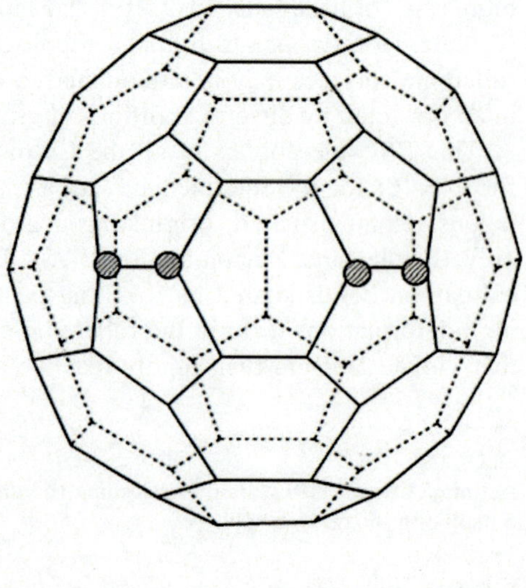

Fig. 11.1 C_{66}–$C_{2v}{}^{\#0011}$ fullerene in the direct space; bonds connecting *dashed atoms* evidence two pairs of touching pentagons

Table 11.1 (continued)

V_i	c_i	G^D Connected vertices	Direct graph nodes
14	6	1 3 28 31 32 35	1 2 9 13 59 60
15	6	7 9 16 21 23 31	29 30 37 41 61 62
16	6	5 9 15 21 25 33	17 21 40 41 63 62
17	6	8 10 18 22 24 32	34 35 43 44 65 66
18	6	6 10 17 22 26 34	22 23 44 45 64 65
19	6	9 20 12 11 23 25	38 39 49 50 52 56
20	6	10 11 12 19 24 26	49 52 48 42 46 53
21	6	1 2 15 16 31 33	4 5 8 63 61 62
22	6	3 4 17 18 32 34	10 11 14 64 65 66
23	6	7 9 11 15 19 27	30 31 37 38 50 51
24	6	8 10 11 17 20 27	35 36 42 43 48 47
25	6	5 9 12 16 19 30	17 18 39 40 55 56
26	6	6 10 12 18 20 30	26 22 53 54 45 46
27	6	7 8 11 23 24 28	27 32 36 47 31 51
28	6	7 8 14 27 31 32	27 28 32 33 59 60
29	6	5 6 13 30 33 34	19 20 24 25 57 58
30	6	5 6 12 25 26 29	19 25 18 55 26 54
31	6	1 7 14 15 21 28	1 5 28 60 29 61
32	6	3 8 14 17 22 28	9 10 33 59 34 66
33	6	2 5 13 16 21 29	7 8 20 58 21 63
34	6	4 6 13 18 22 29	14 15 24 57 23 64
35	6	1 2 3 4 13 14	2 13 12 16 3 6

Table 11.2 *Top*: coordination strings $\{b_{im}\}$ of 66 vertices of C_{66}–$C_{2v}{}^{\#0011}$ mono-chromatic direct graph are grouped in 14 equivalence classes p with cardinality k_p, producing a – wrong – resonance spectra (5×2; 5×4; 3×8; 1×12) with degeneracy of lines $p = 11,12,13,14$ (shaded). *Bottom*: coordination dual strings $\{b_{im}\}^D$ solve this degeneracy reproducing the correct ^{13}C–NMR spectrum of the molecule with 19 lines (*5×2; 14×4*) for isomer C_{66}–$C_{2v}{}^{\#0011}$

p	k_p	Symmetry-equivalent sites V_i	$w_i/2$	$\{b_{im}\}$ $M = 9$
1	2	V_{62} V_{65}	157.5	3 6 9 11 10 11 9 4 2
2	2	V_{41} V_{44}	159	3 6 8 11 11 10 9 6 1
3	2	V_{11} V_4	164	3 6 7 10 11 10 9 7 2
4	2	V_{49} V_{52}	164.5	3 6 8 10 10 10 8 6 4
5	2	V_{12} V_3	170	3 6 7 10 9 9 9 6 6
6	4	V_{61} V_{63} V_{64} V_{66}	158	3 6 9 10 11 11 8 6 1
7	4	V_{21} V_{23} V_{29} V_{34}	158.5	3 6 8 11 11 11 9 4 2
8	4	V_{20} V_{24} V_{28} V_{33}	159.5	3 6 8 11 11 10 9 5 2
9	4	V_{17} V_{22} V_{30} V_{35}	160	3 6 8 10 11 11 9 6 1
10	4	V_2 V_6 V_{13} V_{16}	165.5	3 6 8 9 10 10 9 7 3
11	**8**	V_{19} V_{25} V_{27} V_{32} V_{37} V_{40} V_{43} V_{45}	160.5	3 6 8 10 11 11 9 5 2
12	**8**	V_5 V_8 V_{10} V_{14} V_{57} V_{58} V_{59} V_{60}	161.5	3 6 8 10 11 10 9 6 2
13	**8**	V_{18} V_{26} V_{31} V_{36} V_{38} V_{39} V_{42} V_{46}	162.5	3 6 8 10 10 11 9 5 3
14	**12**	V_1 V_7 V_9 V_{15} V_{47} V_{51} V_{54} V_{55} V_{48} V_{50} V_{53} V_{56}	163.5	3 6 8 10 10 10 9 6 3

Table 11.2 (continued)

p	k_p	Symmetry-equivalent sites V_i	$\{b_{im}\}^D$ $M^D = 5$		
1	2	V_{62} V_{65}	6 10 10 7 1	6 10 10 7 1	6 10 11 6 1
2	2	V_{41} V_{44}	5 10 11 7 1	6 10 10 7 1	6 10 10 7 1
3	2	V_{11} V_4	5 9 10 8 2	5 9 10 8 2	6 10 11 6 1
4	2	V_{49} V_{52}	5 10 10 8 1	6 9 9 7 3	6 9 9 7 3
5	2	V_{12} V_3	5 9 10 8 2	5 9 10 8 2	6 8 10 8 2
6	4	V_{61} V_{63} V_{64} V_{66}	6 10 10 7 1	6 10 11 6 1	6 10 10 7 1
7	4	V_{21} V_{23} V_{29} V_{34}	5 10 11 7 1	6 10 10 7 1	6 10 10 7 1
8	4	V_{20} V_{24} V_{28} V_{33}	5 10 11 7 1	6 10 11 6 1	6 10 10 7 1
9	4	V_{17} V_{22} V_{30} V_{35}	5 10 11 7 1	6 10 10 7 1	6 9 10 7 2
10	4	V_2 V_6 V_{13} V_{16}	5 9 10 8 2	6 10 10 7 1	6 8 10 8 2
11	4	V_{19} V_{25} V_{27} V_{32}	5 10 11 7 1	6 10 11 6 1	6 9 10 8 1
12	4	V_{37} V_{40} V_{43} V_{45}	5 10 11 7 1	6 10 10 7 1	6 9 10 7 2
13	4	V_5 V_8 V_{10} V_{14}	5 9 10 8 2	6 10 11 6 1	6 10 10 7 1
14	4	V_{57} V_{58} V_{59} V_{60}	6 10 10 7 1	6 10 11 6 1	6 10 10 7 1
15	4	V_{18} V_{26} V_{31} V_{36}	5 10 11 7 1	6 9 10 7 2	6 9 10 8 1
16	4	V_{38} V_{39} V_{42} V_{46}	5 10 11 7 1	6 9 9 7 3	6 9 10 7 2
17	4	V_1 V_7 V_9 V_{15}	5 9 10 8 2	6 10 10 7 1	6 10 10 7 1
18	4	V_{47} V_{51} V_{54} V_{55}	5 10 10 8 1	6 9 10 7 2	6 9 10 8 1
19	4	V_{48} V_{50} V_{53} V_{56}	5 10 10 8 1	6 9 9 7 3	6 9 10 7 2

11.2 Heuristic Topological Model

This paragraph shows a simple way for computing the number of molecular independent sites of a fullerene molecule starting just from its connectivity data based on molecular topological invariants.

Molecular graph G with N vertices offer a parade of topological invariants (Todeschini and Consonni 2000); among the most celebrated we have the Wiener index *W(N)*, the integer number defined as the semi-sum of the minimum distances d_{ij} between all couples of vertices V_i and V_j:

$$W(N) = 1/2\ \Sigma_i\, w_i \quad i = 1, \ldots, N \tag{11.1}$$

The invariant w_i expresses the contribution to *W(N)* coming from vertex V_i:

$$w_i = \Sigma_m\ mb_{im} \quad m = 1, \ldots, M \tag{11.2}$$

Wiener coefficient b_{im} gives the number of nodes in the m-coordination shell of site V_i and M is the maximum distance present in the graph $M = max\{d_{ij}\}$. The topological index w_i is called Wiener-weight (WW) of vertex V_i. All nodes in a fullerene direct graph have $b_{i1} = 3$; the dual graph of a fullerene shows the connectivity among its faces then it has $b_{i1} = 5$ for pentagons and $b_{i1} = 6$ in case of hexagons. The ordered string is called the *coordination string* of site V_i:

$$\{b_{im}\} = \{b_{i1} b_{i2} \ldots\ldots b_{im} \ldots\ldots b_{iM-1} b_{iM}\} \quad m = 1, \ldots, M \tag{11.3}$$

of vertex V_i naturally group symmetry-equivalent molecular sites, since equivalent atoms should have the same string. The opposite statement may be invalid since accidental degeneracy may affect this fast sorting method as shown by the earliest applications of topological invariants to C_{76}, C_{78} fullerenes molecules (Ori et al. 1992, 1993) reporting also the first calculations of the Wiener indices of C_{60}–I_h and C_{70}–D_{5h} fullerenes. *Colored* direct graphs represent very good tool to solve the accidental degeneracy (Ori et al. 1992, 1993) and correctly reproduces, from topology, the ^{13}C–NMR experimental patterns for the C_{76}, C_{78} molecules.

According to the scheme given in Fig. 11.3, we introduce a new topological method, still very fast and intuitive, that avoids any graph-coloring by simulating resonance spectra using only a coordination strings $\{b_{im}\}$ and $\{b_{im}\}^D$ coming from direct (G) and dual (G^D) mono-chromatic graphs. The relationship among nodes of G and dual G^D reflects the natural topological correspondence among fullerene vertices and faces. In the fullerene molecular graphs G each atom V_i is represented by a 3-connected node with a given coordination string $\{b_{im}\}$. V_i stays on the corner among three faces A_p, A_q, A_r that, in turn, are also vertices of the dual graph G^D with coordination strings $\{b_{pm}\}^D, \{b_{qm}\}^D, \{b_{rm}\}^D$. This triplet of strings, indicated by the short-hand notation $\{b_{im}\}^D$, represents the *dual coordination string* of V_i.

Coordination strings $\{b_{im}\}$ and $\{b_{im}\}^D$ carry-on a valuable amount of topological information that, in many cases, solves the degeneracy among symmetry-distinct molecular sites. The heuristic validity of this approximated algorithm is shown hereunder by the application to C_{66}–$C_{2v}{}^{\#0011}$ fullerene (Figs. 11.1 and 11.2) characterized by a 19-lines ^{13}C–NMR spectrum with (5×2; 14×4) relative intensities, as observed on metallofullerene $Sc_2@C_{66}$ (Wang et al. 2000). The isomer represented in Fig. 11.1 has 66 carbon atoms, the first 16 of them conventionally placed on the two pairs of fused pentagons, other 40 on the 8 IPR pentagonal faces and the remaining 10 at the interceptions of three hexagons. The topological determination

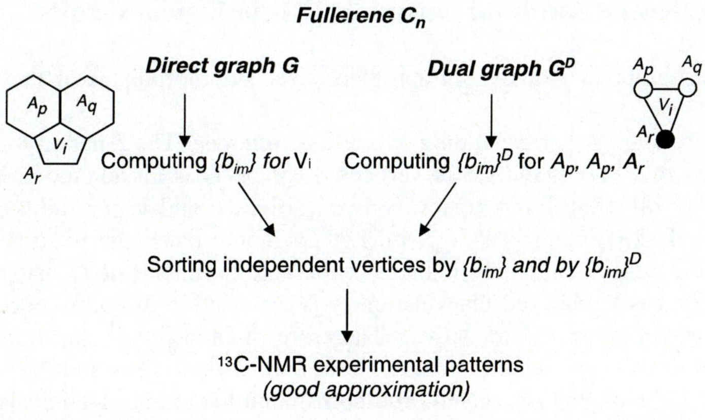

Fig. 11.3 Topological approximate method combines WW strings values from direct and dual graphs and correctly approximates independent molecular sites of a given fullerene just from adjacency data

of WW sets, according to Eq. (11.2), is represented in Table 11.2, where accidental degeneracy of $\{b_{im}\}$ strings is quite evident. WW values in fact group the 66 nodes in 14 distinct sets (called *p, as resonance peak)* of symmetry-equivalent sites V_i with multiplicity k_p ranging from 2 to 12, producing a wrong 14-peaks resonance spectra (5×2; 5×4; 3×8; 1×12) compared to the experimental 19-lines pattern (5×2; 14×4). To solve this error in predicting molecular symmetry, our heuristic method just requires a further fast sorting of the 66 molecular nodes by taking in considerations the *dual coordination strings* $\{b_{im}\}^D$ values, according to the computational scheme given in Fig. 11.2. It is in fact immediately evident from Table 11.2 that shaded lines with multiplicity 8 and 12 get split in quartets by the action of the $\{b_{im}\}^D$ values reproducing the 19-lines (5×2; 14×4) ^{13}C–NMR spectrum experimentally measured. For example, the set of 8 vertices V_{19} V_{25} V_{27} V_{32} V_{37} V_{40} V_{43} V_{45} (see Table 11.2 shaded line 11) are grouped in two independent sets respectively made by V_{19} V_{25} V_{27} V_{32} and V_{37} V_{40} V_{43} V_{45} carbon atoms. In a similar way, the remaining shaded lines are resolved and the ^{13}C–NMR spectrum for this C_{66}–$C_{2v}$$^{\#0011}$ isomer is correctly interpolated, as shown in Table 11.2 bottom. Sum of $w_i/2$ column entries gives the Wiener index $W = 10674$ of the C_{66}–$C_{2v}$$^{\#0011}$ fullerene, whereas its dual graph (Fig. 11.2) has $W^D = 1595$.

This effective topological algorithm exploits the structural information present in the topological coordination strings $\{b_{im}\}$ and $\{b_{im}\}^D$ allowing a quick prediction of molecular symmetry and resonance lines. Possible approximations, originated by accidental degeneracy of $\{b_{im}\}$ and $\{b_{im}\}^D$ numerical values, are however possible. These limitations are resolved by the more evolute, but still purely topological, method which is described in the next paragraph based on the generation of the automorphisms of C_{66} fullerenes. It is nevertheless worth to remark that coordination strings effectively reduce the number of fullerenes for which complete set of automorphism operators will be generated.

11.3 Fullerene Automorphisms and Topological Orbits

In this paragraph, we analyze graph-theoretical automorphisms of the fullerenes C_{66}.

Let G be a graph corresponding to one C_{66} fullerene. The function $f : V(G) \rightarrow V(G)$, where $V(G)$ is the set of vertices of graph G is an automorphism if and only if the following holds for every two vertices u and v: u is adjacent to v if and only of $f(u)$ is adjacent to $f(v)$. Let us denote by $\sim$ the relation on $V(G)$ such that $u \sim v$ if and only if there is the automorphism of G that maps u to v. It can be easily checked that relation $\sim$ is the relation of equivalence (i.e. it is symmetric, reflexive and transitive). Therefore, it induces the partition of $V(G)$ in the set of (disjoint) classes. These classes are called orbits and the vertices of the graph G belonging to a given orbit correspond to symmetry-equivalent carbon atoms.

In chemical studies of the fullerene C_{66}, it is shown that particularly interesting fullerenes are those whose graphs have 19 orbits such that 14 of them consist of four vertices each and 5 of them consist of two vertices each. As mentioned, it is well known that there are 8 such fullerenes (Wang et al. 2000). Our aim is to give the procedure for fast identification of these fullerenes. Since, there are 4478 fullerenes, it would be unpractical to precisely calculate orbits for each of them. Therefore, we propose the solution that operates in two stages which will be explained in more details in the next paragraphs. In the first stage, it calculates the coarser classification of the vertices of the observed fullerene graph. It checks if the fullerene can be immediately eliminated as the possible candidate. If not, then orbits are calculated and the precise check is done. Let us denote by $G_1, G_2, ..., G_{4478}$ corresponding graphs of these 4478 fullerenes and let us observe one of these graphs G_i, where $1 \leq i \leq 4478$. First we use Dijkstra algorithm for calculation of all distances in G. Dijkstra algorithm for finding the distance from one fixed vertex (called initial vertex) to all other vertices is given by the following procedure (Dijkstra 2010):

1. Assign a distance value to every vertex. Set it to zero for the selected initial vertex and to infinity for all other vertices.
2. Mark all vertices as unvisited. Set initial vertex as current.
3. For current vertex, consider its all unvisited neighbors and calculate their tentative distance (from the initial vertex). If this distance is less than the previously recorded distance (infinity in the beginning, zero for the initial vertex), overwrite the distance.
4. When we finish considering all neighbors of the current vertex, mark it as visited. A visited vertex will not be checked ever again; its distance recorded now is final and minimal.
5. If all vertices have been visited, finish. Otherwise, set the unvisited vertex with the smallest distance (from the initial vertex) as the next "current vertex" and continue from step 3.

It can be shown that good implementation of this algorithm works in linear time. Since, we need to use every vertex as initial vertex in order to obtain distance matrix, we can do this in quadratic time. Now, let us assign each vertex its *distance code* $dc\,(v) = (x_1, x_2, ..., x_{65})$, where x_i is the number of vertices on the distance i from v (*distance code* generalizes *coordination string* of Eq. (11.3). Obviously, if vertices u and v are in the same orbit, then $dc\,(u) = dc\,(v)$. Now, we can introduce the equivalence relation $\underset{dc}{\square}$ on $V\,(G_i)$. Let us call its classes of equivalence super-orbits. Each super-orbit is in general the union of orbits, see as an example the shaded lines $p = 7,8,9,10$ in the top of Table 11.2. Let us denote s_{ij} the number of super-orbits of cardinality j corresponding to G_i. If G_i has 19 orbits such that 14 of them consist of four vertices each and 5 of them consist of two vertices each, then it must hold:

$$s_{i1} = 0 \text{ and } s_{i3} = 0 \text{ and } s_{i2} + s_{i6} \leq 5 \tag{11.4}$$

These three simple conditions immediately reduce the number of potential candidates from 4478 to only 13. For these 13, we explicitly construct their orbits. Probably the fastest general purpose algorithm for finding orbits has been developed

(McKay 2010). Also, there are more specialized procedures for analyses of trivalent graphs (Galil et al. 1987). However, we shall use very simple recursive algorithm. Since it solves our problem for each fullerene in less than 0.02 second for each of these 13 fullerenes, it was not necessary to use faster, but more complicate algorithms. Let us assume that vertices in G_i are denoted by $v_1, ..., v_{66}$ and let $d(u, v)$ denotes the distance of vertices u and v. The algorithm for finding automorphisms is presented by the following pseudo code. The following recursion is called by *rec* (1) (it is assumed that initially all vertexes are considered to be *free*):

```
rec(int loc)
If loc = 67 then f is automorphism
Else
    For each free vertex u
        if dist(v_i, v_loc) = dist(f(v_i),u)
        for all i = 1,...,loc − 1
          put f(v_loc) = u and set u is non-free vertex
          rec(loc + 1)
          set u is free vertex
```

After all automorphisms are found, it is relatively simple to determine all orbits. Alternative algorithms for generating automorphism groups of fullerenes are recently reported in literature (Ashrafi and Ahmadi 2005).

11.4 Results and Conclusions

The graph automorphisms generation method presented in this article correctly sieves eight C_{66} molecules with proper orbits among the 4478 fullerene isomers in a very effective way.

All these molecules in Table 11.3 have C_{2v} molecular symmetry and exactly reproduce the ^{13}C–NMR resonance spectrum with the observed 19-lines (19-orbits) and multiplicity (5×2; 14×4). In Table 11.3 they are identified by the values of their topological graph invariants, the Wiener index *W* and the *topological efficiency index* $\rho = w_{ave}/w_{min}$ being the average w_{ave} and minimum w_{min} quantities easily computed from Eq. (11.2). Current computations produces an interesting isomer, molecule *d* in Table 11.3, that is connected to the stable C_{66}–$C_{2v}^{\#0011}$ by a Stone-Wales rotation (SW) of the 4 faces 19,20,11,12 located along graph external border that correspond, one may say, to the south-pole of the actual molecule. Starting from isomer *a* of Fig. 11.2, SW rotates the edge 19–20 and transforms pentagons 11,12 and hexagons 19, 20 in two new hexagons 11,12 and pentagons 19,20 of isomer *d*. In Fig. 11.4 it shows two more pairs of fused pentagons, maintaining C_{2v} symmetry of C_{66}–$C_{2v}^{\#0011}$ fullerene.

We remark that the stable isomer C_{66}–$C_{2v}^{\#0011}$ tends to minimize both the molecular Wiener index $W = 10674$ *and* the topological efficiency index $\rho = 1{,}0268$, pointing out the role of *topological efficiency* in predicting stable isomers of a given

Table 11.3 Topological invariants of the eight C_{66}–C_{2v} isomers with 19-orbits (5×2; 14×4); C_{66}–$C_{2v}$$^{\#0011}$ stable isomer (molecule a), has low values of both indices W and ρ; dual graph indices W^D, ρ^D are also listed

Mol	W^D	ρ^D	W	ρ
a	1592	1.0338	**10674**	**1.0268**
b	1588	1.0803	10687	1.0347
c	1584	1.0776	10696	1.0456
d	1586	1.0538	10713	1.0241
e	1590	1.1217	10723	1.0868
f	1583	1.0769	10723	1.0868
g	1583	1.0518	10745	1.0304
h	1588	1.1343	10761	1.0979

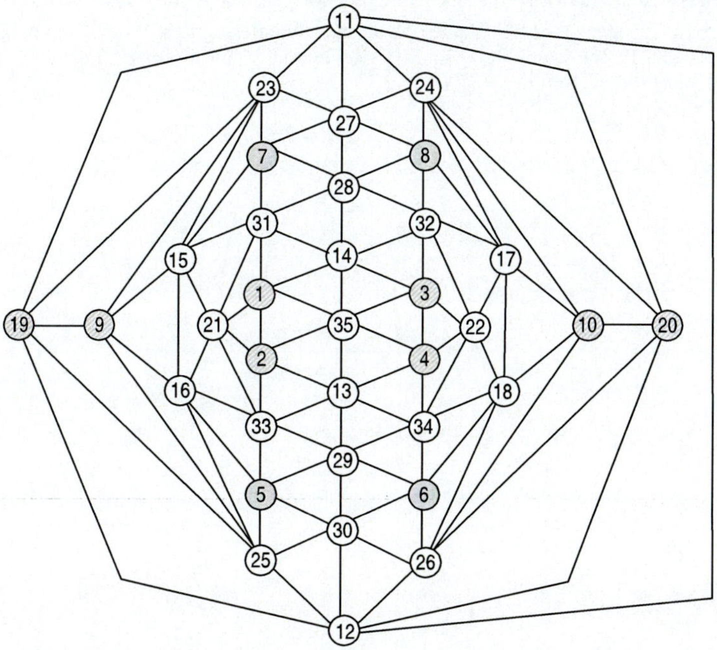

Fig. 11.4 Dual space graph of the C_{66} isomer derived from C_{66}–$C_{2v}$$^{\#0011}$ by Stone-Wale rotation of faces 19,20,11,12; *dashed* 5-edged nodes evidence its four pairs of touching pentagons

C_n fullerene, as studies on C_{28} suggest (Ori et al. 2009). Cataldo et al. (2010) also reported topological efficiency effects to explain the relative chemical stability of graphite lattice portions and fullerene fragments (nanocones) built around a pentagonal face. This original topological concept is currently under deeper investigations and its applications on chemical stability of fullerene will be the matter of forthcoming papers, with the aim to shed new lights on formation mechanism of the interconnected fullerenes and graphene.

References

Ashrafi AR, Ahmadi MR (2005) Internet J Nanotechnol 1(2):1

Cataldo F, Ori O, Iglesias-Groth S (2010) Mol Simul 36(5), 341

Cozzi F, Powell WH, Thilgen C (2005) Pure Appl Chem 77(5):843

Cui Y-H, Tian WQ, Feng J-K, Chen D-L (2010) J Nanopart Res 12:429

Dijkstra E (2010) http://en.wikipedia.org/wiki/Dijkstra%27s_algorithm

Fowler PW, Manolopoulos DE (1995) An atlas of fullerenes. Clarendon, Oxford

Galil Z, Hoffmann CM, Schnorr CP, Weber A (1987) J ACM 34:513

Kroto H (1987) Nature 329:529

McKay BD (2010) http://www.cs.sunysb.edu/~algorith/implement/nauty/implement.shtml

Ori O, Cataldo F, Graovac A (2009) Fuller Nanotub Carbon Nanostruct 17(3):308

Ori O, D'Mello M (1992) Chem Phys Lett 197(1–2):49

Ori O, D'Mello M (1993) Appl Phys A Solids Surf 56:35

Réti T, László I (2009) Acta Polytech Hung 6(5):85

Tan Y-Z, Xie S-Y, Huang R-B, Zheng L-S(2009) Nat Chem 1:450

Todeschini R, Consonni V (2000) Handbook of molecular descriptors. Wiley, Weinheim

Wang C-R, Kai T, Tomiyama T, Yoshida T, Kobayashi Y, Nishibori E, Takata M, Sakata M, Shinohara H (2000) Nature 408:426

Chapter 12
The Topological Background of Schwarzite Physics

Giorgio Benedek, Marco Bernasconi, Eugenio Cinquanta, Luca D'Alessio, and Marzio De Corato

You see things and you say "Why?"
But I dream things that never were and I say "Why not?"
(George Bernard Shaw, Back to Mathusalem)

Abstract About 10 years ago the synthesis of random carbon schwarzites by supersonic cluster beam deposition has endowed the rich sp^2 carbon family with its three-dimensional member. Its reluctance to grow as a three-periodic minimal surface according to topological and physical predictions still prevents schwarzites from being a hot topic, although spongy carbon is already having countless applications. Understanding the links between topology and quantum structure, possibly with the help of large-scale quantum molecular dynamics simulations should trace the route to the synthesis of periodic schwarzites. In this perspective, after a brief account on the growth and characterization of spongy carbon, we review the elementary topology of schwarzites, their stability and growth conditions as derived from pure topological arguments, the electronic structure and the electron-phonon interaction of the smallest periodic schwarzites and what can be learnt by the topological monitoring of quantum molecular dynamics.

12.1 Introduction

The investigation of new sp^2-bonded carbon architectures, marked by the discovery of fullerenes (Kroto et al. 1985) and nanotubes (Iijima 1991), and more recently by the synthesis of spongy carbon (Donadio et al. 1999; Barborini et al. 2002a; Benedek et al. 2003) and the isolation of single graphite layers (graphene) (Novoselov et al. 2004, 2005a, b; Geim and Novoselov 2007; Castro Neto et al. 2009), is opening fascinating perspectives for nanostructured carbon as a novel

G. Benedek (✉)
Donostia International Physics Centre (DIPC), 20018 Donostia-San Sebastián, Spain;
Dipartimento di Scienza dei Materiali, Universitá di Milano-Bicocca, 20125 Milano, Italy
e-mail: giorgio.benedek@mater.unimib.it

F. Cataldo et al. (eds.), *The Mathematics and Topology of Fullerenes*,
Carbon Materials: Chemistry and Physics 4, DOI 10.1007/978-94-007-0221-9_12,

all-purpose material (Benedek and Bernasconi 2004). The early observation of superconductivity in alkali metal-doped fullerenes (Rotter et al. 1992), field-emission (Wang et al. 1998), and supercapacitance (Niu et al. 1997) from arrays of nanotubes, the extraordinary transport (Geim and Novoselov 2007; Castro Neto et al. 2009; Seol et al. 2010), electrical (Stoller et al. 2008) and electro-mechanical (Cadelano et al. 2009; Li et al. 2010) properties of graphene, and the unconventional magnetism of spongy carbon (Rode et al. 2004; Arčon et al. 2006) are just a few examples of the vast areas of application of the most versatile among elemental materials. While fullerenes, nanotubes, and graphite layers aggregate through comparatively weak van der Waals forces, spongy carbon constitutes a fully covalent highly-connected three-dimensional (3D) form of sp^2 carbon, which combines many valuable properties of fullerenes, nanotubes and graphene with a robust 3D architecture. Triply periodic minimal surfaces (Lenosky et al. 1992; Townsend et al. 1992) have been theoretically suggested as possible model structures for spongy carbon, which has since termed *schwarzite*, after the name of the mathematician Hermann Schwarz (Schwarz 1890) who first investigated that class of surfaces.

Schwarzites synthesized by supersonic cluster beam deposition (SCBD) (Barborini et al. 2002; Milani and Iannotta 1999) are characterized by a nanometric porosity and, as suggested by numerical simulations of the TEM images (Benedek et al. 2003), by the structure of a *random schwarzite* (Lenosky et al. 1992) which grows in the form of a *self-affine minimal surface* (Bogana et al. 2001; Benedek et al. 2005). Thus, besides offering appealing technological perspectives, this novel material shows intriguing aspects of differential geometry and topology.

It is somewhat surprising that carbon schwarzites, despite their very interesting structural properties and viable applications in efficient supercapacitors (Diederich et al. 1999) and field emitters (Boscolo et al. 2000; Benedek et al. 2001a; Ferrari et al. 1999), did not receive yet much attention. It is therefore convenient to spend first a few words about the growth method of carbon schwarzites by SCBD (Section 12.2), also because their growth and structural properties appear to be closely related to their topological features. After introducing some elementary concepts on the topology of sp^2 carbon forms, and illustrating the class of three-periodic P- and D-type schwarzites and the effects of self-affine distortion (Section 12.3), it is shown (Section 12.4) that the stability and growth of sp^2 carbon in the form of random schwarzites, rather than as nanotubes or fullerenes, is actually determined by simple initial topological conditions (Benedek et al. 2003). The surface minimality has direct implications on the growth kinetics, which may present a quasi-deterministic character. It is also shown that, unlike fullerenes where abutting five-fold rings are unfavoured, in schwarzites seven-fold rings tend to aggregate thus preventing the formation of crystalline three-periodic structures. A calculation of the free energy of schwarzites including the entropic vibrational and configurational contributions allows to estimate the average porosity in thermal equilibrium as a function of the deposition energy (Section 12.5). The existing calculations of the electronic structure and electron-phonon interaction of the smallest schwarzites, and the possible links between certain topological features and the electronic properties, are discussed in Section 12.6. Finally the predictions of relevant structural, thermal and electronic properties based on quantum molecular dynamics simulations are briefly discussed in Section 12.7. The reader should be advised that the present review is restricted to the small class of schwarzite structures which are accessible to quantum simulations and ab-initio calculations and may help understanding some features observed in spongy carbon. The vast number of sp^2 forms and fantastic structures which can be generated by complex mathematical algorithms and their intriguing topological aspects, though absolutely relevant for the analysis of the spongy forms of carbon and the future construction of regular architectures, are not discussed here, being found in the other chapters of this volume.

12.2 The Birth of Random Schwarzites

Nanostructured sp^2 carbon-based solids represent a class of materials where the surface curvature and the structural organization on the nanometric scale, ranging from less than 1 nm to a few hundreds of nanometres, dramatically influence the mechanical, chemical and physical properties (Rao and Dresselhaus 2001). Porous carbon networks are of great importance in many areas of science and technology including catalysis, energy storage, chromatography, gas and liquid purification and molecular sieving (Lu and Chung 1997; Kyotani 2001). The high specific surface area, chemical inertness and large pore volumes are the important parameters for these applications. The complexity of the carbon structure makes the control of the pore size and structure a difficult technological problem. Many synthetic techniques have been proposed for the production of meso- and macroporous carbon (Lu and Chung 1997; Kyotani 2001; Ryoo et al. 2001; Kajii et al. 2000), but more intriguing is the creation of nanoparticles with a specific surface curvature which can be used to control the porosity of the material.

Total energy calculations show that carbon schwarzites are in general more stable than fullerenes with a similar absolute value of the Gaussian curvature (Lenosky et al. 1992; Vanderbilt and Tersoff 1992). Nevertheless schwarzite-like materials have not been observed during carbon-arc synthesis of fullerenes and nanotubes, suggesting that a new technique should be implemented for the production of such exotic carbon structures. An effective production technique for *random carbon schwarzites* was found to be a bottom-up approach based on the assembling of sp^2 nanometric clusters. This was achieved by means of SCBD of carbon clusters produced by a Pulsed Microplasma Cluster Source (PMCS) (Barborini et al. 1999; Piseri et al. 2001) and assembled onto a substrate.

The source chamber consists of a ceramic cavity hosting along a vertical axis two cylindrical electrodes separated by a gap a few millimeters wide. One of the electrodes, the cathode, is made of graphite and constitutes the target of a pulsed helium beam, which is injected into the source chamber through a solenoid valve along a horizontal axis. On the opposite side of the horizontal axis there is a nozzle which allows for the supersonic molecular beam expansion outside the source chamber into vacuum. The helium pulse directed against the graphite cathode is ionized, after a fixed delay of a few hundreds microseconds, by an intense pulsed discharge (V $\sim$ 750 v). The helium plasma ablates the cathode surface removing carbon atoms via sputtering; then cluster aggregation occurs at low temperature (T $\sim$ 100 K) in the high pressure region in front of the cathode. These particular thermodynamic conditions allow for the formation of full sp^2 fullerene-like carbon clusters with a mass distribution peaked around 600 atoms/cluster (Barborini et al. 2002). Since in the supersonic cluster beam the spread in the kinetic energy per atom is much smaller than the atom binding energy, the clusters reach the substrate with their initial morphology and size distribution, thus leading to the synthesis of nanostructured pure sp^2 carbon films under controlled conditions (Milani et al. 1999; Barborini et al. 2002; Donadio et al. 1999). It should be noted, however, that the carbon atoms impact the substrate surface with a translational kinetic energy of 0.1–0.4 eV/atom.

The film growth via SCBD can be viewed as a random stacking of particles as for ballistic deposition. The resulting material is characterized by a low density as compared to that of films assembled atom by atom and it shows different degrees of order depending on the scale of observation (Milani et al. 1999; Barborini et al. 2002). The characteristic length scales are determined by the deposition energy, the cluster dimensions and by their fate after deposition. Carbon cluster beams are characterized by the presence of a finite mass distribution and by the presence of isomers of different stability and reactivity. Once on the substrate, stable clusters can survive almost intact while reactive isomers can coalesce to form a more disordered phase (Barborini et al. 2002; Milani et al. 1997). A transmission electron microscope (TEM) analysis of nanostructured carbon films shows that, at this scale, the morphology is reminiscent of the precursor clusters (Milani et al. 1999; Lenardi et al. 1999). TEM micrographs show the presence of an amorphous matrix with small closed shell particles and bundles of graphene sheets (Fig. 12.1). Large onion-like and tubular particles have also been observed.

Fig. 12.1 TEM micrograph showing closed graphitic particles and graphene sheets dispersed among amorphous material (adapted from Milani et al. 1999)

In the case of cluster assembling one should recall that, due to finite cluster mass distribution, relatively large clusters are somehow "diluted" among small particles. Large clusters can act as seeds for the formation of nodular defects, which evolve like isolated structures protruding from the average thin film surface. Depending on the initial density of defects, there is a critical film thickness where the nodular structures start to merge. These defects have a profound influence on the evolution of the surface morphology. Roughness, scale invariance, and spatial correlation of the film surface depend on the cluster precursor size and film thickness. In general the surface spatial correlation and its dependence on the film thickness show the characters of a fractal self-affine growth (Barabàsi and Stanley 1995). The surface corrugation can be described by a function $h(\mathbf{r}, t)$ expressing the height of the surface at position $\mathbf{r}$ with respect to the average surface plane, when the film has a thickness t. For a steady deposition rate, t can be interpreted as a time coordinate. At a given t the height-height correlation function is defined as

$$w(\mathbf{r}, t) \equiv \sqrt{\langle [h(\mathbf{r} + \mathbf{r}', t) - h(\mathbf{r}', t)]^2 \rangle_{\mathbf{r}'}}, \tag{12.1}$$

where the average is taken over all the surface positions $\mathbf{r}'$. For an isotropic surface this function only depends on the distance $r \equiv |\mathbf{r}|$ and grows with r as long as it is smaller than a correlation length ξ, then it saturates to a value $w(\infty, t)$ which defines the *surface roughness* for a thickness t. Self-affinity is characterized by two power-laws (Barabàsi and Stanley 1995)

$$w(r, t) \propto r^{\alpha}, \quad r << \xi, \tag{12.2}$$

$$w(\infty, t) \propto t^{\beta}, \tag{12.3}$$

which express a scale invariance in the directions normal and parallel to the growth direction through the *roughness exponent* α and the *growth exponent* β, respectively. The evolution of the roughness during deposition and the extension of spatial correlation on the surface plane turn out to be peculiar of the particular growth mechanism and largely independent of the nature of the physical system. In other words the growth mechanisms can be ascribed to certain universality classes with well defined exponents α and β. In SCBD experiments growing nanostructured carbon described in this chapter the exponents derived with atomic force microscopy (AFM) (Buzio et al. 2000) are

$$\alpha = 0.66 \pm 0.02, \quad \beta = 0.50 \pm 0.03. \tag{12.4}$$

These exponents can be ascribed to the class of processes described by the Kardar-Parisi-Zhang (KPZ) equation with spatially correlated noise (see Barabàsi and Stanley 1995, chapter 22). They roughly agree also with the renormalization-group results for the isotropic growth model within the quenched noise regime (Barabàsi and Stanley 1995, chapter 10).

Scale invariance is seen to extend over up to three decades in the thicker films. This allows to compare the results of molecular dynamics simulations, which necessarily are performed on the nanometric scale, to experiments made on a larger length scale (Fig. 12.2). An example is shown in Fig. 12.3, where the calculated AFM image for a film obtained from a molecular dynamics simulation is compared to the AFM image of a real SCBD carbon film (Bogana et al. 2001). The morphology of the surface looks very similar in the two images, although the are obtained on length scales which differ by more than two orders of magnitude.

The typical correlation lengths ξ for films assembled by small clusters range from 50 to 7 nm. The lowest value refers to film where the density of nodular defects is low and their coalescence has not taken place. These films are very uniform and flat. Apart from isolated nodules, large structures are not present and consequently the correlation length closely reflects the size of the basic morphological units. As long as the morphology develops and larger and larger features appear, correlation length increases. A similar behavior is seen for films assembled by large clusters and it is observed at considerably lower thickness.

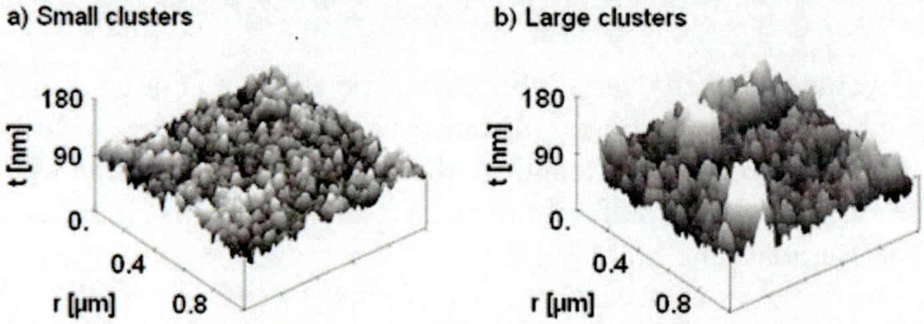

Fig. 12.2 The surface roughness of SCBD carbon films depends on the average size of the precursor clusters. It is defined by the height-height correlation function $w(r,t)$ which grows with the distance r as r^{α} for r much smaller than a correlation length ξ, and reaches a saturation value $w(\infty,t)$ for $r >> \xi$. The latter grows with the film thickness t as t^{β}. Experiment gives a *roughness exponent* $\alpha = 0.66 \pm 0.02 \cong \frac{2}{3}$ and a *growth exponent* $\beta = 0.50 \pm 0.03 \cong \frac{1}{2}$

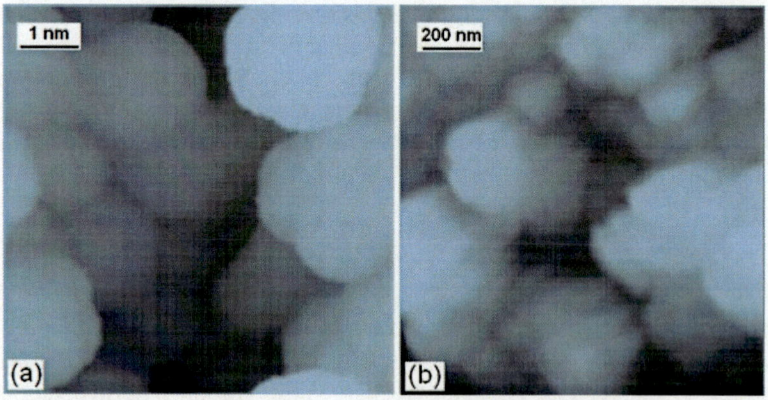

Fig. 12.3 Visual comparison of the simulated AFM image on the nanometric scale (**a**) with an experimental AFM image at the scale of 200 nm (**b**), as reported by Lenardi et al. (1999) (adapted from Bogana et al. 2001)

By adding a metal catalyst during the cluster formation, either by bubbling the He gas stream with a metallic precursors through a liquid metal-organic compound prior to the injection into the PMCS (Fig. 12.4a) or by using mixed cathodes, it is possible to control to some extent the cluster formation inside the source and the composition of the supersonic beam so as to obtain fairly pure schwarzitic structures (Fig. 12.4b). The metal-organic molecules in the buffer gas are cracked by the electric discharge, thus providing metal atoms and highly reactive radicals to the condensing carbon cloud. By using Molybdenum (V) isopropoxide $Mo(OC_3H_7)_5$ in isopropanol and Cobalt (II) methoxyethoxyde $Co(OCH_2CH_2OCH_3)$ in 2-methoxyethanol as catalysts and PMCS in the standard operational mode, it was possible to obtain spongy schwarzite-like carbon films like the one shown in the TEM picture of Fig. 12.5. The material looks like a free-standing film made by several interconnected undulated foils. In high resolution scanning electron micrograph, the contrast created by secondary electrons shows that voids are present in the bulk of the sample indicating that the material has a complex three-dimensional porous structure. The structure of spongy carbon consists of thin carbon layers (1–2 nm thick) interconnected to form a network with overall thickness up to micrometers. A crucial role is played by the metal-organic catalyst whose concentration and dispersion seem to determine the final curvature and morphology of the material.

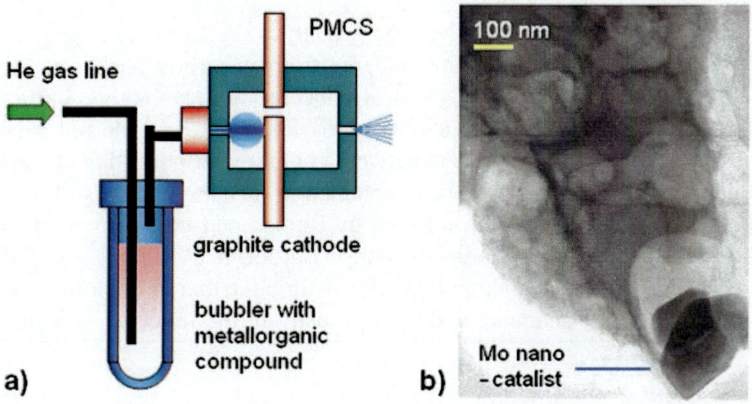

Fig. 12.4 Catalytic growth of carbon schwarzites can be obtained either by passing the carrier He gas through a bubbler containing a metalorganic compound (**a**) or by using a mixed electrode containing the needed amount of catalytic metal nanoparticles. (**b**) A schwarzites grows from a molybdenum catalyst nanoparticle, apparently as a self-affine structure with pore diameters increasing with the distance from the catalyst

The formation of the spongy carbon is assisted by the presence of metal catalyst nanoparticles, as clearly appears in Fig. 12.4b. The size, concentration and dispersion of catalyst nanoparticles are believed to determine the final morphology and curvature of the material. By comparing the material obtained with mixed cathodes with that obtained with metal-organics, it has been verified that the presence of large catalyst concentrations (several percent) in the form of relatively large clusters favours the production of carbon nanoparticles, whereas a finely dispersed

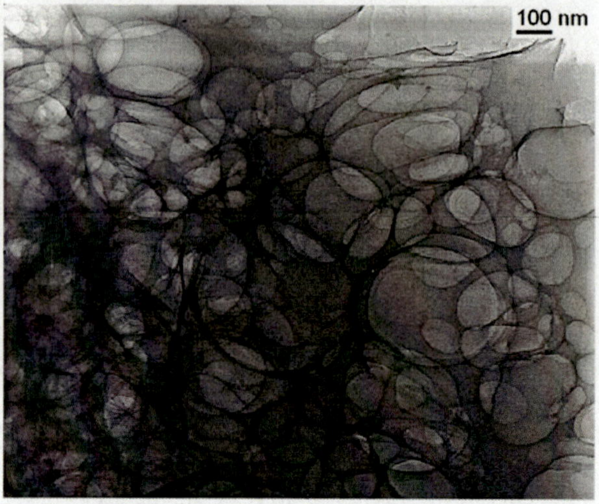

Fig. 12.5 A transmission electron microscope (TEM) image of a random carbon schwarzite obtained by supersonic cluster beam deposition with a deposition energy of 0.1 eV/atom (Barborini et al. 2002a). Raman and near-edge x-ray absorption fine structure (NEXAFS) spectra indicate a pure sp^2 bonding structure, suggesting a single, highly connected graphene sheet with an average pore diameter in the range of 100 nm

catalyst at a low concentration drives the growth to the formation of spongy networks. Generally, the physical vapour deposition of carbon nanostructures uses mixed cathodes, which gives a high local concentration of catalyst particles. By changing the metallic precursors, it is also possible to control the porosity of the material. For example, cobalt leads to networks of narrower pores than those formed with molybdenum. The spongy carbon obtained in this way consists of a fully sp^2 three-dimensional structure, as confirmed by Raman and near-edge x-ray absorption fine structure (NEXAFS) spectroscopy (Barborini et al. 2002b). The TEM images (Fig. 12.5) suggested a topological structure like that of random schwarzites (Lenosky et al. 1992), characterized by a porosity in the range of 10^2 nm and by surface minimality.

12.3 Schwarzite Topology

The aim of this section is to show how topology alone can help one to predict the structure and some relevant physical properties of sp^2 carbon on the mesoscopic scale from parameters which are supposed to be known on the atomic scale, such as the bond strengths and the surface stiffness constants. A basic question is whether a graphite sheet can be transformed into a surface characterized by a negative Gauss curvature everywhere through the creation of a sufficient number of negative disclinations, which occur wherever a 6-membered ring is replaced by a larger ring. A special case of negative Gauss curvature occurs when the mean curvature is zero everywhere, which corresponds to a minimal surface. The conjecture that a minimal surface is particularly stable has stimulated much theoretical work on hypothetical graphite sheets (graphenes) with the structure of a periodic schwarzite (Fig. 12.6)

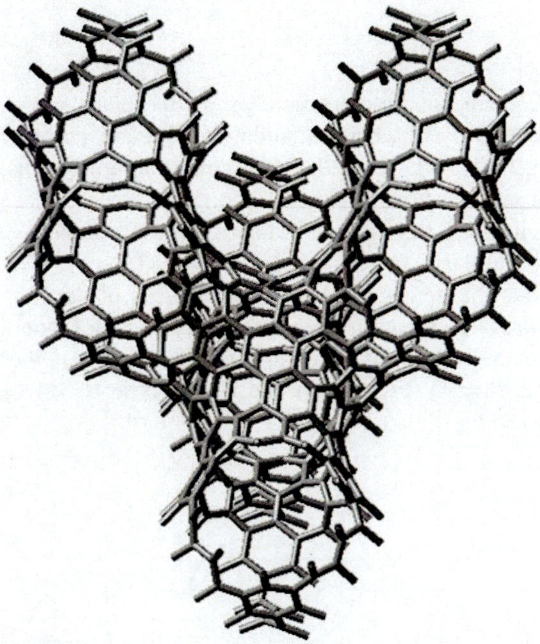

Fig. 12.6 The carbon schwarzite fcc-$(C_{84})_2$ obtained from a tiling with carbon hexagonal and heptagonal rings of a three-periodic D-type minimal surface (Schwarz 1890). The unit elements of a D-type schwarzite are centred at the sites of a diamond lattice. Each unit cell contains two elements and each element is made of 12 heptagons and any number h ($\neq 1$) of hexagons. Here $h = 28$. This is the smallest schwarzite with non-abutting heptagons (equivalent to C_{60} which is the smallest fullerene with non-abutting pentagons and has $h = 20$). Ab-initio calculations show however that, unlike fullerens where abutting pentagons are less stable, in schwarzites abutting heptagons are more stable, which favours non-periodic schwarzites

(McKay and Terrones 1991; Terrones and McKay 1993). Similar theoretical sp^2 carbon structures like polybenzenes (O'Keeffe et al. 1992), hollow graphites (Benedek et al. 1997a, b; Coté et al. 1998) and plumber's nightmares (Vanderbilt and Tersoff 1992), which have been investigated theoretically, can be assigned to the general family of schwarzites.

From the topological point of view graphenes like fullerenes, graphite sheets, nanotubes and schwarzites are described as a polygonal tiling of the surface, where each vertex corresponds to a carbon atom, each edge to a covalent bond and each polygon to a carbon ring. Moreover each atom has a three-fold coordination. The surface covered by the polygonal tiling of carbon rings is characterized by its *connectivity* or order of connection k. According to Hilbert and Cohn-Vossen (1932) the order of connection is the number plus one of the closed cuts which can be made on the given surface without breaking it apart in two pieces. The surface topology can be alternatively characterized either by the *Euler-Poincaré characteristic* χ or by the *genus* g, which are related each other and to k by the equations

$$\chi = 3 - k = 2(1 - g), \quad g = \frac{1}{2}(k - 1) = 1 - \frac{\chi}{2}. \tag{12.5}$$

For example a sphere, or the equivalent projective plane with one point at infinity, is split into two parts when cut along one single closed line and therefore $k = 1$ ($g = 0, \chi = 2$) (Fig. 12.7a, b). A simple (one-hole) torus, or a plane closed by cyclic boundary conditions, can be cut along two closed lines without splitting it in two pieces, whereas a third cut would split it apart, so that $k = 3$ ($\chi = 0, g = 1$) (Fig. 12.7a, c). Similarly for an n-hole torus k=1 + 2n, χ=2(1 − n) and g=n (Fig. 12.7e). Thus the genus represents the number of "holes" (or "handles") of a generalized torus. The notion of connectivity includes one-face surfaces, corresponding to even values of k (semi-integer g, odd χ): the Möbius ring has $k = 2$ ($g = 1/2, \chi = 1$) (Fig. 12.7f), the Klein bottle $k = 4$ ($g = 3/2, \chi = -1$) (Fig. 12.7h), and so on.

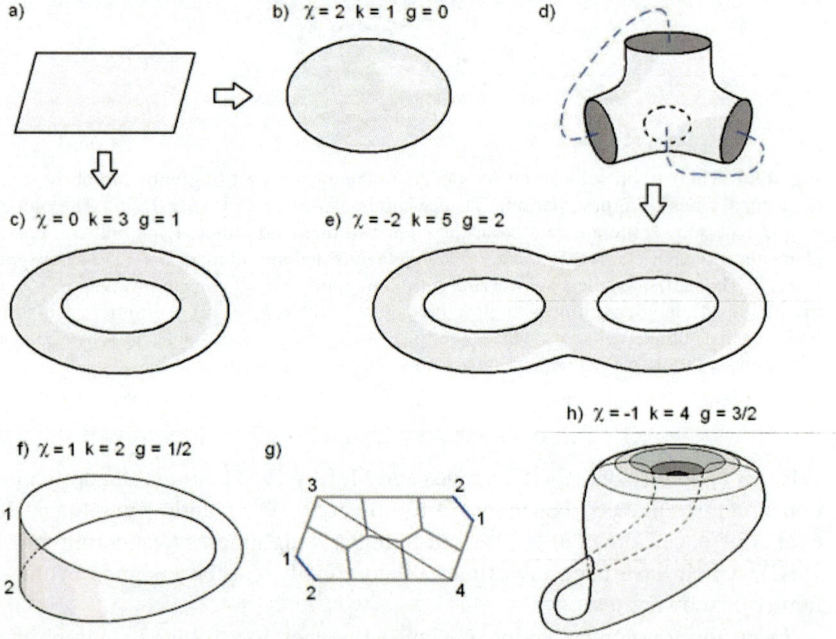

Fig. 12.7 Surfaces of different topology supporting sp^2 carbon. A graphene sheet (**a**) can be closed assuming either the topology of a sphere (fullerenes: (**b**)) or of a one-hole (-handle) torus (a nanotube with a cyclic boundary condition or a toroidal fullerene: (**c**)). The element of D-type schwarzite (**d**) after being closed on itself (*broken lines*) is topologically equivalent to a two-hole (-handle) torus (**e**). In principle a closed sp^2 carbon form can extend over a one-face surface, like a Möbius ring (**f**) or a Klein bottle (**h**). Each form is identified by its Euler-Poincaré characteristic χ or equivalently by the connectivity k or genus g (see text). Within the restricted class of sp^2 carbon with only 6-, 5- and 7-rings, the tiling of these forms require a prevalence of 5- (7-) fold rings for positive (negative) χ. For the Möbius ring the saturation of all bonds implies a minimum tiling with seven 5-rings and one 7-rings (**g**)

While fullerenes are represented by a closed surface topologically equivalent to a sphere ($k = 1$), uncapped nanotubes, graphene sheets and schwarzites are open surfaces with an infinite extension in one, two or three dimensions, respectively. However sp^2 surfaces characterized by a periodic atomic structure can be reduced to a closed surface by applying cyclic boundary conditions. In this way uncapped nanotubes and graphene become topological equivalent to an ordinary (one-hole) torus ($k = 3$) (Fig. 12.7a, c). On the other hand periodic infinite surfaces like schwarzites would have an infinite connectivity. However, similarly to the procedure for crystalline lattices in solid state physics, cyclic boundary conditions may be applied on a finite portion of the periodic surface, so as to make k, g and χ finite and linearly dependent on the actual number of unit cells. It is therefore convenient to define the corresponding parameters for the unit cell, k_{cell}, g_{cell} and χ_{cell}. Their values are obtained by closing the portion of surface contained in the unit cell on itself as implied by the cyclic boundary conditions, and the number g_{cell} of handles generated by the closure operation gives $\chi_{cell} = 2(1 - g_{cell})$ and $k_{cell} = 2\,g_{cell} + 1$.

For a three-periodic surface having the periodicity of the simple cubic lattice (P-type gyroid [21], Fig. 12.8a) the unit cell contains one element, whereas the one having the structure of the diamond lattice, (D-type gyroid (Lenosky et al. 1992), Fig. 12.8b) has two elements per unit cell. The two elements may differ just for an inversion operation (as in diamond) or have a different size (as in the sphalerite lattice), but they are topologically identical since they coordinate the same number (four) of neighbor elements. In this case the closure operation may be applied to a single element (Fig. 12.7d), which gives $g_{el} = 2$ ($\chi_{el} = -2$ and $k_{el} = 5$), equivalent to that of a two-handle torus (Fig. 12.7e). Note that for both D- and P-type gyroids $g_{cell} = 3$ ($\chi_{cell} = \chi_{el} = -4$ and $k_{cell} = k_{el} = 7$). It is important to remark that the genus of a cell (e.g., a simulation cell, which may contain several unit cells of the lattice) is always less than the genus per element times the number of elements contained in the cell due to the internal connections.

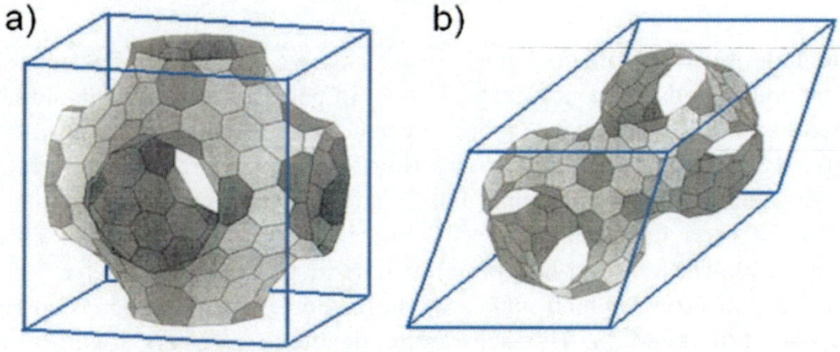

Fig. 12.8 The tiling with 6- (*light grey*) and 7- (*dark grey*) rings of the unit cell of a P-type (**a**) and D-type (**b**) schwarzite, both having 216 atoms per unit cell. The 7-rings are 24 per unit cell in both cases. The unit cell of the D-type schwarzite is made of two identical but inequivalent elements, containing twelve 7-rings each, joined in the staggered position as atoms are in the diamond lattice (Lenosky et al. 1992)

The polygonal tiling of a surface is subject to Euler's theorem linking the numbers of atoms (v), of bonds (e) and of rings (f) to the connectivity by the equation (Hilbert and Cohn-Vossen 1932), also known as Poincaré's formula:

$$\begin{aligned} v - e + f &= \chi \\ &= 3 - k = 2(1 - g). \end{aligned} \tag{12.6}$$

Note that for the three-fold coordination implied by sp^2 hybridization

$$e = \frac{3}{2} v \tag{12.7}$$

For periodic schwarzites it may be convenient to refer all quantities in Eq. (12.6) to the unit cell, or to the unit element, when a comparison is needed to closed sp^2 forms like fullerenes and capped nanotubes, whereas for the statistical arguments of Section 5, Eq. (12.6) shall refer to a large number of unit cells with cyclic boundary conditions. By calling f_n the number of n-membered rings (hereafter called n-rings) and inserting Eq. (12.7) into (12.6) it is found

$$6\chi = \sum\nolimits_n (6 - n) f_n. \tag{12.8}$$

Note that this conditions on the numbers of different rings is *independent of the number of 6-rings*, which can therefore be any natural number (except 1 (Meija 2006)).

Hereafter we shall restrict to the class of sp^2 structures with only 5-, 6- and 7-rings (5-6-7 class), unless the cases of schwarzites with 8- or 9-rings are explicitly stated. From the pure topological standpoint the extension to other possible structures with larger or smaller rings is indeed straightforward. For 5-6-7 structures it is

$$f_7 - f_5 = -6\chi. \tag{12.9}$$

For fullerenes ($\chi = 2$) with no 7-rings the well known result $f_5 = 12$ is obtained. For open nanotubes and graphene sheets $f_5 = f_7$: in perfect structures this number can be zero, whereas in defective graphene or nanotubes 5- and 7-rings always occurs in pairs, e.g., through the Stone-Wales transformation which converts four adjacent 6-rings into two 5–7 ring pairs. For D- and P-type schwarzites with no 5-rings $f_7 = 24$ in each unit cell ($f_7 = 12$ per element for the D-type schwarzite) (Fig. 12.8a, b). The smallest D-type schwarzite in this class has twelve 7-rings per element and no 6-ring, which makes 28 atoms per element, to be compared with the smallest fullerene C_{20}. This schwarzite, denoted *fcc*-$(C_{28})_2$, together with the fullerene C_{20}, are examples of *platonic tiling*, made of only one kind of polygons. Larger D-type schwarzites of the 6–7 class are obtained by adding 6-rings and have the formula *fcc*-$(C_m)_2$ where $m = 28 + 2f_6$ ($f_6 \neq 1$) is the number of atoms per element, while the simple-cubic P-type schwarzites of the 6–7 class shall be denoted by *sc*–C_m with $m = 56 + 2f_6$. Schwarzites as well all fullerenes with two kinds of polygons are examples of *archimedean tiling*.

Equation (12.8) holds also for odd values of χ, i.e., for a steric distribution of bonds and rings lying on a one-face surface, e.g., a Möbius ring ($\chi = 1$, Fig. 12.7f). In this case $f_5 - f_7 = 6$. An example of tiling is shown in Fig. 12.7g, where the two edges 1–2 are supposed to be joined and a bond 3–4 is added. With the rule that adjacent rings have only one edge in common, the bond distribution of Fig. 12.7g yields $f_5 = 7$ and $f_7 = 1$.

D-type schwarzites of the 6–8 class have, according to Eq. (12.7), six 8-rings per element and the general formula fcc-$(C_l)_2$ with $l = 16 + 2f_6 \quad (f_6 \neq 1)$. The smallest (platonic) form has 16 atoms per element. The polybenzenes studied by O'Keeffe et al. (1992) belong to this class and are obtained by inserting one 6-ring in each plane normal to a three-fold {111} axis. Another example, obtained by the insertion of three 6-rings in each plane normal to a {111} axis – altogether 24 6-rings to give fcc-$(C_{64})_2$ – is briefly discussed in Section 12.6. According to Eq. (12.7), also schwarzites of the 6–9 class can exist, the smallest of which is made of four 9-rings and 12 atoms per element. The general formula is fcc-$(C_j)_2$ with $j = 12 + 2f_6 \quad (f_6 \neq 1)$. To our best knowledge no specific study on the stability of these structures is available.

The minimal gyroid surfaces which support the above schwarzite structures can be analytically described by the Weierstrass-Enneper representation in the complex plane (Hyde 1999; Hoffman 1996). The P- and D-type gyroids are just two special cases, where one can be continuously transformed into the other by the Bonnet transformation (Hyde 1999). Thus the simulations of TEM images with deformed P- or D-type gyroids, discussed in the next section, can be easily extended to any intermediate case obtained by a Bonnet transformation. The shapes of P- and D-type gyroids (Fig. 12.8a, b) are well approximated by the lowest terms of a Fourier expansion as

$$\cos x + \cos y + \cos z = 0, \tag{12.10}$$

$$\cos x \cos y \cos z + \sin x \sin y \sin z = 1, \tag{12.11}$$

respectively, where the coordinates x, y, z are in units of some conventional length, say $a_0 = 1$ nm. The spongy carbon structures like those shown in Figs. 12.4b and 12.5) are examples of random schwarzites, which can be obtained from a numerical simulation, as done, e.g., by Lenosky et al. (Lenosky et al. 1992; Townsend et al. 1992). Another method to generate images of apparently random schwarzites is to simulate their growth processes by applying to a three-periodic gyroid surface a continuous scale change along the growth direction (z axis), so as to mimic the observed self-affinity. This can be done for a P-type schwarzite by introducing in Eq. (12.10) a scaling factor z^{β} so as to give a distorted surface obeying the equation

$$\cos(xz^{-\beta}) + \cos(yz^{-\beta}) + \cos\left(\frac{z^{1-\beta}}{1-\beta}\right) = 0, \tag{12.12}$$

with $\beta = 1/2$ taken from experiment. A comparison of a portion of the TEM image of Fig. 12.5 with a portion of the surface given by Eq. (12.10) plotted with a contrast and field depth similar to that of the TEM image is shown in Fig. 12.9; a similar comparison between another portion of the TEM image and a small part of two compenetrating distorted D-type schwarzites is shown in Fig. 12.10. A visually similar image could be obtained with a single D-type surface with a longer field depth. There is a clear visual resemblance between the experiment and the distorted P-type simulation, whereas the comparison with the image of the compenetrating D-type surfaces is less convincing. The latter, however, shows certain quasi-circular features which are seen in the TEM image but not in the P-type simulation.

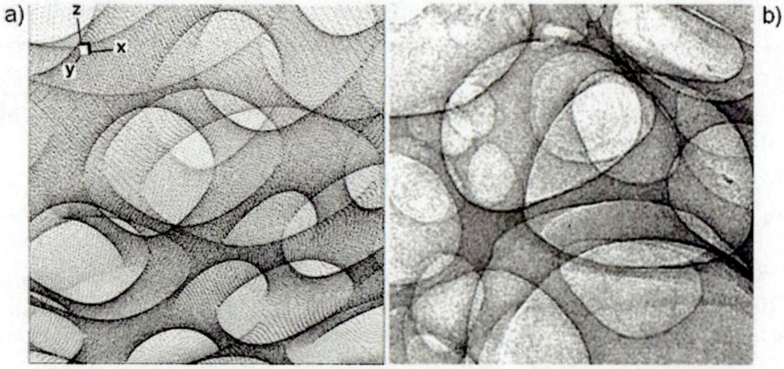

Fig. 12.9 Comparison between a portion of a distorted ($\beta = 1/2$) P-type schwarzite, Eq. (12.12) (**a**) and of a random carbon schwarzite as observed by TEM (**b**). The contrast of the simulated image has been chosen so as to give a field depth comparable to that of the TEM image (Barborini et al. 2002a)

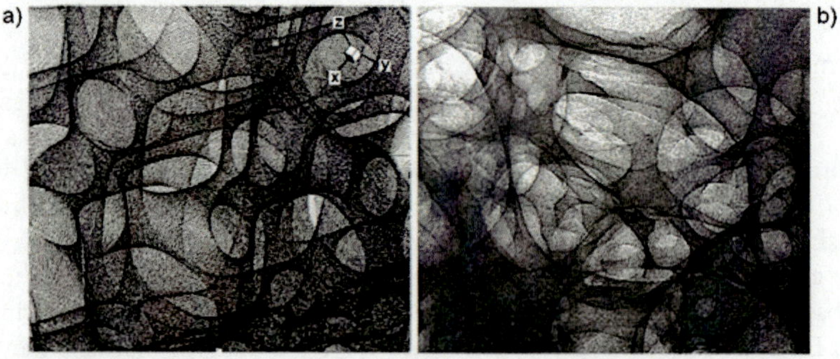

Fig. 12.10 Comparison between a portion of two compenetrating distorted ($\beta = 1/2$) D-type schwarzites (**a**) and another portion of a random carbon schwarzite as observed by TEM (**b**). The contrast of the simulated image has been chosen so as to give a field depth comparable to that of the TEM image. A similar correspondence would be obtained with a single distorted D-type surface with either longer field depth

A brief discussing on whether the scale distortion preserves minimality is in order. The minimality condition for a surface represented by the equation $x = x(y, z)$ is fulfilled when (Osserman 1986)

$$\left[1+\left(\frac{\partial x}{\partial z}\right)^2\right]\frac{\partial^2 x}{\partial y^2} - 2\frac{\partial x}{\partial y}\frac{\partial x}{\partial z}\frac{\partial^2 x}{\partial y \partial z} + \left[1+\left(\frac{\partial x}{\partial y}\right)^2\right]\frac{\partial^2 x}{\partial z^2} = 0, \qquad (12.13)$$

which corresponds to a vanishing mean curvature at any point of the surface. The transformation in Eq. (12.12), defined by

$$x' = xz^{-\beta}, \quad y' = yz^{-\beta}, \quad z' = z^{1-\beta}/(1-\beta) \qquad (12.14)$$

is seen to violate Eq. (12.13) by terms of order β/z because

$$\frac{\partial x}{\partial y} = \frac{\partial x'}{\partial y'}, \quad \frac{\partial x}{\partial z} = \frac{\partial x'}{\partial z'} + o(\beta z^{-1}), \quad \frac{\partial^2 x}{\partial y^2} = \frac{\partial^2 x'}{\partial y'^2}z^{-\beta}, \qquad (12.15)$$

$$\frac{\partial^2 x}{\partial z^2} = \frac{\partial^2 x'}{\partial z'^2}z^{-\beta} + o(\beta z^{-1})z^{-\beta}, \quad \frac{\partial^2 x}{\partial y \partial z} = \frac{\partial^2 x'}{\partial y' \partial z'}z^{-\beta} + o(\beta z^{-1})z^{-\beta}, \qquad (12.16)$$

whereas the principal curvatures decrease like $z^{-\beta}$. Thus for $\beta < 1$ the minimality condition is slowly recovered in the distorted schwarzite model at sufficiently large z.

12.4 Schwarzite Stability

Once established the forms of pure sp^2 carbon which are allowed by topology, the second general question is under which conditions schwarzites, rather than nanotubes or fullerenes, are produced in a catalyzed SCBD experiments. To answer this question one needs first to consider the total energy of a curved single-walled sp^2 carbon as a function of its geometry. In a previous study (Benedek et al. 2003) it has been suggested that a good approximation to the total energy of sp^2 carbon surfaces is provided by the Helfrich's form for membranes and foams (Helfrich 1973; Oguey 1999; Sullivan 1999):

$$E = \int_A dA(\gamma + \kappa H^2 - \overline{\kappa} K), \qquad (12.17)$$

where A is the (portion of the) surface which the total energy refers to,

$$H = \frac{1}{2}\left(\frac{1}{R_1} + \frac{1}{R_2}\right) \quad K = \frac{1}{R_1 R_2}, \qquad (12.18)$$

are the mean and gaussian curvatures, respectively, with R_1 and R_2 the principal radii of curvature, $\gamma = 2.82$ eV/Å^2 is the energy for unitary *flat* surface (a free-standing graphene sheet) (Sullivan 1999), κ and $\bar{\kappa}$ are stiffness constants associated with cylindrical and elliptical/hyperbolic deformations of the surface, respectively.

Since both H^2 and K are inversely proportional to a square length, only the term in γ in Eq. (12.17) is extensive, whereas the two curvature terms depend on the *aspect ratio* and the *connectivity*, respectively, and not on the length scale. The latter property is a consequence of the Gauss-Bonnet theorem (Osserman 1986). A corollary of this theorem states that *for any closed orientable surface S of genus g for which a representation of class r > 3 exists the surface integral of the Gaussian curvature K is given by*

$$\int_A K\,dA = 2\pi\chi. \tag{12.19}$$

Minimal surfaces are characterized by $R_1 = -R_2$ at all positions, and therefore the total energy of a schwarzite is readily obtained from Eq. (12.19) as

$$E_{schw} = \gamma A - 2\pi\chi\bar{\kappa}. \tag{12.20}$$

Density functional (DF) calculations for single-walled nanotubes of variable radius (Sullivan 1999; White et al. 1993), and C_{60} (Sullivan 1999) yield for all structures about the same stiffness constants: $\kappa \cong 3.1$ eV and $\bar{\kappa} \cong 1.7$ eV. Other calculations for graphene (Cadelano et al. 2009; Lu et al. 2009) and nanotubes (Kudin et al. 2001) give κ ranging from 2.80 to 2.92 eV. Consistently a value of $\bar{\kappa} = 1.5$ eV can be extracted by means of Eq. (12.20) from the available calculations of the cohesive energy of schwarzites (Lenosky et al. 1992; Vanderbilt and Tersoff 1992; O'Keeffe et al. 1992; Gaito et al. 2001). This shows that the Helfrich's form for the total energy approximately holds also for all forms of pure sp^2 carbon with universal values of the stiffness constants.

The total energy expressed by Eq. (12.17) has the important property, if $\bar{\kappa}$ is constant, of having *a stable local minimum for a minimal surface*, since for $\kappa > 0$ the integral on H^2 is always positive unless $H = 0$, while the integral over the Gaussian curvature K is, according to Gauss-Bonnet theorem, independent of any small continuous deformation of the surface. Thus sp^2 carbon taking the shape of a minimal surface like schwarzites are stable forms (up to effects of the contour where $\bar{\kappa}$ may change, as discussed below). If the negative disclinations yielding a negative Gauss curvature are exclusively due to heptagons, the number of disclinations N_d is fixed by the Euler-Poincaré characteristic as $N_d = 6(2 - \chi)$ independently of the length scale of the surface (Sadoc 1997).

For a free-standing surface the stiffness constants κ and $\bar{\kappa}$ exclusively depend on the electronic structure associated with the sp^2 hybridization, but they can be modified locally at the edges of the carbon surface or along the line where it docks at the substrate or, more significantly, at a catalyst nanoparticle. It is argued that such local values of κ and $\bar{\kappa}$, and the initial values of the curvature radii as determined by any local geometric constrain may give general indications about whether the growth process of sp^2 carbon will preferentially lead to fullerenes, nanotubes or schwarzites. A local energy minimum with respect to the local aspect ratio, expressed by R_1/R_2, implies a relationship between the ratios of the curvature radii and of the stiffness constants:

$$\frac{R_1}{R_2} = \frac{2\bar{\kappa}}{\kappa} - 1. \tag{12.21}$$

The surface deformation energy densities for (spherical) fullerenes ($R_1 = R_2 \equiv R$), straight nanotubes ($R_1 \equiv R, R_2 \to \infty$) and schwarzites ($R_1 = -R_2 \equiv R$) are $(\kappa - \bar{\kappa})/R^2, \kappa/4R^2$ and $\bar{\kappa}/R^2$, respectively, and therefore for any given R_1 the values of κ and $\bar{\kappa}$ define three different topological domains: schwarzites are favored for $\bar{\kappa} < {}^1\!/_4\kappa$, nanotubes for ${}^1\!/_4\kappa < \bar{\kappa} < {}^3\!/_4\kappa$ and fullerenes for $\bar{\kappa} > {}^3\!/_4\kappa$ (Fig. 12.11). From Eq. (12.21) it is seen that in each domain the surface energy minimum lines are $\bar{\kappa} = \kappa$ for spherical fullerenes, $\bar{\kappa} = \kappa/2$ for straight nanotubes (Fig. 12.11, broken lines) and $\bar{\kappa} = 0$ for minimal schwarzites. Arguably away from the broken lines or $\bar{\kappa} = 0$ non-spherical fullerenes, bent nanotubes or non-minimal schwarzites are favored. The calculated values of $\bar{\kappa}$ and κ (Fig. 12.11, red cross) fall in the nanotube/graphene domain close to the $\bar{\kappa} = \kappa/2$ line, thus explaining why it is relatively easy to grow straight nanotubes.

It should be remarked that the local values of $\bar{\kappa}$ and κ, either at the surface termination into vacuum, where the growth takes place by cluster addition, or at the contact with a catalyst, are likely to be different from the calculated values for the free-standing structures. One should consider that the local change in the electronic structure, e.g., a π bond-charge depletion or accretion, can substantially modify $\bar{\kappa}$. The charge redistribution produced by a catalyst depends on the actual size of catalyst nanoparticles, which may explain why the growth of schwarzites supersedes that of nanotubes when metallorganic precursors are used. In this case the metallic particles are in general very small and highly dispersed, which can preserve over an extended region the value of $\bar{\kappa}$ appropriate to grow schwarzites. Another important remark is that once the growth has started in one domain it is very unlikely that the system jumps into another domain since this would require, for *topological reasons*, a prohibitive reshuffling of bonds. For example, jumping from the fullerenes (schwarzite) to the nanotube domain implies the annihilation of twelve 5-(7-) membered rings, which makes the energy hills between valleys rather high.

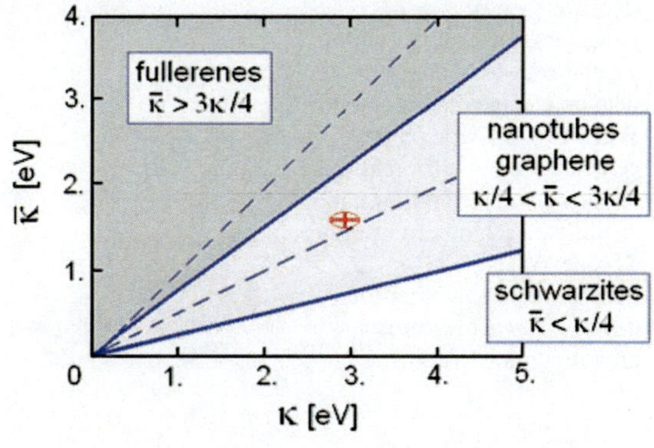

Fig. 12.11 The stability regions of sp^2 carbon surfaces as functions of the two stiffness constants. The *broken lines* indicate the minima of the surface energy for quasi-spherical fullerenes and nanotubes. The minimum surface energy for schwarzites would occur at $\bar{\kappa} = 0$

According to this picture an initial $\overline{\kappa} < {}^1\!/_4\kappa$ at the catalyst-carbon contact line determines the growth of a schwarzite, which continues despite the gradual change of the ratio $\overline{\kappa}/\kappa$. It may be thought that during the sequential addition of fragments to the edge of the growing schwarzite atoms adjust so as to fulfill as much as possible the minimality condition $R_1 = -R_2$, which means to keep as close as possible to the bottom of the surface energy valley. Would the local property $H = 0$ be exactly fulfilled at each addition of material, the growth could be viewed in this particular case as a deterministic process with the minimum production of entropy. This descends directly from the beautiful Euler theorem linking an extensive property, such as the area of a surface, to an intensive one, the differential conditions $H = 0$ (i.e., Eq. (12.13)), and may apply to any other physical problem whose total (free) energy minimum can be represented by the area of a minimal surface.

These arguments only tell about the topology of the growing structures with no information about shape, order and symmetries. This requires a thermodynamic approach where the appropriate thermodynamic potential is considered. The ordering depends very much on the mutual interactions between rings of different sizes. In fullerenes abutting 5-rings are not favored; it may be said that they repel each other, which makes the single isomer of C_{60} with no abutting 5-rings particularly stable, beautifully ordered and highly symmetrical. On the other hand total energy calculations on schwarzites (Gaito et al. 2001; D'Alessio 2007) show that the bond energy between two 7-rings is $\varepsilon_{77} = -5.107$ eV, between a 6- and a 7-ring $\varepsilon_{67} = -5.181$ eV and between two 6-rings $\varepsilon_{66} = -5.587$ eV. Thus the separation of two abutting 7-rings costs 0.332 eV and therefore abutting 7-rings are favored (D'Alessio 2007). Figure 12.12 shows a small region of a schwarzite where some re-shuffling of bonds

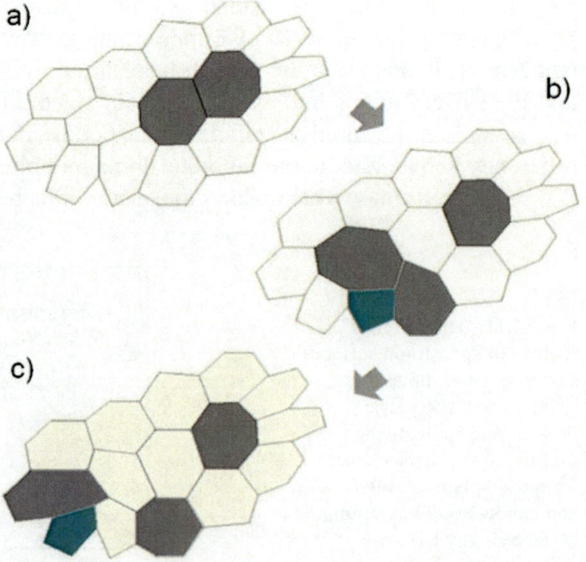

Fig. 12.12 Two abutting 7-rings (*a, dark polygons*) are separated by a re-shuffling of bonds at the cost of the creation of a 5–7 ring pair (*b,c; dark-green pair of polygons*). This occurs within a small schwarzite region with no change of its contour. When two 5–7 ring pairs generated in two different regions get adjacent they can annihilate each other through a Stone-Wales transformation, leaving four 6-rings. The net result is a migration of a 7-ring

leads to the separation of two initially abutting 7-fold rings with no change of the contour. This costs the formation of a 5–7 ring pair. However two 5–7 ring pairs formed in two different regions of the sample which become adjacent can annihilate through a Stone-Wales transformation leaving four 6-rings. The net result is a separation of two 7-rings. This process, however, is energetically less favored than the coalescence of two separated 7-rings, which may explain why random schwarzites seem to be more likely than well ordered three-periodic structures. The bond reshuffling implies however a change of configurational entropy, which requires some further discussion.

12.5 Thermodynamics

As a simple example, the configurational entropy associated with all possible distributions of a fixed number f_7 of 7-rings in a schwarzite of given χ can be estimated for a restricted class of schwarzites, for example for the D-type ($\chi_{el} = -2$). The number of possible configurations is given by the number of isomers within this restricted class. The calculation for the smallest D-type schwarzites ($(C_{28})_2$ to $(C_{40})_2$) has been approached (D'Alessio 2007) by adapting to schwarzite elements the spiral sequencing method originally developed for fullerenes by Manolopoulos, Fowler et al. (Yoshida and Fowler 1992; Manolopoulos and Fowler 1992; Fowler et al. 1995; Manolopoulos and Fowler 1997; Achida et al. 1998; Làszlò et al. 2001) for the enumeration of isomers and spectral analysis. The results for small schwarzites can be extrapolated to larger samples by means of some simple combinatorial argument subject to the further restriction that the D-type elements are connected by six-atom necks, so that an isolated element only contains 6- and 7-rings. This is clearly a crude approximation which retains however a tutorial value and is worth discussing here. For a D-type element of fcc-$(C_m)_2$ with $f_{6,el} \equiv (m-28)/2$ and $f_{7,el} = 12$, the number $W_{el}(f_{6,el})$ of isomers per element grows like the number of combinations of 7- and 6-rings (the latter being $f_{6,el}+2$ for including one half of the four necks), divided by the number of permutations of the four necks, by the number (3) of possible ways of closing the element on itself (Fig. 12.7d) and by 2, since the distinction between the internal and external surfaces of a schwarzite (extroversion isomery) is irrelevant. This gives

$$W_{el}(f_{6,el}) \approx \frac{1}{144}\binom{14+f_{6,el}}{12} \tag{12.22}$$

For example, for $f_{6,el} = 2, 4, 6$ Eq. (12.22) gives $W_{el}(f_{6,el}) = 12,\ 128,\ 874$ which compare fairly well with the exact figures 11, 125 and 893 (D'Alessio 2007).

Consider now a D-type schwarzite of N elements closed by cyclic boundary conditions. It has a Euler-Poicaré characteristic $\chi = -2(N-1)$ and a total area $A = N(f_{6,el}A_6 + 12A_7)$, where A_n is the area of an n-ring. The total number of configurations is then $W = (W_{el})^N$, which gives an entropic contribution

$$S_c/k_B = \ln W = \left(1 + \frac{1}{2}|\chi|\right) \ln W_{el}(f_{6,el}). \tag{12.23}$$

By introducing via the Gauss-Bonnet theorem, Eq. (12.19), the average Gauss curvature

$$\overline{K} \equiv 2\pi\chi/A \tag{12.24}$$

and assuming a constant area and temperature T, the corresponding (configurational) Helmholtz free energy per unit area can be written via Eq. (12.20) as

$$\frac{F_c}{A} = \gamma + \overline{\kappa}\,|\overline{K}| - \frac{k_B T}{f_{6,el}A_6 + 12A_7} \ln W_{el}(f_{6,el}) \tag{12.25}$$

Since the mean pore radius $\overline{R} = \overline{K}^{-1/2}$ depends on the number of 6-rings, one can obtain $\overline{R}$ at thermal equilibrium by setting $(\partial F_c/\partial f_{6,el})_{A,T} = 0$, which gives for large $f_{6,el}$ and $\beta \equiv 1/k_B T$:

$$f_{6,el} \approx \exp(\beta\pi\overline{\kappa}/3), \quad \overline{R} \approx (A_6/4\pi)^{1/2} \exp(\beta\pi\overline{\kappa}/6). \tag{12.26}$$

These equations show that for increasing temperature $\overline{R}$ decreases which means that the porosity increases. The present equilibrium description can hardly adapt to the SCBD process, unless it is assumed that at the spot hit by the beam there is a defined average temperature proportional to the flux and to the energy per atom. Although the travelling clusters in the supersonic beam are very cold, at the impact on the surface their translational kinetic energy shares among all degrees of freedom. Thus the vibrational contribution to the free energy has to be added. This amounts to (Horton and Maradudin 1975)

$$\frac{F_v}{A} = \frac{k_B T}{f_{6,el}A_6 + 12A_7} \sum\nolimits_j [x_j \coth x_j - \ln(2\sinh x_j)], \quad x_j \equiv \frac{1}{2}\beta\hbar\omega_j, \tag{12.27}$$

where $\hbar\omega_j$ are the phonon energies of an element. Since only sums over the whole phonon spectrum are of interest in the present discussion, the same average set of frequencies is used for all elements, and the four elements coordinated by the central one are assumed to be rigid. The last condition ensures a finite energy for all the acoustic modes of the whole schwarzite, whose branches are replaced by the respective top energies. This treatment of the acoustic modes is similar to a Debye approximation, which works quite well in the calculation of thermodynamic functions at comparably high temperatures as in the present case.

The phonon spectrum has been calculated for the smallest D-type schwarzites with a simple nearest-neighbour force constant model with radial and shear force constants, which are taken the same for all bonds and equal to those of graphite (Benedek and Onida 1993; Benedek et al. 1993). In this case the dynamical matrics can be block-diagonalized into three adjacency matrices (Manolopoulos et al. 1991; Manolopoulos and Fowler 1992; Fowler et al. 1995; Achida et al. 1998) whose

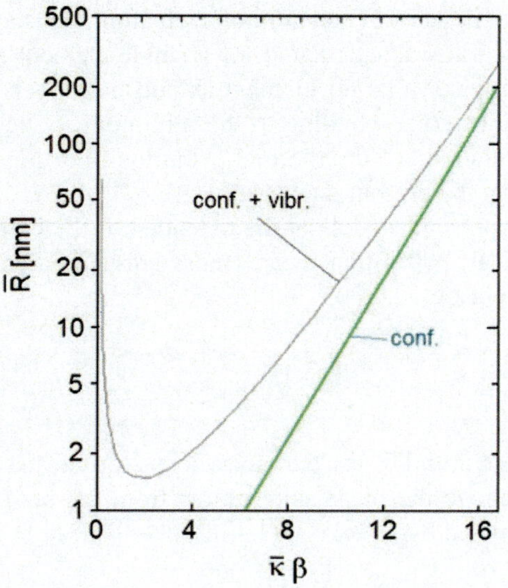

Fig. 12.13 Mean radius (*on a logarithmic scale*) of the pores of a d-type schwarzite with a random distribution of the twelve 7-rings per element calculated as a function of the dimensionless parameter $\bar{\kappa}\beta = \bar{\kappa}/k_B T$ temperature from the Helmholtz free energy including both configurational and vibrational contributions (*curve*) or the configurational contribution only (*straight line*). When phonon contributions are included the mean radius shows a minimum of about 15 nm for $T \cong \bar{\kappa}/2k_B$ and a rapid increase above this temperature. According to molecular dynamics simulations (Section 12.7) this behaviour can be interpreted as due to a rapid graphitization that schwarzite undergoes above 3500 K, before melting. This temperature corresponds to $\bar{\kappa} \cong 0.6$ eV which is less than one half of the ab-initio value but falls into the schwarzite domain (Fig. 12.11)

eigenvalues give the phonon energies for the different polarizations. Model force constants for sp^2 carbon given as functions of the bond length are available for a more precise calculation of the phonon energies (Benedek et al. 1993). This however is not needed at the level of approximation adopted in this discussion.

Figure 12.13 displays the mean pore radius calculated as a function of the dimensionless parameter $\bar{\kappa}\beta = \bar{\kappa}/k_B T$ from the Helmholtz free energy including both configurational and vibrational contributions (curve) or the configurational contribution only (straight line). When the phonon contributions are included the mean radius shows a minimum for $T \cong \bar{\kappa}/2k_B$ corresponding to about 15 nm, and a rapid increase above this temperature. According to molecular dynamics simulations described below this behaviour can be interpreted as due to a rapid graphitization that schwarzite undergoes above 3500 K, before melting. This temperature corresponds to $\bar{\kappa} \cong 0.6$ eV which is less than one half of the ab-initio value but falls into the schwarzite domain (Fig. 12.11). It is also noted that the experimental schwarzite shown in Fig. 12.5 has pore radii in the range of 100 nm, which corresponds to a formation temperature of 500 K for $\bar{\kappa} \cong 0.6$ eV or 1170 K for $\bar{\kappa} \cong 1.4$ eV. Both values are reasonable in view of the fact that the incident energy per atom, initially associated with one translational degree of freedom, is subsequently distributed over the three vibrational degrees of freedom of each atom. For the present example the vibrational contribution was calculated for a single element size corresponding to $m = 36$; for consistency a larger element should have been used so as to have a mean curvature radius corresponding to the minimum $\overline{R}$. Nevertheless the molecular dynamics simulations of the thermal evolution of a three-periodic schwarzite qualitatively confirm the present analysis.

In principle the present model can be extended to the case of self-affinity. It is found (Benedek et al. 2005) that mean Gauss curvature decreases for increasing thickness t as (Benedek et al. 2005)

$$|\overline{K}| = |\overline{K}_o| \frac{1+\beta}{1-\beta} t^{-2\beta}, \tag{12.28}$$

and $\overline{K}_o$ is the average initial Gauss curvature. Thus for the quasi-equilibrium growth regime discussed above the mean pore radius from the pure configurational part, Eq. (12.26), is corrected by a factor $\sqrt{(1-\beta)/(1+\beta)} t^{\beta} \approx t^{\beta}/\sqrt{3}$.

12.6 Electronic Structure and Electron-Phonon Interaction

The band structure of electrons freely moving on a periodic minimal surface has been investigated by Aoki et al. (2001). When the actual sp^2-bonded atomic structure is considered, schwarzites, like nanotubes, are either metals or insulators, depending on the topological structure and element size. Tight-binding calculations of the band structure have been performed for the smallest D-type schwarzites fcc-$(C_m)_2$ of tetrahedral symmetry ($m = 28$, 36 and 40) (Gaito et al. 2001; Benedek et al. 1997, 2001). More recent *ab-initio* calculations of the electronic band structure of *fcc*-$(C_{28})_2$ are also available (Spagnolatti et al. 2003). The structure of *fcc*-$(C_{28})_2$ exists in two enantiomers of opposite chirality (Fig. 12.14a, b), having however the same band structure (Fig. 12.14c). As also seen in the density of electron states (DOS) (Fig. 12.14d), rather large gaps occur between the valence as well as between the conduction bands. The Fermi level (E_F) crosses the lowest conduction band, which confers to *fcc*-$(C_{28})_2$ a metallic character. The DOS's of the next tetrahedral schwarzites *fcc*-$(C_{36})_2$ and *fcc*-$(C_{40})_2$ as obtained by tight-binding calculations (Gaito et al. 1998) are displayed in Fig. 12.15. The occurrence of many sharp peaks in the DOS of both schwarzites is indicative of rather flat bands due to the existence of localized electronic states within each element, notably at the 7-rings, due to the absence of conjugation.

There is an interesting alternation in the conducting properties: while *fcc*-$(C_{28})_2$ is metallic, the next one, *fcc*-$(C_{36})_2$, having four 6-rings per element, is an insulator and *fcc*-$(C_{40})_2$, with six 6-rings per elements, is metallic. For increasing m larger and larger portions of the surface acquire a graphene-like structure with, however, a slight negative Gaussian curvature, similar to graphene subject to a shear strain. The effects of a shear strain on the electronic structure of graphene have been theoretically investigated in a recent paper by Cocco et al. (2010), who show that a

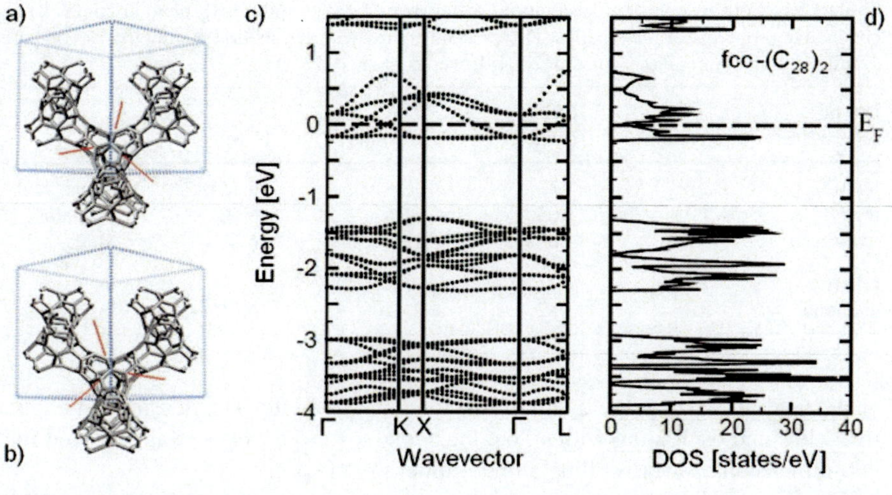

Fig. 12.14 (**a,b**) The two enantiomers of the D-type schwarzite fcc-$(C_{28})_2$ made of only 7-rings, here shown in the conventional cubic cell with four formula units. The chirality is evidenced by the directions of three bonds and the lack of mirror symmetry with respect to the (110) plane. The two enantiomers have the same electronic band structure (**c**) obtained from an ab-initio calculation (Spagnolatti et al. 2003). Comparatively large gaps occur between valence as well between conduction bands, as also seen in the density of electron states (DOS). The Fermi level (E_F) cuts the lowest conduction band, which confers to fcc-$(C_{28})_2$ a metallic character

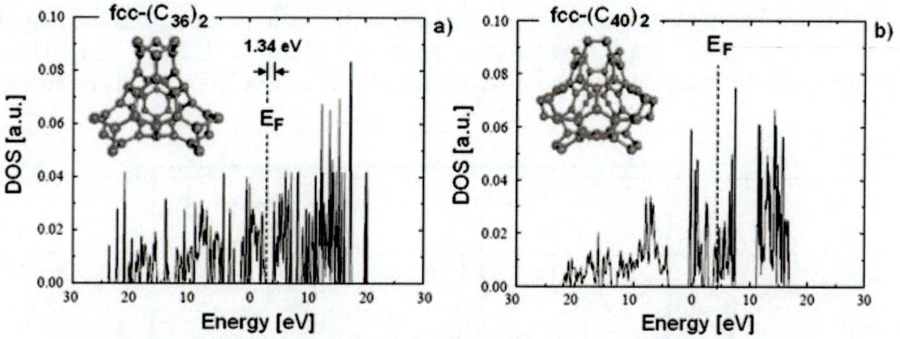

Fig. 12.15 Density of the electronic states (DOS) of the D-type schwarzites fcc-$(C_{36})_2$ (**a**) and fcc-$(C_{36})_2$ (**b**) a tight-binding calculation (Gaito et al. 2001). As seen from the position of the Fermi level (E_F), the former schwarzite is an insulator with a gap of 1.34 eV, the latter is a metal. The presence of many sharp peaks in the DOS is indicative of bands of states strongly localized within each element. The insets show the element atomic structures

shear strain opens a gap at the Dirac points, with the intriguing consequence that a band-gap engineering based on the application of suitable stress field would be possible (Cocco et al. 2010). The data of Table 12.1 show that schwarzites are particularly stable, with a cohesive energy per atom which rapidly increases in absolute

Table 12.1 Cohesive energy per atom (E_{coh}), density, bulk modulus (**B**), bond strength (**b**) and conductive property for the smallest D-type schwarzites with tetrahedral symmetry, as compared to fullerite and diamond (Gaito et al. 2001; Benedek et al. 1997, 2001)

D-type schwarzite	E_{coh} (eV/atom)	Density (g/cm^3)	B (Mbar)	b (Mbar Å^3)	
fcc-$(C_{28})_2$	−7.66	1.33	1.58	16.12	Metal
fcc-$(C_{36})_2$	−7.71	1.05	1.26	16.20	Insulator
fcc-$(C_{40})_2$	−7.92	1.60	1.92	16.25	Metal
fullerite	−7.99	1.71	0.14	–	Insulator
diamond	−8.36	3.52	4.42	16.71	Insulator

value, tending for large *m* to that of diamond and graphite. The density is comparatively low and oscillating with *m*, but must tend to zero for $m \rightarrow \infty$ as expected for a two-dimensional surface filling a three-dimensional space.

The search of superconductivity in exotic carbon forms, e.g., in clathrates (Blase et al. 2010), has stimulated an *ab-initio* study of the vibrational structure and of the electron-phonon coupling in *fcc*-$(C_{28})_2$ of the 6–7 class and *fcc*-$(C_{64})_2$ of the 6–8 class (Spagnolatti et al. 2003). Figure 12.16 shows the calculated phonon density of states at the Γ-point of *fcc*-$(C_{28})_2$ with indications of the even-symmetry optical modes which mostly contribute to the electron-phonon interaction. Their electron-phonon coupling is explicitly indicated in meV units (if larger than 2 meV). It appears that the largest contribution comes from phonons which deform the narrow necks joining two neighbour elements. Here the Gauss curvature is the largest, which supports the conjecture made for clathrates (Blase et al. 2010) that a larger Gauss curvature should favour superconductivity. The calculation yields however a

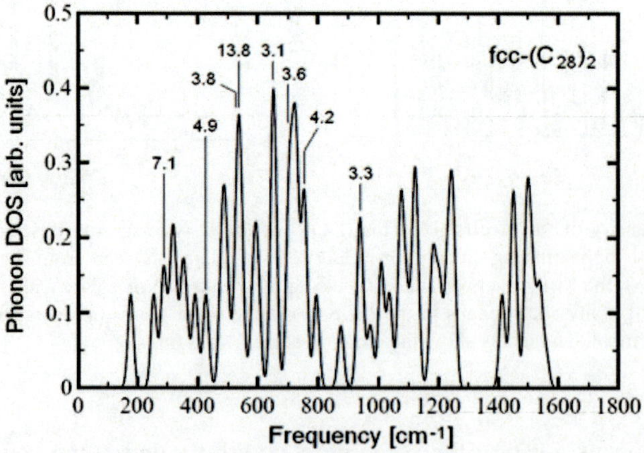

Fig. 12.16 The *ab-initio* density of phonon states at the Γ-point (zone center) of *fcc*-$(C_{28})_2$. Some of the phonon peaks are labelled by the corresponding calculated values of the electron-phonon coupling (in meV units), if larger that 2 meV (adapted from (Spagnolatti et al. 2003))

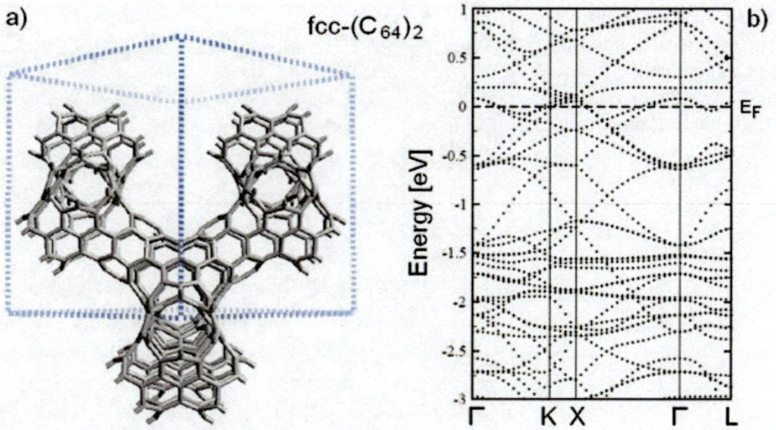

Fig. 12.17 The fcc-$(C_{64})_2$ belongs to the class of schwarzites made of 8- and 6-rings and has, according to Eq. (12.8), six 8-rings per element (**a**). The appreciable conjugation which is allowed by the even-ring structure removes the localization effects seen in schwarzites of the 6–7 class: no gap is found in the band structure (**b**), the resulting solid is metallic with a comparatively low density at the Fermi level (E_F), and the electron-phonon interaction is small

discouraging $\lambda = 0.116$ for the overall electron-phonon coupling parameter, which is quite smaller that that of doped fullerenes. Nevertheless it is conjectured that doping, by shifting E_F to regions of much higher density of states, could increase λ up to a factor five (Spagnolatti et al. 2003). For comparison a similar analysis has been carried out for the schwarzite *fcc*-$(C_{64})_2$ of the 6–8 class, which has larger and much less curved necks due to the insertion of 6-rings. The structure and the *ab-initio* electronic bands are shown in Fig. 12.17a, b (Spagnolatti et al. 2003). The first important difference with respect to the schwarzites of the 6–7 class is the disappearance of gaps. This is attributed to the appreciable conjugation which is allowed by the even-ring structure and removes the localization effects seen in 6–7 class schwarzites. The resulting metallic solid has however a comparatively low density at the Fermi level (E_F), and the resulting electron-phonon interaction is even smaller than in *fcc*-$(C_{28})_2$.

12.7 Quantum Molecular Dynamics Simulations

Thanks to the development of efficient tight-binding molecular dynamics (TBMD) methods (Colombo 2005), there have been also a few TBMD simulations of the growth and temperature evolution of low-coordinated carbon structures from cluster assembling, aiming at clarifying the conditions for schwarzite formation (Spadoni et al. 1997; Benedek et al. 1998; Yamaguchi et al. 2007; Bogana and Colombo 2007; Rosato et al. 1999, 2001; Donadio et al. 1999). In particular it has been investigated how the size distribution of clusters, their density and kinetic energy are effective in the growth of sp^2 schwarzitic material rather than mixed sp^2-sp^3 carbon or less-coordinated forms like carbynes (Rosato et al. 2001; Bongiorno et al. 2005; Agarwal

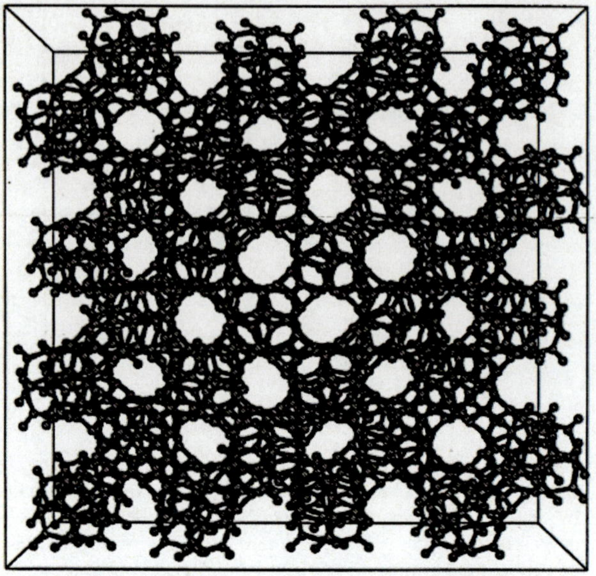

Fig. 12.18 The cell with periodic boundary conditions made of 32 unit cells of the D-type schwarzite *fcc*-$(C_{36})_2$ for a tight-binding molecular dynamics simulation (Rosato et al. 1999, 2001)

et al. 2010). Here we briefly discuss just one particular simulation which shows the useful information which topology can provide in the molecular dynamics of a complex structure.

A D-type *fcc*-$(C_{36})_2$, represented in Fig. 12.18 inside a simulation cell of 32 unit cells (64 elements) with cyclic boundary conditions, is gradually heated from room temperature to 4250 K. The evolution is monitored through the topological connectivity (Fig. 12.19a), evaluated over a subunit of five unit cells (10 elements). As shown in Fig. 12.19b, the initial connectivity of the subunit is $k = 19$ and is slowly reduced by increasing the temperature down to 16 due to some bond breaking and reshuffling. Just above 3800 K the connectivity drops to 3, thus signalling a rapid graphitization of the schwarzite. It would be quite hard to visualize what looks to be a topological phase transition by just examining the simulation snapshots. On the contrary the change of connectivity, as derived from Eq. (12.6) by counting at each time the numbers of bonds and rings, constitutes a sort of *topological order parameter* which allows to monitor the phase change from an ordered schwarzite to a disordered graphite-like material.

Further examples of TBMD simulations illustrating the dependence of the output on the temperature variation protocol can be found in Refs. (Spadoni et al. 1997; Benedek et al. 1998). A mixture of large clusters (C_{39} fragments) dispersed in a gas of C_2 dimers, under a gradual increase of temperature from 1500 to 3500 K starts coalescing until they form a single connected cluster. The structure, reminding of a random schwarzite, shows the formation of some 7-ring associated with a negative Gaussian curvature. For comparison the cluster coalescence at constant

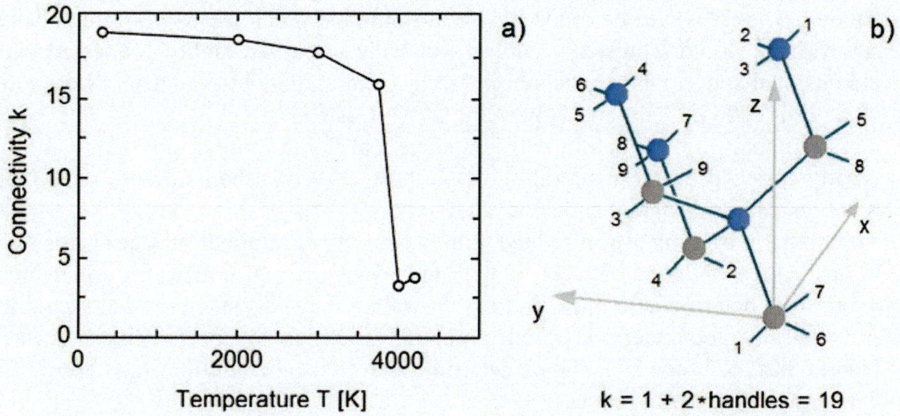

Fig. 12.19 (**a**) Tight-binding molecular dynamics simulation of the graphitization of the schwarzite fcc-$(C_{36})_2$ (Fig. 12.14). The evolution for increasing temperature is monitored by the connectivity of a subunit of eight elements (four molecules) (**b**). Since nine handles are required to close the eight-element subunit on itself (by joining the corresponding numbers in (**b**)), the low-temperature connectivity is 19. The graphitization occurs slightly below 4000 K, where the connectivity suddenly drops from $k = 16$ to the graphene value $k = 3$, thus depicting a topological phase transition (adapted from Rosato et al. 1999, 2001)

temperature of 3500 K leads instead to a graphite-like structure dominated by 6-membered rings. Other simulations starting from a gas of only carbon dimers yield open tubular structures with the corresponding connectivity $k = 3$. It is hoped that larger and larger-scale quantum simulations will help finding a viable route to the new world of crystalline schwarzites.

12.8 Conclusion

Although schwarzitic carbon sponges did not know the glamour of the ordered forms of sp^2 carbon, they have nevertheless led to important applications, some of which have been mentioned in the introduction. Countless examples may be found in the literature, ranging from the engineering of SCBD carbon-based composites (Bongiorno et al. 2005) to biological applications, one for all the recent demonstration of interfacing live cells with nanocarbon substrates (Agarwal et al. 2010). The path towards low-dimensional carbon for nanotechnologies has now reached the still poorly known world of pure carbon chains, carbynes, which were looked for since the time of fullerene discovery (Kroto et al. 1985). The recent production of carbynes by SCBD (Ravagnan et al. 2002, 2007) and from graphene (Jin et al. 2009) is calling for new theoretical investigations (Ravagnan et al. 2009). However the real challenge for possible developments on more fundamental questions is, in our opinion, the synthesis of ordered three-periodic schwarzites or even supported planar architectures formed by two-periodic schwarzites, eventually obtained by joining nanotubes, as it would be made by a nano-plumber. Such a new class of ordered sp^2 carbons would represent a natural extension to curved highly connected two-dimensional spaces of what has been learnt from and about graphene (Novoselov et al. 2004, 2005; Geim and Novoselov 2007; Castro Neto et al. 2009). The fascinating mutual implications linking graphene and Dirac fermion physics would be greatly enriched by the exploration of periodic curved carbon surfaces once their topology can be designed and controlled.

Another intriguing aspect related with the growth of minimal surfaces is the one-to-one correspondence between global and local minimal conditions which may allow for a deterministic growth along the valley floor of the energy landscape. In principle any global thermodynamic potential which can be represented as a surface integral has minima which can be determined from local conditions, and allows for a deterministic growth process.

Carbon, as the most versatile element of the periodic table, keeps stimulating the ingenuity of versatile scientists. The invention of new *possible* sp^2 structures by means of powerful mathematical tools, like, e.g., those recently investigated by Diudea (2005) (Fig. 12.20), as well as the great excitement started with the isolation of a single graphene sheet (Novoselov et al. 2004, 2005; Geim and Novoselov 2007; Castro Neto et al. 2009), now raised to the rank of paradigm linking distant areas of physics, suggest that fantasy joined to the rigorous thought are the lifeblood of science.

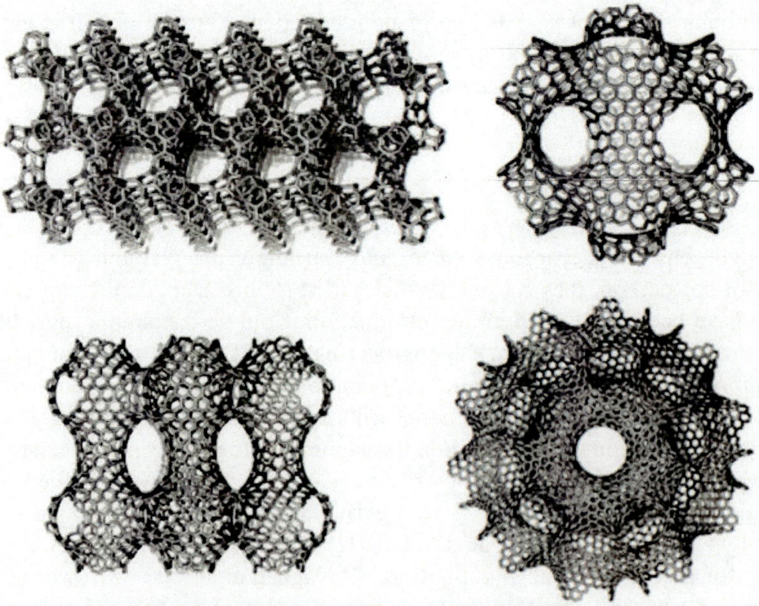

Fig. 12.20 A few examples of nanoporous carbon allotropes designed by Mircea Diudea (2005) by septupling map operations

Acknowledgement One of us (GB) gratefully acknowledges the Ikerbasque Foundation for support within the project ABSIDES. We acknowledge many stimulating discussions with Prof. Paolo Milani (University of Milano).

References

Achida Y, Fowler PW, Mitchell D, Zerbetto F (1998) J Phys Chem A 102:6835
Agarwal S, Zhou X, Ye F, He Q, Chen GCK, Soo J, Boey F, Zhang H, Chen P (2010) Langmuir Lett 26:2244
Aoki H, Koshino M, Takeda D, Morise H, Kuroki K (2001) Phys Rev B 65, 035102
Arčon D, Jagličič Z, Zorko A, Rode AV, Christy AG, Madsen NR, Gamaly EG, Luther-Davies B (2006) Phys Rev B 74:014438
Barabàsi A-L, Stanley HE (1995) Fractal concepts in surface growth. Cambridge University, Cambridge
Barborini E, Piseri P, Milani P (1999) J Phys D 32: L105
Barborini E, Piseri P, Milani P, Benedek G, Ducati C, Robertson J (2002a) Appl Phys Lett 81:3359; see also E. Gerstner, Nature materials update (2002), http://www.nature.com/materials/news/news/021107/portal/m021107-1.html
Barborini E, Siviero F, Vinati S, Lenardi C, Piseri P, Milani P (2002b) Rev Sci Instrum 73:2060
Benedek G, Bernasconi M (2004) Encyclopaedia of nanoscience and nanotechnology. Marcel Dekker, New York, NY, p 1235
Benedek G, Bernasconi M, Donadio D, Colombo L (2001b) In: Benedek G, Milani P, Ralchenko VG (eds) Nanostructured carbon for advanced applications. Kluwer, Dordrecht, p 89
Benedek G, Colombo L, Gaito S, Galvani E, Serra S (1997a) J Chem Phys 106:2311
Benedek G, Colombo L, Gaito S, Serra S (1997b) In: Paoletti A, Tucciarone A (eds) The physics of diamond. IOS, Amsterdam pp 575–598
Benedek G, Colombo L, Spadoni S, Gaito S, Milani P (1998) In: Turchi PEA, Gonis A, Colombo L (eds) Tight-binding approach to computational materials science. MRS Symposium Proceedings vol 491. MRS, Warrendale, p 529
Benedek G, Milani P, Ralchenko VG (eds) (2001a) Nanostructured carbon for advanced applications. Kluwer, Dordrecht.and papers therein
Benedek G, Onida G (1993) Phys Rev B 47:16471
Benedek G, Onida G, Righetti M, Sanguinetti S (1993) Nuovo Cim 15D:565; see also In: Bortignon PF, Broglia RA, Schrieffer JR (eds) (1994) Perspectives in many-particle physics. North-Holland, Amsterdam
Benedek G, Vahedi-Tafreshi H, Barborini E, Piseri P, Milani P, Ducati C, Robertson J (2003) Diamond Rel Mater 12:768
Benedek G, Vahedi-Tafreshi H, Milani P, Podestà A (2005) In: Beck C et al (ed) Complexity, metastability and non-extensivity. World Scientific, Singapore, pp 146–155
Blase X, Benedek G, Bernasconi M (2010) In: Colombo L, Fasolino AL (eds) Computer-based modeling of novel carbon systems and their properties, carbon materials: chemistry and physics 3, Fasolino. Springer, Berlin Heidelberg, chapter 6
Bogana M, Colombo (2007) L Appl Phys A 86:275
Bogana M, Donadio D, Benedek G. Colombo L (2001) Europhys Lett 54:72
Bongiorno G, Lenardi C, Ducati C, Agostino RG, Caruso T, Amati M, Blomqvist M, Barborini E, Piseri P, La Rosa S, Colavita E, Milani P, Nanosci J (2005) Nanotechnol 10:1
Boscolo I, Milani P, Parisotto M, Benedek G, Tazzioli F (2000) J Appl Phys 87:4005
Buzio R, Gnecco E, Boragno C, Valbusa U, Piseri P, Barborini E, Milani P (2000) Surf. Sci. 444: L1
Cadelano E, Palla PL,Giordano S, Colombo L (2009) Phys Rev Lett 102:235502
Castro Neto AH, Guinea F, Peres NMR, Novoselov KS, Geim AK (2009) Rev Mod Phys 81:109
Cocco G, Cadelano E, Colombo L (2010) Phys Rev B 81:241412 (**R**)

Colombo L (2005) Riv Nuovo Cimento 28:1
Coté M, Grossman JC, Cohen ML, Louie SG (1998) Phys Rev B 58:664
D'Alessio (2007) Thesis, Univ. Milano-Bicocca (unpublished)
Diederich L, Barborini E, Piseri P, Podestà A, Milani P, Scheuwli A, Gallay R (1999) Appl Phys Lett 75:2662
Diudea MV (2005) J Chem Inf Model 451002; see also Vizitiu AE, Diudea MV, chapter 3 of the present volume
Donadio D, Colombo L, Milani P, Benedek G (1999) Phys Rev Lett 84:776
Ferrari AC, Satyanarayana BS, Robertson J, Milne WI, Barborini E, Piseri P, Milani P (1999) Europhys Lett 46:245
Fowler PW, Manolopoulos DE, Orlandi G, Zerbetto F (1995) J Chem Soc Faraday Trans 91:1421
Gaito S, Colombo L, Benedek G (1998) Euro Phys Lett 44:525; erratum: 81:559 (2001)
Geim AK, Novoselov KS (2007) Nat Mater 6:183
Helfrich W (1973) Z Naturforsch 28:768
Hilbert D, Cohn-Vossen S (1932) Anschauliche Geometrie. Springer, Berlin
Hoffman D (1996) Nature 384:28
Horton GK, Maradudin AA (1975) Dynamical properties of solids – vol 2: crystalline solids. North Holland, Nort-Holland Amsterdam p33
Hyde ST (1999) In: Sadoc JF, Rivier N (eds) Foams and emulsions. Kluwer, Dordrecht, p 437
Iijima S (1991) Nature 324:56
Jin C, Lan H, Peng L, Suenaga K, Iijima S (2009) Phys Rev Lett 102:205501
Kajii H, Kawagishi Y, Take H, Yoshino K, Zakhidov AA, Baughman RH (2000) J Appl Phys 88:758
Kroto HW, Heath JR, O'Brien SC, Curl RF, Smalley RE (1985) Nature 318:162
Kudin KN, Scuseria GE, Yakobson BI (2001) Phys Rev B 64:235406
Kyotani T (2001) Carbon 38:269
Làszlò I, Rassat A, Fowler PW, Graovac A (2001) Chem Phys Lett 342:369
Lenardi C, Piseri P, Briois V, Li Bassi A, Bottani CE, Milani PJ(1999) J Appl Phys 85:7159
Lenosky T, Gonze X, Teter M, Elser V (1992) Nature 355:333
Li G, Luican A, Lopes des Santos JMB, Castro Neto AH, Reina A, Kong J, Andrei EY (2010) Nat Phys 6:109
Lu Q, Arroyo M, Huang R (2009) J Phys D Appl Phys 42:102002
Lu W, Chung DDL (1997) Carbon 35:427
Manolopoulos DE, Fowler PW (1992) J Chem Phys 96:7603
Manolopoulos DE, Fowler PW (1997) J Chem Soc Faraday Trans 93:3289
Manolopoulos DE, May JC, Down SE (1991) Chem Phys Lett 181:105
Meija J (2006) Analytic Bioanalytic Chem 386:4
McKay AL, Terrones H (1991) Nature 352:762
Milani P, Barborini E, Piseri P, Bottani CE, Li Bassi A (1999) Eur Phys J D 9:63
Milani P, Ferretti M, Piseri P, Bottani CE, Ferrari A, Li Bassi A, Guizzetti G, Patrini MJ (1997) Appl Phys 82:5793
Milani PE, Iannotta S (1999) Cluster beam synthesis of nanostructured materials. Springer, Berlin
Niu C, Sichel EK, Hoch R, Moy D, Tennent H (1997) Appl Phys Lett 70:1480
Novoselov KS, Geim AK, Morozov SV, Jiang D, Zhang Y, Dubonos SV, Grigorieva IV, Firsov AA (2004) Science 30:666
Novoselov AK, Geim AK, Morozov SV, Jiang D, Katsnelson MI, Grigorieva IV, Dubonos SV, Firsov A (2005a) Nature 438:197
Novoselov KS, Jiang D, Booth T, Khotkevich VV, Morozov SM, Geim AK (2005b) Proc Natl Acad Sci USA 102:10451
Oguey C (1999) In: Sadoc JF, Rivier N (eds) Foams and emulsions. Kluwer, Dordrecht, p 417
O'Keeffe M, Adams GB, Sankey OF (1992) Phys Rev Lett 68:2325
Osserman R (1986) A survey of minimal surfaces. Dover, New York, NY
Piseri P, Podestà A, Berborini E, Milani P (2001) Rev Sci Instrum 72:2261

Rao AM, Dresselhaus MS (2001) In: Benedek G, Milani P, Ralchenko VG (eds) Nanostructured carbon for advanced applications, NATO science series II, vol 24. Kluwer, Dordrecht, p 3
Ravagnan L, Manini N, Cinquanta E, Onida G, Sangalli D, Motta C,Devetta M, Bordoni A, Piseri P, Milani P (2009) Phys Rev Lett 102:245502
Ravagnan L, Piseri P, Bruzzi M, Miglio S, Bongiorno G, Baserga A, Casari CS, Li Bassi A, Lenardi C, Yamaguchi Y, Wakabayashi T, Bottani CE, Milani P (2007) Phys Rev Lett 98:216103
Ravagnan L, Siviero F, Lenardi C, Piseri P, Barborini E, Milani P, Casari CS, Li Bassi A, Bottani CE (2002) Phys Rev Lett 89:285506
Rode AV, Gamaly EG, Christy AG, Fitz Gerald JG, Hyde ST, Elliman RG, Luther-Davies B, Veinger AI, Androulakis J, Giapintzakis J (2004) Phys Rev B 70:054407; highlighted by R. F. Service, Science 304:42
Rosato V, Celino M, Benedek G, Gaito S (1999) Phys Rev B 60:16928
RosatoV, Celino M, Gaito S, Benedek G (2001) Comp Mater Sci 20:387
Rotter LD, Schlesinger Z, McCauley JP, Coustel N, Fisher JE, Smith AB (1992) Nature 355:532
Ryoo P, Joo SH, Kruk M, Jaroniek M (2001) Adv Mater 13:677
Sadoc JF (1997) In: Sadoc JF, Rivier N (eds) Foams and emulsions. Kluwer, Dordrecht, p 511
Schwarz KHA (1890) Gesammelte mathematische abhandlungen. Springer, Berlin
Seol JH, Jo I, Moore AL, Lindsay L, Aitken ZH, Pettes MT, Li X, Yao Z, Huang R, Broido D, Mingo N, Ruoff RS, Shi L (2010) Science 328:231
Spadoni S, Colombo L, Milani P, Benedek G (1997) Eur phys Lett 39:269
Spagnolatti I, Bernasconi M, Benedek G (2003) Eur Phys J B 32:181
Stoller MD, Park S, Zhu Y, An J, Ruoff RS (2008) Nano lett 8:3498
Sullivan JM (1999) In: Sadoc JF, Rivier N (eds) Foams and emulsions. Kluwer, Dordrecht p 379
Terrones H, McKay AL (1993) In: Kroto HW, Fisher JE, Cox JE (eds) The fullerenes. Pergamon Press, Oxford, p 113
Townsend SJ, Lenosky T, Muller DA, Nichols CS, Elser V (1992) Phys Rev Lett 69:921
Vanderbilt D, Tersoff J (1992) Phys Rev Lett 68:511
Wang H, Setlur AA, Lauerhaas JM, Dai JW, Seelig EW, Chang RPH (1998) Appl Phys Lett 72:2912
White CT et al (1993) In: Billups WE, Ciufolini MA (eds) Buckminsterfullerenes. VCH, New York, NY, p 125
Yamaguchi Y, Colombo L, Piseri P, Ravagnan L, Milani P (2007) Phys Rev B 76:134119
Yoshida M, Fowler PW, (1992) J Chem Phys 96:7603

Chapter 13
High π-Electronic Stability of Soccer Ball Fullerene C_{60} and Truncated Octahedron C_{24} Among Spherically Polyhedral Networks

Haruo Hosoya

Abstract By using the technique of topological symmetry the characteristic polynomials of highly symmetrical π-electron carbon networks of regular and semiregular polyhedra were factored out, and their stability on MO basis was analyzed and discussed. Besides the soccer ball-shaped C_{60} fullerene, a π-electron system of truncated octahedron-shaped C_{24} was suggested to be a candidate for a stable spherical carbon network. Mathematical basis and explanation for the high stability of these two substances among other polyhedral carbon networks with highly degenerate molecular orbitals were obtained.

13.1 Introduction

It has been established that irrespective of the grade of accuracy of the theory adopted, the soccer ball-shaped fullerene C_{60} has exceedingly high stability of conjugated π-electron network among the 1812 spherical isomers composed of 20 hexagons and 12 pentagons with all the same vertex-degree of three (Fowler and Manolopoulos 1995; Cioslowski 1995; Osawa et al. 1998). However, this conclusion has been derived from the results of a huge number of calculations. On the other hand, very few discussions have been given not only on the possibility of other types of polyhedral π-electron networks of carbon (Glukhovtsev et al. 1990; Liu et al. 1991; Sokolov and Stankevich 1993; King 1998; Ceulemans et al. 2002), but also on the mathematical and chemical ground for the stability of these types of molecules.

With regard to these problems, the present author has developed the theory of topological symmetry by which the secular determinant of a highly symmetrical π-electron network can be factored out according to a simple recipe (Hosoya and

H. Hosoya (✉)
Ochanomizu University (Emeritus), Bunkyo-ku, Tokyo 112-8610, Japan
e-mail: hosoya@is.ocha.ac.jp

F. Cataldo et al. (eds.), *The Mathematics and Topology of Fullerenes*,
Carbon Materials: Chemistry and Physics 4, DOI 10.1007/978-94-007-0221-9_13,

Tsukano 1994). This theory can be extended to the density of states of infinitely large network of graphite, or graphene, easily by solving 2 by 2 determinant followed by simple manipulation of taking its limit to infinity (Hosoya et al. 1995). In this paper, although giant fullerenes and nanotube structures are not treated, this technique is applied to various spherically polyhedral π-electron networks mostly of regular and semi-regular polyhedra. Those networks were also treated as in formal discussion, in which the degrees of the vertices of the networks exceed three.

Although the high stability of the soccer ball-shaped C_{60} fullerene was ascertained as has been expected, truncated octahedron-shaped C_{24} was suggested also to be a candidate for a stable spherical π-electron network of carbon. The results of this study might be helpful for designing new types of stable spherical π-electronic networks.

13.2 Spherically Polyhedral Networks

Among the five Platonic solids, or regular polyhedra, the following two pairs, cube 4^3 and octahedron 3^4, and dodecahedron 5^3 and icosahedron 3^5, are duals with each other, while the dual of tetrahedron 3^3 is tetrahedron itself, where the symbols for these polyhdera are taken from Cundy and Rollett (1952); Williams (1979). From these regular polyhedra almost all the Archimedean solids, or semi-regular polyhedra, can be derived and classified into three groups as in Figs. 13.1, 13.2, and 13.3.

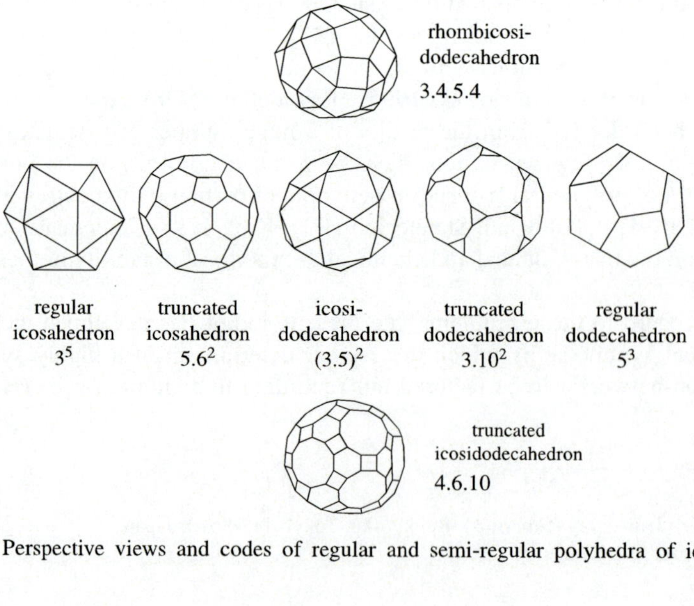

Fig. 13.1 Perspective views and codes of regular and semi-regular polyhedra of icosahedral symmetry

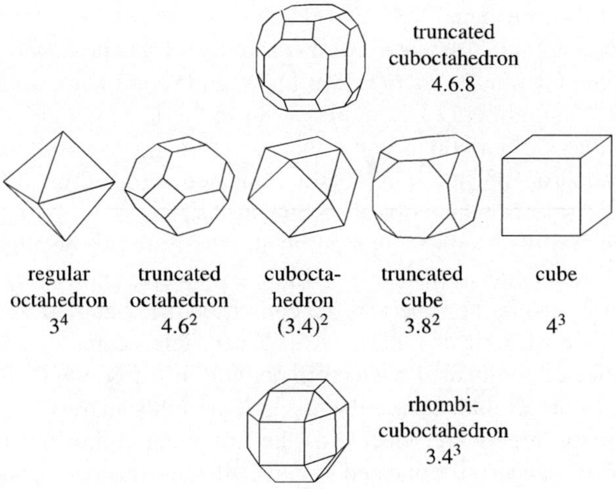

Fig. 13.2 Perspective views and codes of regular and semi-regular polyhedra of octahedral symmetry

Fig. 13.3 Perspective views and codes of regular and semi-regular polyhedra of tetrahedral symmetry

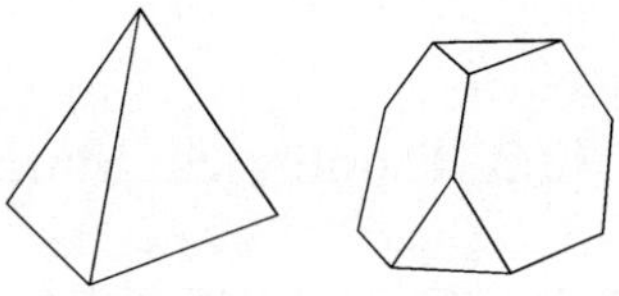

The first group polyhedra belong to I_h (icosahedral) symmetry and can be generated by symmetrical truncation of either of dodecahedron and icosahedron. The most popular one is truncated icosahedron, or soccer ball-shape, 5.6^2. Namely, a regular pentagon and a pair of consecutive hexagons are surrounding each vertex of this polyhedron, which can be obtained by truncating all the vertices of regular icosahedron in such a manner that each edge is trisected. Icosidodecahedron $(3.5)^2$ can be obtained either by bisecting each edge of 3^5 or 5^3. Truncated dodecahedron 3.10^2 can be obtained from proper truncation of 5^3. Truncated icosidodecahedron 4.6.10 and rhombicosidodecahedron 3.4.5.4 are generated from the intersection of three polyhedra, i.e., 3^5, 5^3, and rhombic tricontahedron, which is not shown here but is the dual of cuboctahedron $(3.4)^2$ (See later). Among these five semi-regular polyhedra of icosahedral symmetry, the three truncated polyhedra, i.e., 5.6^2, 3.10^2, and 4.6.10 are cubic graphs, the degrees of whose vertices are all three, and are potentially capable of forming π-electron networks of carbon atoms. On the other hand, the degrees of all the vertices of 3.4.5.4 and $(3.5)^2$ are four, thus suggesting rare chance of forming a π-electron network of carbon.

Similarly from the pair of 3^4 and 4^3, five semi-regular networks of octahedral symmetry (O_h), i.e., truncated octahedron 4.6^2, truncated cube 3.8^2, truncated cuboctahedron 4.6.8, and cuboctahedron $(3.4)^2$ and rhombicuboctahedron 3.4^3 can be generated. Among them $(3.4)^2$ and 3.4^3 have the least possibility of forming a π-electron network of carbon, because their vertex-degrees are all four.

From tetrahedron 3^3 one can obtain truncated tetrahedron 3.6^2, which is a candidate of a π-electron network of carbon atoms.

Among the family of semi-regular polyhedra there are two peculiar entities, snub cube $3^4.4$ and snub dodecahedron $3^4.5$, where six squares and twelve pentagons are, respectively, isolated in the spherical sea composed of 32 and 80 triangles. Further, the vertex-degree of both of them is five. Thus there seems to be no possibility for the existence of conjugated π-electron network of carbon with these polyhedral structures. They are excluded from the discussion in this analysis.

Then the main targets are selected as the following eight semi-regular polyhedra: truncated icosahedron, truncated dodecahedron, truncated icosidodecahedron, rhombicosidodecahedron, truncated octahedron, truncated cube, truncated cuboctahedron, and truncated tetrahedron. Beside them regular polyhedra were also supplementarily discussed.

13.3 Topological Symmetry

The general theory of the analysis using topological symmetry has already been given in our earlier paper with the soccer ball-shaped C_{60} fullerene as an example (Hosoya and Tsukano 1994). Then only the results of this fullerene will be given later. In this paper truncated dodecahedron is chosen as an example to demonstrate how the topological symmetry is obtained and taken into consideration for an efficient factorization of the secular determinant of the Hückel molecular orbitals.

13.4 Icosahedral Symmetry

The Schlegel diagram of truncated dodecahedron of D_{5h} geometrical symmetry is given in Fig. 13.4b, where one can easily find eleven decagons of two different sizes and also a larger icosagon by tracing the circle loop joining the central five small decagons. The periphery of this diagram forms the twelfth decagon. Suppose that all the edges in Fig. 13.4b are made of flexible rubber bands. Then one can pick up the above-mentioned icosagon, and enlarge it to form a large periphery as in Fig. 13.4c. Then the ten figures as in Fig. 13.4d composed of a pair of triangles are automatically pendent to the inward of the big icosagon as seen in Fig. 13.4c with a high D_{10h} topological symmetry.

This means that the 60 by 60 determinant of this network can be factored out into the product of ten six by six determinants of the cyclic unit of Fig. 13.4d, where *a* denotes

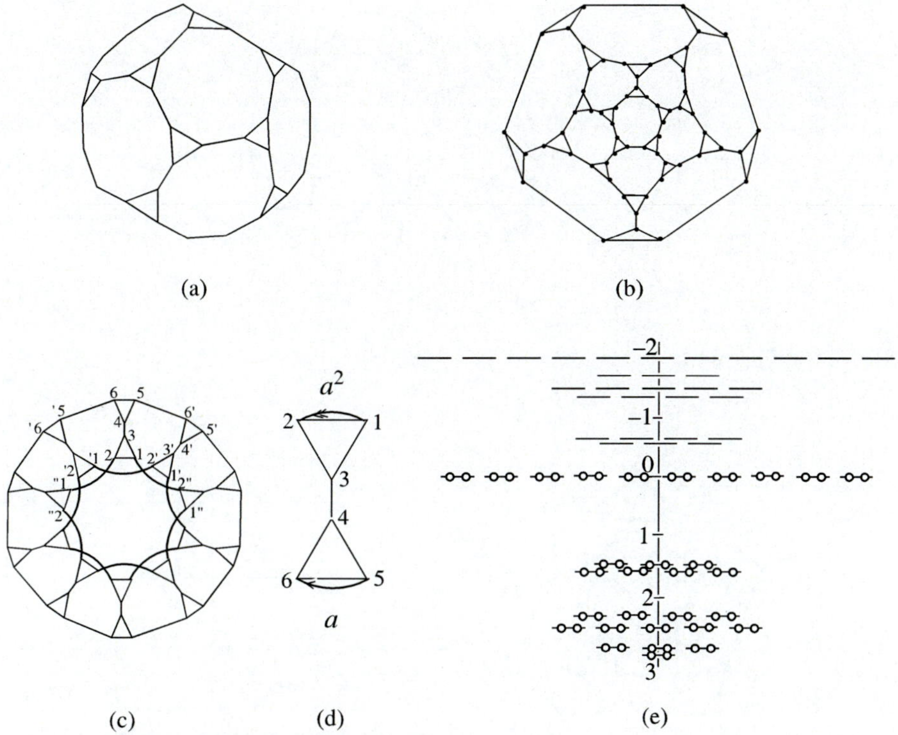

Fig. 13.4 (**a**) Perspective view, (**b**) Schlegel diagram, (**c**) topological symmetry diagram, (**d**) cyclic unit, (**e**) and HMO energy level diagram of truncated dodecahedron

$$a = \exp(2k\pi \mathrm{i}/10) = \exp(k\pi \mathrm{i}/5)\ (k = 1,2,\ldots,10). \tag{13.1}$$

Then according to the standard recipe (Hosoya et al. 1987) for solving a determinant of cyclic symmetry the following six by six determinant can be obtained:

$$\mathrm{P_G}(k,x) = \begin{vmatrix} -x & 1+a^2 & 1 & 0 & 0 & 0 \\ 1+a^{*2} & -x & 1 & 0 & 0 & 0 \\ 1 & 1 & -x & 1 & 0 & 0 \\ 0 & 0 & 1 & -x & 1 & 1 \\ 0 & 0 & 0 & 1 & -x & 1+a \\ 0 & 0 & 0 & 1 & 1+a^* & -x \end{vmatrix}, \tag{13.2}$$

where a^* denotes the complex conjugate of a, or simply a^{-1}.

Then one can obtain the HMO energy level diagram as shown in Fig. 13.4e in a highly degenerate fashion. By putting 60 electrons into the lowest 30 orbitals one gets a closed shell structure. However, the HOMO's are NBMO's with tenfold degeneracy, suggesting a highly reactive electronic structure.

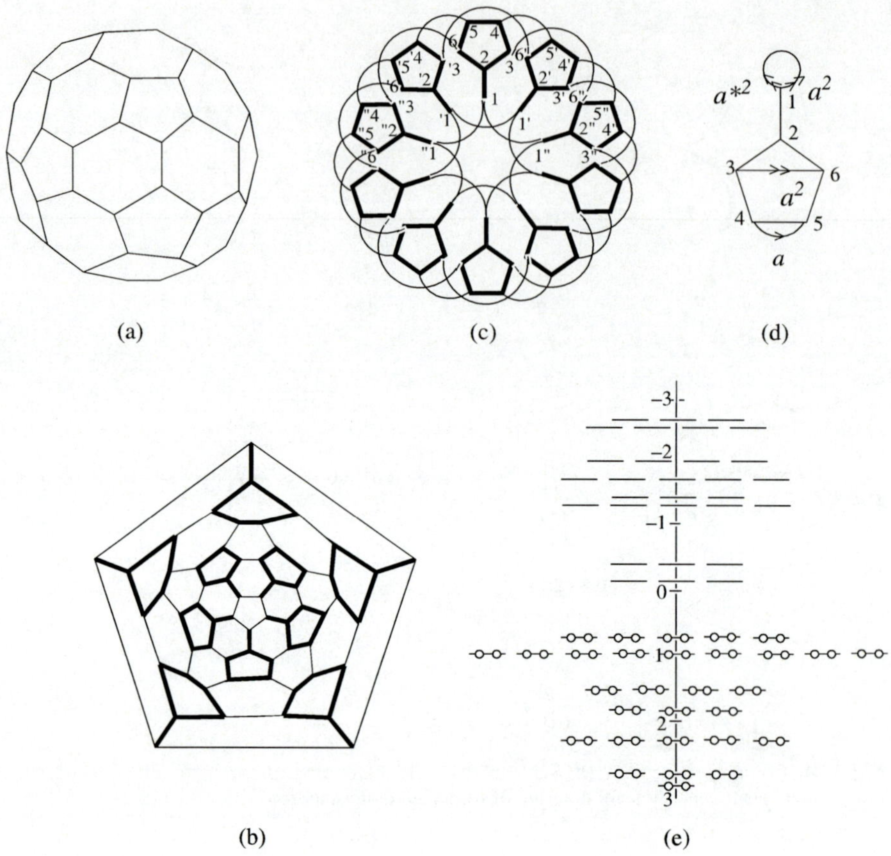

Fig. 13.5 Truncated icosahedron. Captions are the same as that of Fig. 13.4

In contrast to this network the soccer ball-shaped C_{60} fullerene has the highest stability among all the regular and semi-regular polyhedron networks (Hosoya and Tsukano 1994). In Fig. 13.5 the perspective view (a), Schlegel diagram (b), topological symmetry diagram (c) of D_{10h} symmetry, cyclic unit (d), and HMO energy level diagram (e) of this network are shown. In what follows except for a few cases, (a)~(e) are assigned to the figures in the above categories.

From Fig. 13.1 the third icosahedral network of truncated icosidodecahedron can be chosen as a candidate for a possible π-electron network with a semi-regular polygonal skeleton. The Schlegel diagram with fivefold symmetry can be drawn as Fig. 13.6b, where eleven decagons of three different sizes can be seen and the periphery is found to be the twelfth decagon. In this case it is not difficult to find the tenfold rotational "topological symmetry" by noticing the topological structure of Fig. 13.6b', which can be obtained by deforming Fig. 13.6b in a plane so that

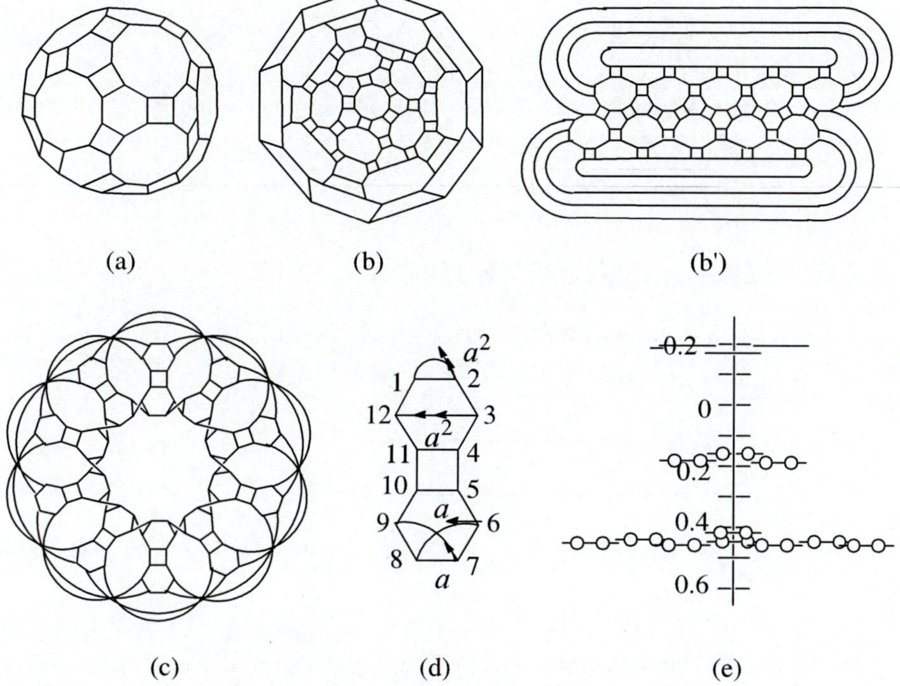

Fig. 13.6 Truncated icosidodecahedron. Captions are the same as that of Fig. 13.4

the pair of "five octagon cycles" symmetrically face with each other. Then one can easily draw Fig. 13.6c with D_{10h} topological symmetry, where a unit structure is composed of a pair of hexagons jointed through a square as shown in Fig. 13.6d. Then the 120 by 120 determinant can be factored into a product of ten 12 by 12 determinants. Since it is not a difficult task to write its explicit expression by following the unit structure of Fig. 13.6d with the same expression of (1) for a, the determinant is not given here. In this case, although the obtained HMO energy levels are generally highly degenerate, HOMO and LUMO are not degenerate but with a very narrow energy gap of $x = \pm 0.1569$ in β unit. In Fig. 13.6e only the energy level diagram in the region near HOMO and LUMO is given. From this diagram subtle possibility of closed shell structure of this π-electron system is suggested.

In this way the order of the characteristic polynomials of the three π-electronic systems, 5.6^2, 3.10^2, and 4.6.10, of icosahedral symmetry could be reduced by a factor of ten. However, the most dramatic simplification by the technique of topological symmetry can be demonstrated in the case of regular octahedron.

See the topological structure of the Schlegel diagram of octahedron in Fig. 13.7. The bold lines form a hexagonal cycle, each of whose vertex, say number n (=1~6), is connected to $n-1$ and $n+1$. Further, each n is also connected to $n-2$ and $n+2$.

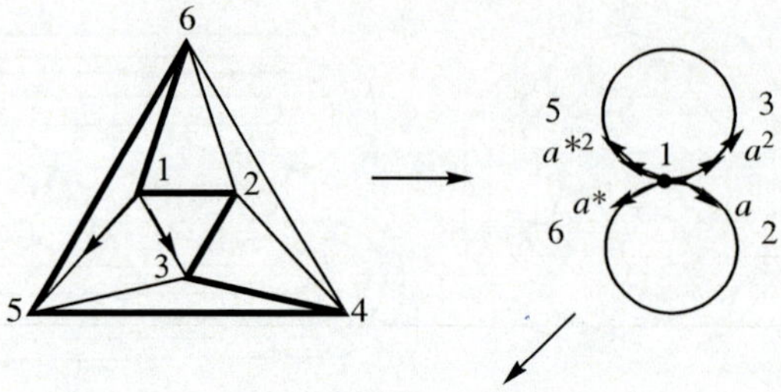

$a = \exp(2k\pi i/6)$
$= \exp(k\pi i/3)$
$a^* = a^{-1}$
$k = 1,2,...,6$

$P_G(k,x) = (a+a^*) + (a^2 + a^{*2}) - x = 0$

$\therefore \quad x = 2(\cos k\pi/3 + \cos 2k\pi/3)$

n	1	2	3	4	5	6
x	0	−2	0	−2	0	4

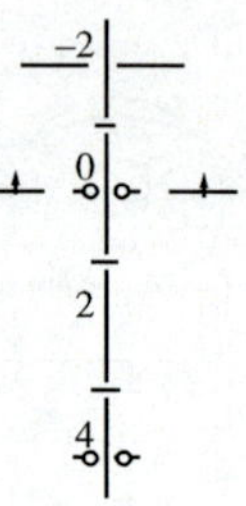

Fig. 13.7 Schlegel diagram, cyclic unit, characteristic polynomial, and HMO energy level diagram of regular octahedron

Then by running through from n=1 to 6, one gets the complete structure of the Schlegel diagram of regular octahedron. This bonding mode at each vertex is drawn as in the upper right of the figure.

The reduced characteristic polynomial $P_G(k,x)$ of a vertex of unit order gives the rigorous expression of x, and accordingly the general expression of the energy level of regular octahedron can be obtained as

$$x = 2(\cos k\pi/3 + \cos 2k\pi/3) \; (k = 1,2.\ldots,6), \tag{13.3}$$

leading to all the integer eigen values as in the table in Fig. 13.7 without solving the secular determinant of the order of six. The HMO energy level diagram is obtained as shown there, but HOMO's are open at NBMO's. Moreover the degrees of the vertices of this network are four. Thus there is no possibility for the existence of regular octagonal π-electronic system.

On the other hand, the regular dodecahedron network is a cubic graph, whose vertex-degrees are all three, as shown in Fig. 13.8, which is a candidate for a spherical unsaturated π-electronic system. In this case the 20 by 20 determinant can be factored into a product of ten 2 by 2 determinants by using the tenfold rotational symmetry as shown in Fig. 13.8c with the cyclic unit of Fig. 13.8d. In this case deformation of Fig. 13.8b into 8b' is helpful for obtaining Fig. 13.8c. Also in this case the π-electronic system becomes open at quadruply degenerate NBMO's as shown in Fig. 13.8e, suggesting again no possibility for its existence.

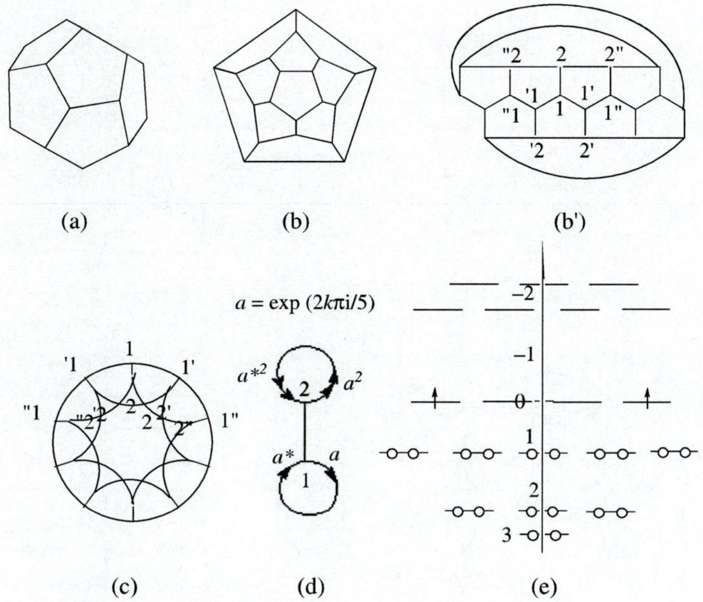

Fig. 13.8 Dodecahedron. Captions are the same as that of Fig. 13.4

13.5 Octahedral and Tetrahedral Symmetry

Now turn to the networks of octahedral symmetry picked out from the three candidates, i.e., truncated octahedron, truncated cube, and truncated cuboctahedron, as given in Fig. 13.2.

All the materials regarding truncated octahedron are given in Fig. 13.9. The Schlegel diagram of Fig. 13.9b has D_{3h} symmetry. However, notice that the arrangement of the six isolated squares is topologically the same as that of the six vertices of the octahedron shown in Fig. 13.7. Then by numbering the vertices in each square as in Fig. 13.9b, one can rearrange the squares so that they have D_{6h} topological symmetry as in Fig. 13.9c, where the six unit structures of Fig. 13.9d are circularly joining with each other. By solving the 4 by 4 determinant the energy level diagram can be obtained as in Fig. 13.9e, predicting a stable closed shell structure with $x = \pm 0.4142$ in β unit for HOMO and LUMO.

The HOMO-LUMO gap of 0.8284β is a little larger than 0.7566β of soccer ball-shaped C_{60} fullerene. Although the torsional energy of the σ-skeleton of truncated octahedron seems to be larger than that of C_{60} fullerene, this substance needs to be studied as a candidate for a stable spherical carbon allotrope.

For truncated cube two different Schlegel diagrams can be drawn as Figs. 13.10b and b'. In this case rather than the conventional octagon-shaped one the triangular

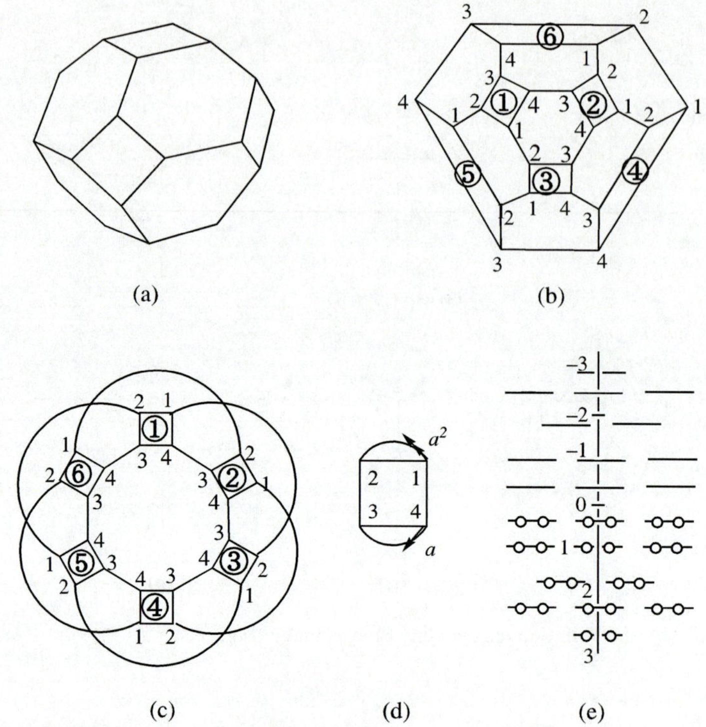

Fig. 13.9 Truncated octahedron. Captions are the same as that of Fig. 13.4

diagram of Fig. 13.10b' is superior for obtaining the topological symmetry diagram of Fig. 13.10c with higher symmetry (D_{6h}), where the unit structure of Fig. 13.10d has a key role in representing the topological symmetry of the whole network. By putting 24 electrons into the energy level diagram of Fig. 13.10e, HOMO's are found to be open at NBMO's, leading negative conclusion for the existence of this π-electron network.

The next candidate is truncated cuboctahedron (Fig. 13.11). By deforming the octagonal Schlegel diagram of Fig. 13.11b into b', it is not so difficult to draw the topological symmetry diagram of Fig. 13.11c with D_{6h} symmetry using the octagon in Fig. 13.11d as a cyclic unit. However, by putting 24 π-electrons into the energy level diagram one gets open HOMO's at NBMO's (Fig. 13.11e). Then truncated cuboctahedron drops out.

One can enjoy the three D_{6h} diagrams of topological symmetry, namely, Figs. 13.9c, 13.10c, and 13.11c of the topological symmetry for these networks of O_h symmetry, which effectively factor out the determinant of the whole graph by a factor of ten yielding highly degenerate MO's for these π-electron networks.

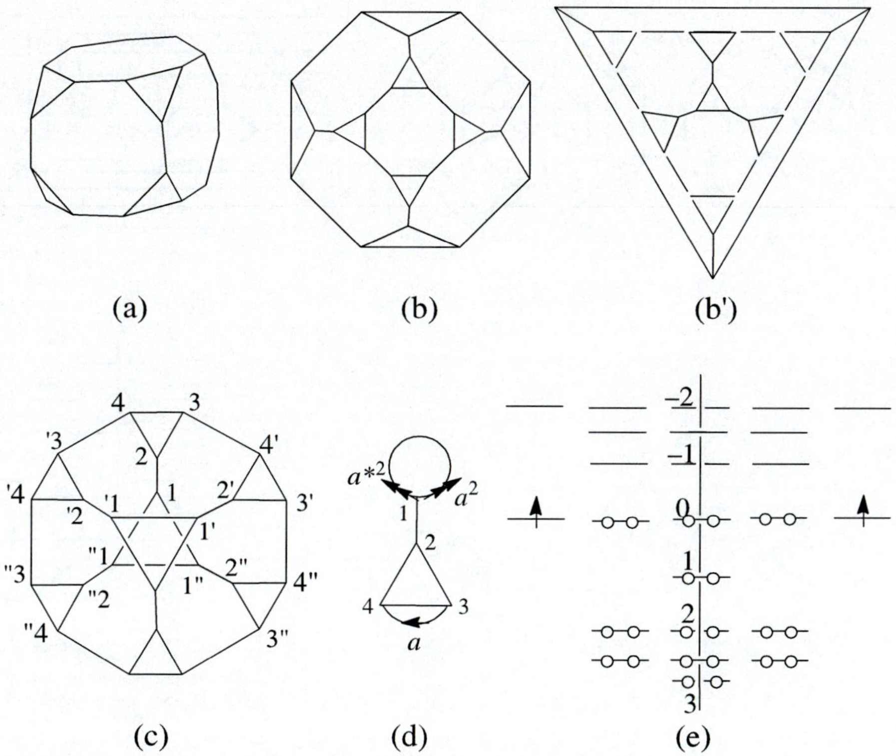

Fig. 13.10 Truncated cube. Captions are the same as that of Fig. 13.4

The situation is almost the same for the case with I_h symmetry. Namely, the diagrams of topological symmetry in Figs. 13.4c, 13.5c, 13.6c, and 13.8c have D_{10h} symmetry.

There exists only one entity for semi-regular polyhedral networks of tetrahedral (T_d) symmetry. Either from the two kinds of Schlegel diagrams as shown in Figs. 13.12b and b'. One can draw the D_{4h} topological symmetry diagram of Fig. 13.12c, which yields a closed shell ground state as in Fig. 13.12e. However, HOMO's are doubly degenerate NBMO's, and stable ground state cannot be expected.

13.6 Networks with High Vertex-Degree

In Figs. 13.1, 13.2, and 13.3 four semi-regular polyhedra, $(3.5)^2$, 3.4.5.4, $(3.4)^2$, and 3.4^3, are found to be 4-regular graphs, where the vertex-degrees are all four. Among the regular polyhedra, octahedron 3^4 also belongs to this group. Although these

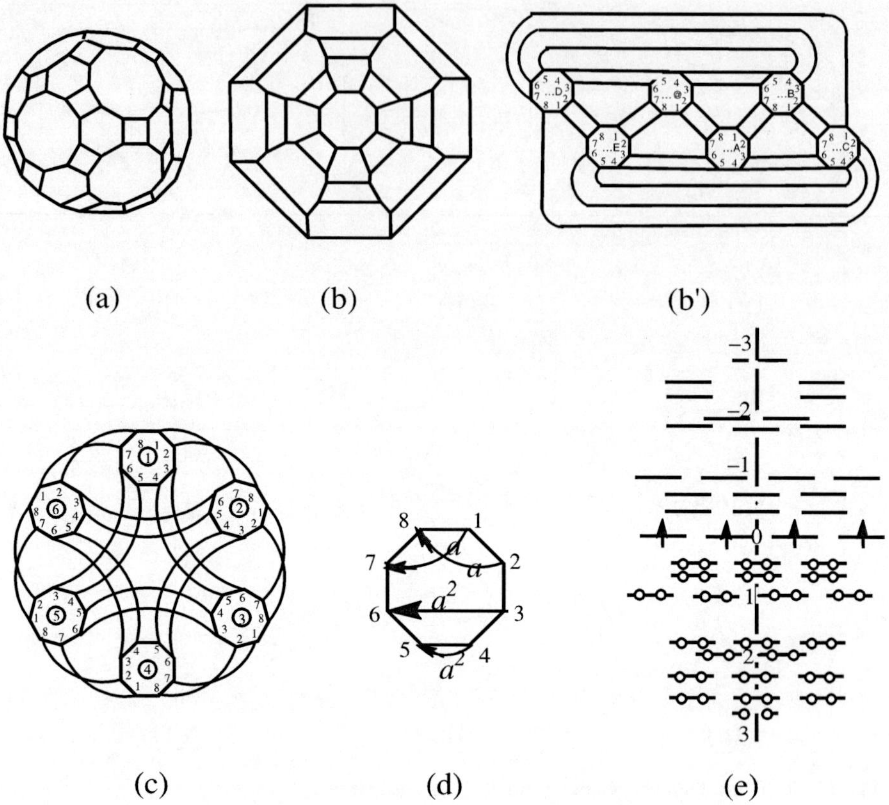

Fig. 13.11 Truncated cuboctahedron. Captions are the same as that of Fig. 13.4

polyhedra are not qualified to become a candidate for a stable π-electron network, let us obtain their orbital energy level diagrams as a formal discussion. The essence of the results are given in Fig. 13.13, where N denotes the number of vertices, or number of π-electrons, and n the degree of degeneracy of topological symmetry. That is, N/n gives the size of the cyclic unit. The value of a is expressed by

$$a = \exp(2k\pi i/n), \qquad (13.4)$$

and k runs from 1 to n.

If N electrons are put into the MO's from the bottom, the (formal) ground state will be open for all these networks as shown in Fig. 13.13. The least stable "π-electron network" will be icosidodecahedron, because its HOMO's are open anti-bonding MO's, while the HOMO's of the rest are open NBMO's. Further, it is interesting to observe that the two pairs of cyclic units, i.e., icosidodecahedron and cuboctahedron, and rhombicuboctahedron and rhombicosidodecahedron, are, respectively, very similar with each other.

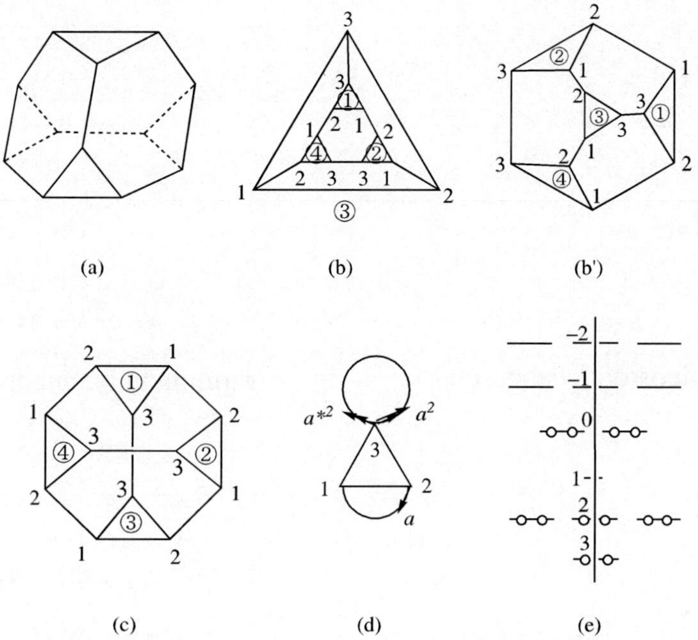

Fig. 13.12 Truncated tetrahedron. Captions are the same as that of Fig. 13.4

13.7 Summary of the Results

Remember that HOMO's of all the networks with the vertex-degree of four are open, and cannot be expected to have stable ground state. Then the results of only the ten cubic graphs are compared in Table 13.1. By seeing the fourth and fifth columns those three networks are selected whose marks are both ○, namely, truncated icosahedron 5.6^2, truncated icosidodecahedron 4.6.10, and truncated octahedron 4.6^2. They have closed-shell structure with HOMO's of bonding character. Their HOMO-LUMO gap is in the following order, $4.6^2 > 5.6^2 >> 4.6.10$. The first inequality sign will be reversed if some strain energy for making their ball-shaped structures are duly taken into consideration. Among all the polyhedral networks, stable or unstable, 4.6.10 has such a peculiar property that its HOMO is not degenerate. Due to their high topological symmetry, the MO's of all the polyhedral networks, irrespective of their stability, are highly degenerate. Then their high degeneracy is not the cause of high stability of the two networks of 4.6^2 and 5.6^2. On the other hand, these two stable networks have such a common property that polygons other than a hexagon is systematically scattered in the spherical sea of hexagons. Then it is plausible that they have a high possibility to be ejected from graphene sheets when exposed by a strong impulse of radiation or particles.

Although this analysis cannot give quantitative conclusion, the direction of the discussion is believed to be qualitatively correct. Of course, there must be the effect

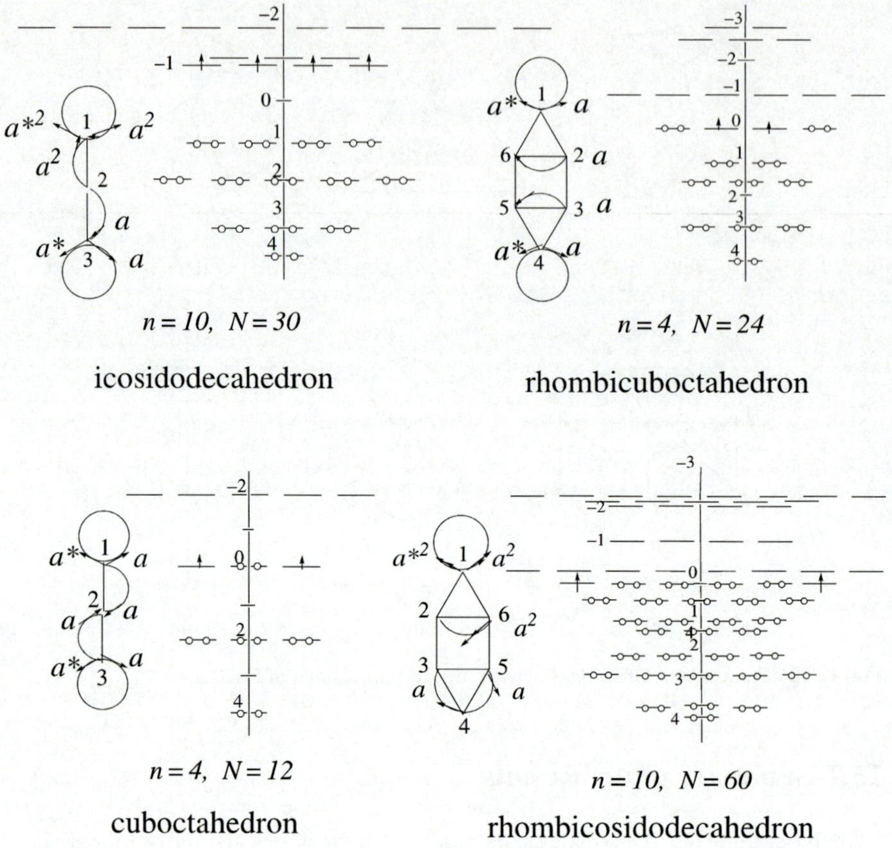

Fig. 13.13 Cyclic units and HMO energy level diagrams of icosidodecahedron, cuboctahedron, rhombicuboctahedron, and rhombicosidodecahedron

of destabilization by six squares in the conjugated π-electron network, and the torsion energy caused by the tetragonal skeleton is also expected to be large to some extent. However, if some optimistic discussion might be allowed to be developed, among the 169 Kekulé structures of this π-electron network (Hosoya 1986) the following one is expected to play dominant contribution to well pay off the above-mentioned destabilization factors. More sophisticated calculation is needed to be performed.

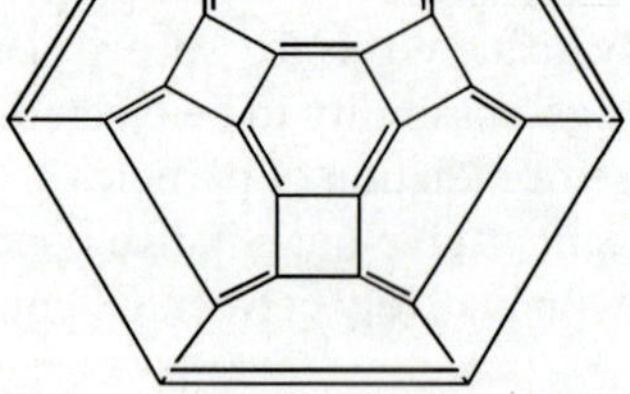

Table 13.1 Comparison of the various properties of regular and semi-regular polyhedra whose vertex degrees are all three

Polyhedron	Code	$\|V\|$	Open(×) Closed(○)	HOMO character	HOMO degeneracy	cf
I_h symmetry						
Dodecahedron	5^3	20	×	△	4	Fig. 13.8
Truncatede icosahedron	5.6^2	60	○	○	5	Fig. 13.5
Truncated dodecahedron	3.10^2	60	○	△	10	Fig. 13.4
Truncated icosidodecahedron	4.6.10	120	○	○	1	Fig. 13.6
O_h symmetry						
Cube	4^3	8	○	○	3	
Truncated cube	3.8^2	24	×	△	5	Fig. 13.10
Truncated octahedron	4.6^2	24	○	○	3	Fig. 13.9
Truncatede cuboctahedron	4.6.8	48	×	△	4	Fig. 13.11
T_d symmetry						
Tetrahedron	3^3	4	×	×	3	
Truncated tetrahedron	3.6^2	12	○	△	2	Fig. 13.12

HOMO character: × anti-bonding, △ non-bonding, ○ bonding
Degeneracy: degree of degeneracy of HOMO

References

Ceulemans A, Compernolle S, Delabie A, Somers K, Chibotaru LF, Fowler PW, Marganska MJ, Szopa M (2002) Phys Rev B 65:115412

Cioslowski J (1995) Electronic structure calculations on fullerenes and their derivatives. Oxford University Press, Oxford

Cundy M, Rollett AP (1952) Mathematical models. Clarendon Press, Oxford

Fowler PW, Manolopoulos DE (1995) An atlas of fullerenes. Clarendon Press, Oxford

Glukhovtsev MN, Simkin B Ya, Yudilevich IA (1990) Theor Exper Chem 26:212

Hosoya H (1986) Comp Math Appl 12B:271

Hosoya H, Aida M, Kumagai R, Watanabe K (1987) J Comput Chem 8:358

Hosoya H, Okuma Y, Tsukano Y, Nakada K (1995) J Chem Inf Comput Sci 35:351

Hosoya H, Tsukano Y (1994) Fullerene Sci Technol 2:381

King RB (1998) J Math Chem 23:197

Liu X, Klein DJ, Schmalz TG, Seitz WA (1991) J Comput Chem 12:1252

Osawa E, Ueno H, Yoshida M, Slanina Z, Zhao X, Nishiyama M, Saito H (1998) J. Chem Soc Perkin Trans 2:943

Sokolov VI, Stankevich IV (1993) Russ Chem Rev 62:419

Williams R (1979) The geometrical foundation of natural structure. Dover, New York, NY

Chapter 14
The Estrada Index and Fullerene Isomerism

Patrick W. Fowler and Ante Graovac

Abstract Estrada Index, $EE(G)$, defined as the sum of exponentials of the eigenvalues of the adjacency matrix of graph G, is calculated for sets of general cubic polyhedra, and for general and isolated-pentagon fullerenes. Amongst small cubic polyhedra, the "near-fullerenes" and fullerenes minimise EE. Amongst fullerenes, the isolated-pentagon fullerenes minimise EE. The preference for fullerenes over non-fullerenes is significant, but the relative variation of EE with fullerene isomer is tiny (parts per million for general fullerenes, parts per billion for isolated-pentagon fullerenes) and is essentially tracking the number of pentagon adjacencies (and hence overall stability).

14.1 Introduction

The Estrada Index of G, a graph with n vertices and m edges, is defined as the sum of exponentials

$$EE(G) = \sum_{i=1}^{n} e^{\lambda_i}, \tag{14.1}$$

where $\{\lambda_{i,}\ i = 1, \ldots n\}$ are the eigenvalues of $\mathbf{A}(G)$, the adjacency matrix of G (Estrada 2000, 2002, 2004). In the bio-informatics context for which it was originally devised, the index is applied to a vertex-weighted graph G, constructed to incorporate 3D-structural information relevant to protein folding. Weighted graphs are also employed in an application of $EE(G)$ to characterisation of molecular branching (Estrada et al. 2006). However, in applications of the Estrada Index and

P.W. Fowler (✉)
Department of Chemistry, University of Sheffield, Sheffield S3 7HF, UK
e-mail: p.w.fowler@sheffield.ac.uk

F. Cataldo et al. (eds.), *The Mathematics and Topology of Fullerenes*,
Carbon Materials: Chemistry and Physics 4, DOI 10.1007/978-94-007-0221-9_14,

related centrality (Estrada et al. 2005a), bipartivity (Estrada et al. 2005b; Došlić 2005) and thermodynamic (Estrada and Hatano 2007) indices to networks and chemical graphs, G is usually taken to be unweighted. We will be concerned here exclusively with unweighted chemical graphs capable of representing conjugated networks of carbon atoms, i.e., G will be simple, connected, and of maximum degree $\Delta(G) \geq 3$.

For a simple graph, the adjacency matrix has entries $A_{ij} = 1$ if there is an edge between vertices i and j, and $A_{ij} = 0$ otherwise. The eigenvalues of $\mathbf{A}(G)$ for a chemical graph G defined as above lie in the range from +3 to –3 and are conventionally ordered such that

$$+3 \geq \lambda_1 > \lambda_2 \geq \lambda_3 \ldots \geq \lambda_n \geq -3. \tag{14.2}$$

In view of the well known series expansion, convergent for all x,

$$e^x = \sum_{j=0}^{\infty} \frac{x^j}{j\,!}, \tag{14.3}$$

an alternative expression for the Estrada Index is

$$EE(G) = \sum_{j=0}^{\infty} \frac{\mu_j(G)}{j\,!}, \tag{14.4}$$

where $\mu_j(G)$ is the j-th moment of the eigenvalue spectrum of $\mathbf{A}(G)$,

$$\mu_j(G) = \sum_{i=1}^{n} \lambda_i^j = Tr\left[\mathbf{A}(G)^j\right]. \tag{14.5}$$

The moments count the self-returning walks of each length j in the graph, and so provide a useful link between the structure of the graph and its spectrum. The coefficients of the characteristic polynomial, whose roots are the eigenvalues of $\mathbf{A}(G)$, depend on the counts of structural components of G, and can be reconstructed from the moments (Schwenk 1979).

It has been shown that the Estrada Index is well approximated for several important classes of chemical graphs by simple analytical expressions (Gutman and Graovac 2007c; Ginosar et al. 2008). For the cycle on n vertices, C_n, the index is remarkably well represented by

$$EE(C_n) \approx nI_0, \tag{14.6}$$

and for the path on n vertices, P_n, by

$$EE(P_n) \approx (n+1)I_0 - \cosh(2), \tag{14.7}$$

where

$$I_0 = \sum_{k=0}^{\infty} \frac{1}{(k\,!)^2} = 2.27958530\ldots \tag{14.8}$$

An approximation for general graphs, derived with the help of simple assumptions about the distribution of eigenvalues (Gutman et al. 2007f), is

$$EE \approx \frac{n}{k}\ \sinh(k), \tag{14.9}$$

where $k = \sqrt{6m/n}$. This approximation works well for benzenoid graphs (See also Gutman and Radenković 2007d). Upper and lower bounds on $EE(G)$ have been established for various classes of simple graphs (Gutman et al. 2007a; de la Peña et al. 2007). Approximations connecting $EE(G)$ with spectral radius (Gutman et al. 2007b) and other topological indices (Gutman et al. 2007e) have also been investigated.

The related quantity known as the bipartivity index $\beta(G)$ is a measure of how closely the spectrum of G approaches the paired structure characteristic of a bipartite graph. It is defined as

$$\beta(G) = \sum_{i=1}^{n} (e^{\lambda_i} + e^{-\lambda_i}) / \sum_{i=1}^{n} 2e^{\lambda_i} = \sum_{i=1}^{n} \cosh(\lambda_i)/EE(G) \tag{14.10}$$

with $\beta(G) = 1$ for a bipartite graph (Došlić 2005).

The present survey is mainly concerned with the Estrada Index $EE(G)$ for graphs belonging to a well-known class of chemical graphs, the fullerenes. A fullerene is an all-carbon molecule for which the graph is the skeleton of a cubic polyhedron on n vertices with exactly 12 pentagonal faces and all other faces hexagonal. Fullerene graphs exist for $n = 20$, and for all even $n > 22$ (Grünbaum and Motzkin 1963), and typically have large numbers of isomeric forms. Complete sets of fullerene graphs for chemically interesting values of n are readily generated using the face-spiral algorithm (Manolopoulos et al. 1991; Fowler and Manolopoulos 1995, 2006) or the pent-hex puzzle (Brinkmann and Dress 1997, 1998) algorithm embodied in the fullgen program. Complete sets of graphs of general cubic polyhedra, without restriction on face size, can be generated for small values of n using the plantri program. (Both fullgen and plantri programs are available from http://cs.anu.edu.au/~bdm/plantri/).

Two natural questions with reference to $EE(G)$ of fullerenes are: Does the Estrada Index have a simple approximation also for these graphs? Does it distinguish well between isomers (and so have some possible correlation with physical stability or properties of these molecules)? As examples of "near-bipartite" graphs, fullerenes have been studied with the aid of $\beta(G)$, and similar questions can be asked (Došlić 2005). It is also of interest to place the trends for fullerenes in the wider context of cubic polyhedral graphs, as this family provides candidates for small carbon cages below the fullerene threshold of $n = 20$.

14.2 Calculations

Estrada indices were calculated by diagonalisation of adjacency matrices constructed using the previously mentioned programs for three sets of graphs: (i) the 149,960,273 cubic polyhedra with $4 \leq n \leq 30$, (ii) the 30579 general fullerenes with $20 \leq n \leq 70$, (iii) the 167299 isolated-pentagon fullerenes with $60 \leq n \leq 130$. Tables 14.1, 14.2, and 14.3 show statistical summaries of the results.

14.2.1 Cubic Polyhedra

The values in Table 14.1 illustrate several points. The Estrada Index grows essentially linearly with n for cubic polyhedra, though with a significant spread ($\pm 5\%$ of the mean, or more) between maximum and minimum at each n. Equation (14.9), with $m = 3n/2$ (and hence $k = 3$) for cubic graphs, predicts $EE \approx (n/3)\ \sinh(3) \approx 3.339n$. This estimate is based on a two-term approximation to the eigenvalue distribution that implies eigenvalues spread from +3 to −3, an approximation well suited to cubic graphs. Taking further terms does not appear to improve the quality of the estimate (Gutman et al. 2007f). As Fig. 14.1 shows, the plots of maximum and minimum Estrada indices for the cubic polyhedra, after initial curvature settle down to approximately linear growth with slopes bracketing the estimate of 3.339.

Table 14.1 Estrada Index of cubic polyhedral graphs with up to 30 vertices

n	N	EE_{min}	EE_{max}	EE_{mean}	ΔEE
4	1			21.18918	–
6	1			25.07449	–
8	2	29.39381	30.97135	30.18258	1.57754
10	5	35.47908	38.01880	36.97973	2.53972
12	14	41.66461	45.76235	43.77900	4.09774
14	50	47.86854	52.82970	50.70732	4.96116
16	233	54.11469	60.49693	57.66026	6.38224
18	1249	60.64827	68.50431	64.64812	7.85604
20	7595	66.60833	75.62884	71.60423	9.02051
22	49566	73.43659	83.30237	78.57197	9.86577
24	339722	79.69093	91.31446	85.53704	11.62353
26	2406841	86.23227	98.44361	92.50600	12.21134
28	17490241	92.77361	106.14180	99.47755	13.36819
30	129664753	99.31596	114.15836	106.45206	14.84240

n is the number of vertices, N is the number of non-isomorphic graphs of order n in the class, EE_{min}, EE_{max} and EE_{mean} are the minimum, maximum and mean values of the Estrada Index for the N graphs, and ΔEE is the difference $EE_{\text{max}} - EE_{\text{min}}$

Table 14.2 Estrada Index of fullerene graphs with up to 70 vertices

n	N	N_{min}	EE_{min}	EE_{max}	EE_{mean}	ΔEE
20	1	1			66.60833	–
24	1	1			79.69093	–
26	1	1			86.23227	–
28	2	2	92.77361	92.77461	92.77411	0.00100
30	3	3	99.31596	99.31747	99.31663	0.00151
32	6	6	105.85782	105.85930	105.85879	0.00149
34	6	5	112.40016	112.40155	112.40071	0.00139
36	15	14	118.94203	118.94479	118.94312	0.00277
38	17	17	125.48438	125.48755	125.48555	0.00317
40	40	38	132.02672	132.03121	132.02803	0.00449
42	45	45	138.56907	138.57214	138.57033	0.00307
44	89	89	145.11142	145.11579	145.11294	0.00437
46	116	116	151.65417	151.65764	151.65536	0.00347
48	199	199	158.19652	158.20128	158.19796	0.00476
50	271	271	164.73847	164.74496	164.74047	0.00649
52	437	422	171.28122	171.28677	171.28308	0.00555
54	580	540	177.82357	177.82863	177.82561	0.00506
56	924	913	184.36632	184.37228	184.36827	0.00596
58	1205	1205	190.90867	190.91412	190.91082	0.00545
60	1812	1812	197.45023	197.45870	197.45348	0.00848
62	2385	2378	203.99416	204.00052	203.99608	0.00636
64	3465	3451	210.53651	210.54326	210.53875	0.00675
66	4478	4348	217.07926	217.08511	217.08136	0.00585
68	6332	6073	223.62201	223.62876	223.62405	0.00676
70	8149	8149	230.16396	230.17245	230.16667	0.00849

Column headings as in the note of Table 14.1. N_{min} is the fullerene isomer (labelled by position in the spiral order) that has $EE = EE_{\mathrm{min}}$

In the range of our calculations, the spread ΔEE therefore also grows roughly linearly with n. Published bounds for $EE(G)$ are not of practical use for estimating ΔEE for the cubic polyhedra: for example, with $n = 20$, $m = 30$, Theorem 1 (de la Peña et al. 2007) gives $22.80 < EE < 2332$, which as the graphs are regular can be improved by Theorem 2 to $39.17 < EE < 1302$, whereas direct calculation (Table 14.1) gives the far tighter spread of $66.60 < EE < 71.60$.

An intriguing feature of the data for small n is that the cubic polyhedron achieving the minimum value of $EE(G)$, in the range $4 \leq n \leq 30$, also has the minimum value of the parameter F (Domene et al. 1997) where

$$F(G) = \sum_r (6 - r)^2 f_r \geq 12, \tag{14.10}$$

and f_r is the number of faces of size r in the given polyhedral graph (see Fig. 14.2). This simple invariant has been used to quantify the proximity of a cubic polyhedron to a fullerene (amongst cubic polyhedra, $F = 12$ if and only if G is a fullerene graph) and is found to pick out structures that are good candidates for stable polyhedral

Table 14.3 Estrada Index of isolated-pentagon fullerene graphs with up to 130 vertices

n	N	EE_{min}	EE_{max}	EE_{mean}	ΔEE
60	1			197.450227717	–
70	1			230.163963285	–
72	1			236.706711578	–
74	1			243.249457368	–
76	2	249.792204338	249.792205412	249.792204875	0.000001074
78	5	256.334951308	256.334953509	256.334952490	0.000002200
80	7	262.877697417	262.877700820	262.877699285	0.000003403
82	9	269.420445785	269.420447540	269.420446679	0.000001755
84	24	275.963193579	275.963195939	275.963194345	0.000002360
86	19	282.505940762	282.505942660	282.505941687	0.000001898
88	35	289.048688019	289.048690666	289.048689253	0.000002647
90	46	295.591435526	295.591438620	295.591436694	0.000003093
92	86	302.134182890	302.134185680	302.134183976	0.000002791
94	134	308.676930540	308.676932794	308.676931488	0.000002254
96	187	315.219677579	315.219680800	315.219679024	0.000003221
98	259	321.762425464	321.762427967	321.762426334	0.000002503
100	450	328.305172487	328.305176420	328.305173878	0.000003933
102	616	334.847920281	334.847923230	334.847921378	0.000002949
104	823	341.390667501	341.390671040	341.390668826	0.000003539
106	1233	347.933415295	347.933418101	347.933416304	0.000002806
108	1799	354.476162658	354.476166160	354.476163820	0.000003502
110	2355	361.018910022	361.018914220	361.018911286	0.000004198
112	3342	367.561657725	367.561661280	367.561658807	0.000003555
114	4468	374.104405142	374.104408840	374.104406305	0.000003698
116	6063	380.647152543	380.647156400	380.647153798	0.000003857
118	8148	387.189900050	387.189903461	387.189901324	0.000003411
120	10774	393.732647663	393.732652019	393.732648828	0.000004357
122	13977	400.275395064	400.275398581	400.275396331	0.000003517
124	18769	406.818142571	406.818146640	406.818143859	0.000004069
126	23589	413.360890184	413.360893701	413.360891376	0.000003517
128	30683	419.903637691	419.903641760	419.903638888	0.000004069
130	39393	426.446385092	426.446389819	426.446386417	0.000004727

Column headings are as in the note of Table 14.1

carbon cages in that they have low total energies according to quantum-mechanical estimates. For $n = 20$, 24 and 26, the cubic polyhedron of minimum *EE* is the unique fullerene of that order, and the fullerene isomer(s) at $n = 28$ and 30 have lower *EE* than all other cubic polyhedra. For $n = 22$, the cubic polyhedron of minimum *EE* is the edge-truncated dodecahedron ($f_4 = 1,\ f_5 = 10,\ f_6 = 2,\ F = 14$), the closest approximation to a fullerene at this vertex count, and the skeleton of the C_{22} carbon cage of lowest energy (Domene et al. 1997). Figure 14.3 shows a scatter-plot of *EE* and *F* for $n = 20$, illustrating the way that minimisation of *EE* picks out the fullerene from the pack of cubic polyhedra.

It seems a plausible conjecture that the cubic polyhedron of minimum Estrada Index is also of minimal *F* index (a "near fullerene"), and in particular, therefore, for $n = 20$ and $n \geq 24$, is a fullerene. The rationale underlying the conjecture

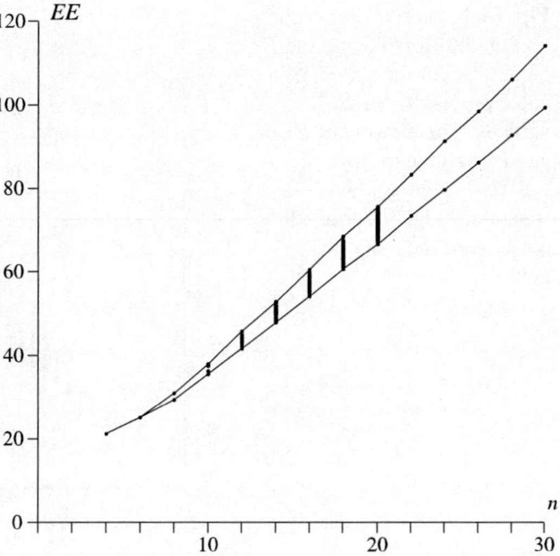

Fig. 14.1 Trends in the Estrada Index (*EE*) for cubic polyhedral graphs with $n \leq$ 30 vertices. The envelope of maximum and minimum values is shown, and, for $n \leq$ 20, all *EE* values in-between are shown; for higher values of *n*, the distribution of *EE* values appears continuous on this scale

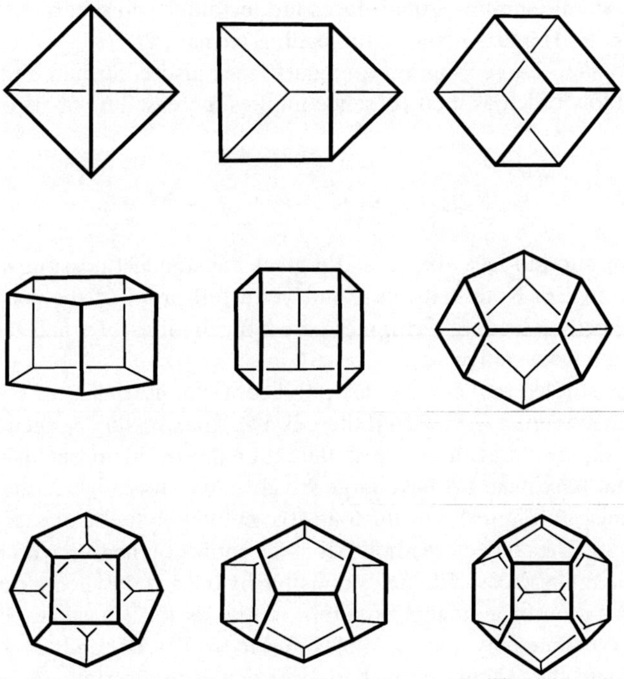

Fig. 14.2 Cubic polyhedra with minimal Estrada Index for $n \leq 22$ vertices. These are all "near-fullerenes" in that they also minimise the F function, which describes the summed-square deviation of the face recipe from that of an idealised all-hexagonal polyhedron. The polyhedra can be identified by labels *n*:*p*, where *n* is the vertex number, and *p* is the order of generation by the plantri program. They are: 4:1, 6:1, 8:2, 10:4, 12:14, 14:50, 16:233, 18:746 and 22:25920. At $n = 20$ and $n \geq 24$, the minimising polyhedra are fullerenes

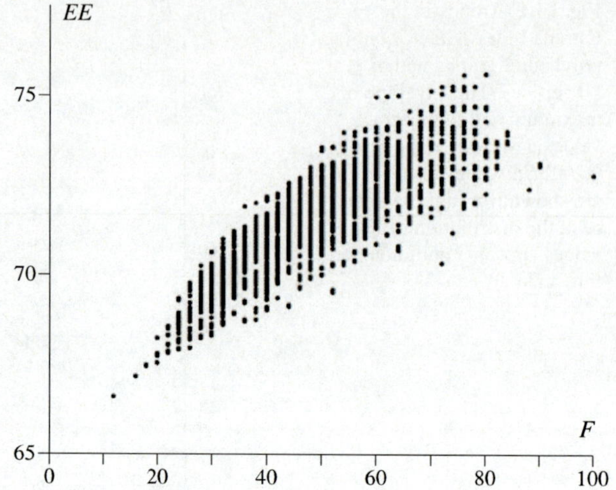

Fig. 14.3 Scatter-plot of the Estrada Index (*EE*) and the *F* parameter for the set of 7595 cubic polyhedra on 20 vertices. The minima of *EE* and *F* coincide in the C_{20} fullerene, seen as a well separated single dot outside the apex of the cloud of points

is that some small (sub-hexagonal) faces are inevitably present in cubic polyhedra ($f_3 + f_4 + f_5 \geq 4$) and so contribute leading terms $+2f_r/(r-1)!$ to the moment expansion for *EE*. Large (super-hexagonal) faces also contribute indirectly to the expansion at low order, as their presence implies increased numbers of small faces, through

$$3f_3 + 2f_4 + f_5 = 12 + f_7 + 2f_8 + 3f_9 + \ldots \tag{14.11}$$

Thus, restriction to face sizes 5 and 6 gives the smallest possible n-independent leading term in *EE*. Note that this is not yet a full proof of the conjecture since, beyond the positive leading term, there are contributions of small faces to higher moments that have negative signs.

Within the studied range, the cubic polyhedra that *maximise* $EE(G)$ can be predicted by first maximising f_3, then if there is a tie, maximising f_4, then f_5, and so on; the moment expansion again provides the rationale for this observation. The cubic polyhedra that maximise *EE* have large *F* values but, for $n \geq 12$, they do not precisely maximise *F*. Figure 14.4 illustrates the cubic polyhedra of maximum $EE(G)$ for $4 \leq n \leq 30$. The conjecture implies that the cubic polyhedron with maximum *EE* has $\lfloor n/3 \rfloor$ triangular faces. When n is divisible by 6, such polyhedra can be obtained by omni-truncation of a smaller structure (which is itself a cubic polyhedron for $n > 6$), as illustrated by the examples for $n = 12$, 18, and 24 in the figure.

Clearly, then, the Estrada Index has some power to discriminate between cubic polyhedra, even though, as the graph parameters n and m determine a large part of $EE(G)$, it cannot be expected to be sensitive to the finer details of structure (Gutman et al. 2007f). From a practical viewpoint, it should be noted that the information about face signatures provided by the extrema of *EE* could of course also be obtained by direct inspection without calculation of eigenvalues.

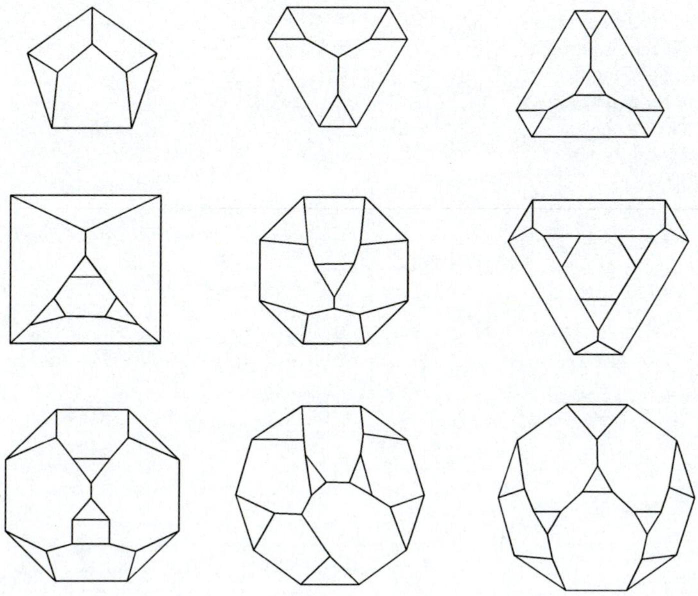

Fig. 14.4 Cubic polyhedra with maximal Estrada Index for $8 \leq n \leq 24$ vertices. All have the maximum possible number of triangular faces. The polyhedra are shown as Schlegel diagrams. In the *n*:*p* labelling used in Fig. 14.2 they are 8:1, 10:3, 12:8, 14:26, 16:104, 18:489, 20:2521, 22:14019 and 24:81880

14.2.2 Fullerenes

When the class of target graphs is restricted to the fullerenes (See Tables 14.2 and 14.3 for general and isolated-pentagon fullerenes, respectively), the variability in $EE(G)$ all but disappears. Figure 14.5 shows the variation of Estrada index for respectively the general fullerenes with up to 70 vertices and the isolated-pentagon fullerenes with up to 130 vertices. All fit smoothly on the same curve. It is impossible, on the scale of the plot, to see the isomer variations in $EE(G)$, even though all isomers within the respective classes are included as separate data-points. So, for example, 8149 isomers, all with slightly different *EE*, are comprehended within the single dot shown for $n = 70$, and 10774 isomers lie behind the single dot for $n = 120$. At fixed *n*, the range ΔEE is a few parts per million of the mean *EE* for general fullerenes, and a few parts per billion for isolated-pentagon fullerenes. The plot of *EE* against *n* is therefore dominated by the linear term, and has a slope only about 1% smaller than the predicted $\sinh(3)/3 \approx 3.339$ obtained from Eq. (14.9). this gives an answer to our first question of finding a simple approximation for *EE* of fullerenes.

Variation with fullerene isomer, though small, is systematic. At each *n*, the minimum value of *EE* is associated with an isomer that also has the smallest possible number of pentagon adjacencies, N_p. The isomer numbers given in the $N_{\min}$

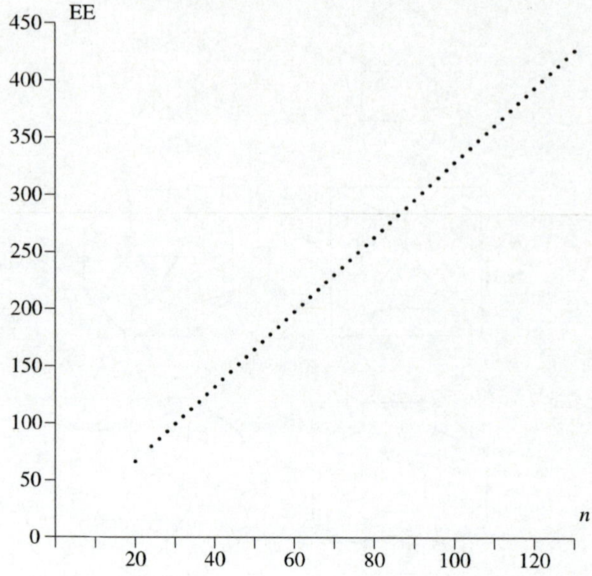

Fig. 14.5 Consolidated plot of the near-linear variation of Estrada Index (*EE*) with vertex count for general fullerenes and isolated-pentagon fullerenes with $n \leq 130$

column of Table 14.2 can be compared with those in published tabulations of minimal-adjacency fullerenes (e.g. Table 14.1 in the Atlas of Fullerenes (Fowler and Manolopoulos 1995)). In some cases (n = 30, 32, 34, 38, 42, 50, 52, 54, 58, 60, 70), the minimum-N_p isomer is uniquely defined and in others there are several minimal-N_p isomers, but in all cases the isomer with minimum *EE* has minimum N_p. Furthermore, as Fig. 14.6 shows for n = 40, 60 and 70, plots of the Estrada Index versus the number of pentagon adjacencies, are approximately linear, with slopes of $\sim 4 \times 10^{-4}$. This observation is readily rationalised on the basis of the expression for *EE* in terms of moments of the eigenvalue spectrum. The first few moments for a general fullerene on n vertices are linear functions of n alone, and hence independent of isomer. They are:

$$\mu_0 = n,\ \mu_1 = \mu_3 = 0,\ \mu_2 = 3n,\ \mu_4 = 15n,\ \mu_5 = 120,\ \mu_6 = 93n - 120,\ \mu_7 = 1680. \quad (14.12)$$

The presence of the pentagonal defects gives rise to constant corrections of the moments μ_5 and μ_7 that would vanish for an (unrealisable) all-hexagonal tessellation of the sphere. The influence of the degree of aggregation of the pentagonal defects appears first at μ_8. Pentagon pairs introduce 8-cycles, and fused pentagon triples introduce extra 9-cycles, and these influence μ_8 and μ_9 in the obvious ways. Pentagon pairs also affect μ_9 indirectly as they reduce the number of possible pentagon-hexagon contacts and hence remove possible 9-cycles. The resulting expressions are

$$\mu_8 = 639n - 1920 + 16N_P \quad (14.13)$$

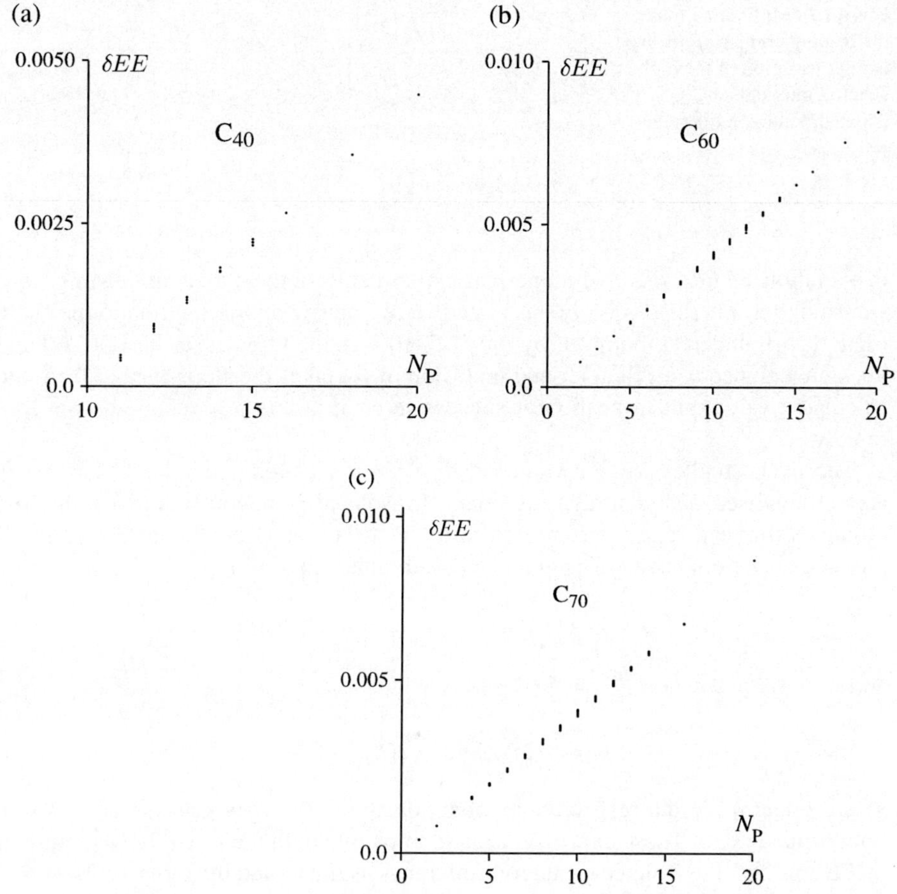

Fig. 14.6 Variation of Estrada Index with pentagon-adjacency count in fullerenes (**a**) C_{40}, (**b**) C_{60} and (**c**) C_{70}. The excess Estrada Index δEE is defined with respect to the isomer that minimises both EE and pentagon adjacency count, N_P

and

$$\mu_9 = 18360 - 36N_P + 18N_T, \qquad (14.14)$$

where N_P is the number of edges common to two pentagons and N_T is the number of vertices common to three pentagons, i.e., the numbers of (possibly overlapping) fused pentagon pairs and fully fused pentagon triples. These moments make small and opposed contributions of $+16/8! \sim +4 \times 10^{-4}$, and $-36/9! \sim -1 \times 10^{-4}$, respectively, to the slope $(\partial EE/\partial N_P)$, and there are further contributions of each sign from higher moments.

Although the moment expansion (4) is not well converged, in the sense that the absolute error incurred by truncation at μ_8 is not small compared to the range

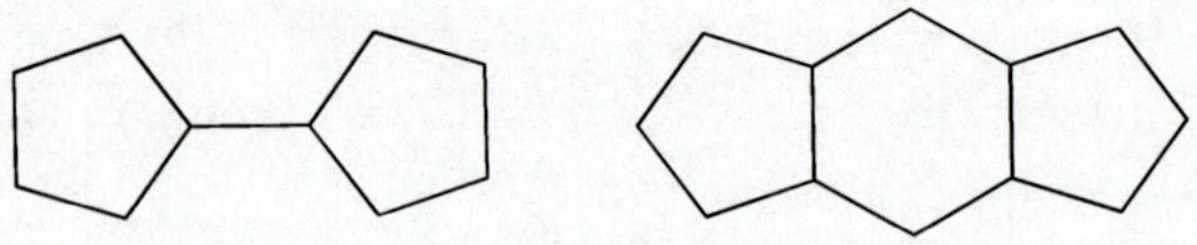
I II

Fig. 14.7 Motifs that make the leading contributions to isomer variation of the higher spectral moments of isolated-pentagon fullerenes

of variation of *EE* with N_P, numerical experiments suggest that the leading non-vanishing term $(1/8!)(\partial\mu_8/\partial N_P)$ is already a reasonable approximation to the slope $(\partial EE/\partial N_P)$, underestimating it by only ~5–10% in the three cases $n = 70$, 60 and 40. Convergence is oscillatory, and inclusion of μ_9 takes the slope further from the calculated value, but truncation at successive even orders μ_{2r} seems to converge rapidly.

The even smaller isomer variation of *EE* for isolated-pentagon fullerenes is also rationalised with a similar argument. In isolated-pentagon fullerenes, the first moments that can reflect differences in the distribution of pentagons are μ_{12} and μ_{13}, which are quoted as (Zhang and Balasubramanian 1993)

$$\mu_{12} = 35169n + 120p + 24q - 240120 \tag{14.15}$$

and

$$\mu_{13} = 1790880 - 260p - 52q, \tag{14.16}$$

where p and q are the respective numbers of motifs of types **I** and **II** (Fig. 14.7), both capped at 30. These expressions are compatible with the tiny relative variation of *EE* amongst the isolated-pentagon fullerenes as illustrated by Table 14.3.

Although the discrimination amongst fullerenes is weak, it seems still true at the larger values of n that the Estrada Index separates fullerenes from other cubic polyhedra. For example, at $n = 60$, the truncated dodecahedron, with the maximum 20 triangles, has an *EE* of 228.1096, some 16% greater than the C_{60} fullerene value.

The bipartivity index (10) does not show such regular behaviour for the small cubic polyhedral graphs. Maximum $\beta(G)$ is achieved for bipartite polyhedra, and when $\beta(G)$ is restricted to non-bipartite graphs it is not well correlated with *EE* or *F*, although it does show some intriguing clustering patterns. (Fig. 14.8). Amongst small fullerenes, the spread of values of $\beta(G)$ is again of the order of ppm (Table 14.4), but there is a general tendency of $\beta(G)$ to rise with N_P (Fig. 14.9). Minimisation of $\beta(G)$ selects minimum-N_P fullerenes, though not always the same isomers as chosen by minimisation of *EE* (compare the N_{min} columns of Tables 14.2 and 14.4). The (very small) positive gradient can be understood in terms of the contribution of μ_8 and the intuition that fullerenes with more 8-cycles are "more nearly bipartite". This correlation with overall stability has been noted before (Došlić 2005), though again it is dependent on a very small part of the total $\beta(G)$ function, and duplicates information that could be obtained in other ways.

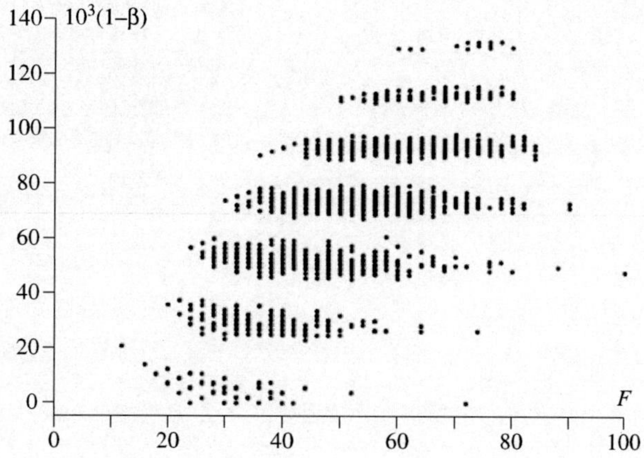

Fig. 14.8 Scatter-plot of the bipartivity index (β) and the F parameter for the set of 7595 cubic polyhedra on 20 vertices. The bipartivity is shown as a deviation from unity. The C_{20} fullerene corresponds to the leftmost dot

Table 14.4 Bipartivity Index of fullerene graphs with up to 60 vertices

n	N_{min}	β_{min}	β_{max}	$10^8 \Delta\beta_{max}$
20	1		0.9791852	
24	1		0.9825995	
26	1		0.9839181	
28	2	0.9850507	0.9850509	19
30	3	0.9860343	0.9860346	26
32	6	0.9868962	0.9868976	13
34	5	0.9876579	0.9876591	124
36	14	0.9883357	0.9883380	228
38	17	0.9889429	0.9889460	312
40	39	0.9894898	0.9894948	499
42	45	0.9899852	0.9899889	372
44	89	0.9904359	0.9904412	530
46	99	0.9908485	0.9908527	423
48	171	0.9912262	0.9912318	565
50	271	0.9915731	0.9915809	774
52	422	0.9918950	0.9919017	668
54	540	0.9921925	0.9921989	643
56	843	0.9924696	0.9924764	689
58	1205	0.9927270	0.9927336	663
60	1812	0.9929660	0.9929757	964

n is the number of vertices, β_{min}, β_{max} and $\Delta\beta_{max}$ are the minimum, maximum and range of the index within the set of fullerenes. N_{min} is the fullerene isomer (labelled by position in the spiral order) that has $\beta = \beta_{min}$. For 62 $n \leq 70$, the isomers of minimum β are 62:2194, 64:3451, 66: 4169, 68: 6073, respectively

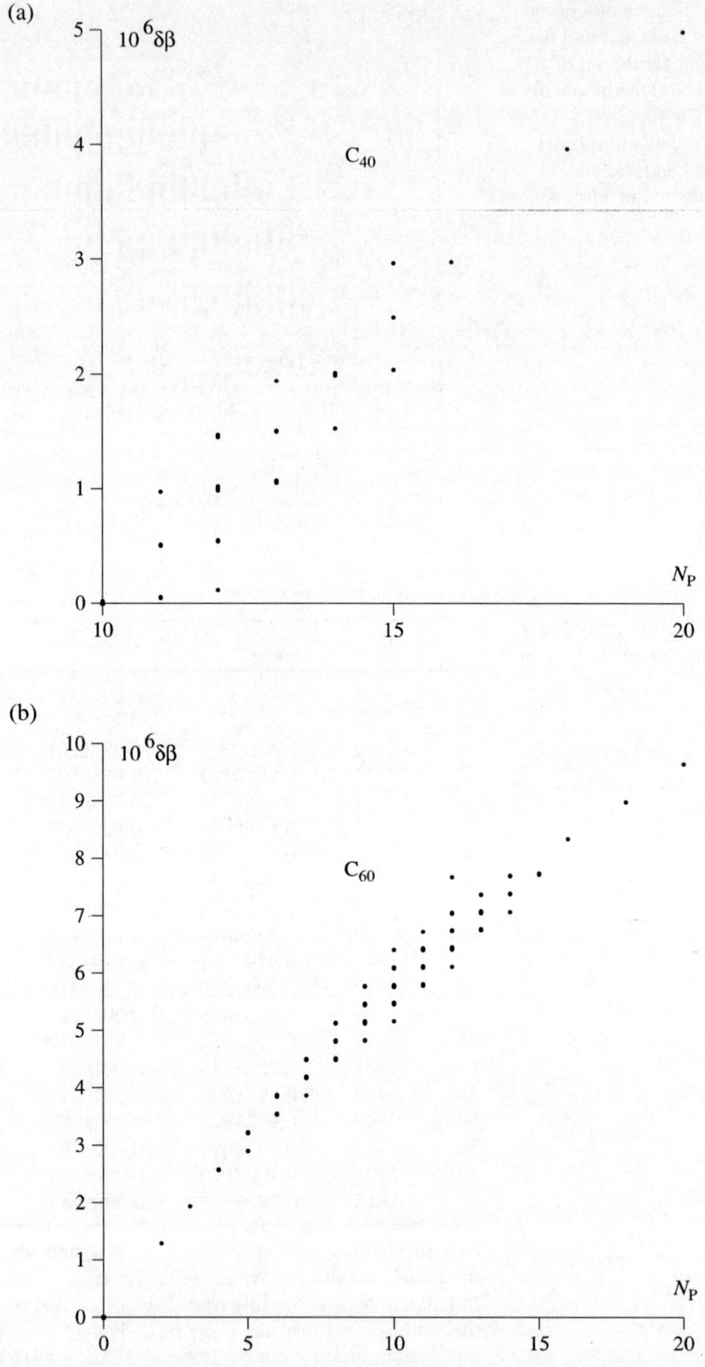

Fig. 14.9 Variation of bipartivity index with pentagon-adjacency in fullerenes (**a**) C_{40} and (**b**) C_{60}. The excess, $\delta\beta$, is defined with respect to the fullerene isomer that the minimises β and the pentagon adjacency count, N_P

14.3 Conclusion

As a potential index of carbon cage stability, the Estrada Index does show some selectivity in picking out fullerenes and near-fullerenes from the general cubic poyhedra, and its extremal values can be interpreted in terms of the face recipes of the polyhedra. This discriminatory power arises from the link between eigenvalues and circuit structure, through the spectral moments. The same link gives rise to the (very small) variation with fullerene isomer, which edges the Estrada Index into the company of indices such as Resistance Distance, Wiener Index, Balaban Index and Complexity/spanning tree count (Fowler 2002, 2003). All these quantities show trends with the number of pentagon adjacencies, a simple invariant which remains the most straightforward indicator of fullerene overall stability.

References

Brinkmann G, Dress AWM (1997) J Algorithms 23:345–358 A constructive enumeration of fullerenes

Brinkmann G, Dress AWM (1998) Adv App Math 21:473–480 PentHex Puzzles: A Reliable and Efficient Top-Down Approach to Fullerene-Structure Enumeration

de la Peña JA,. Gutman I, Rada J (2007) Linear Alg Appl 427:70–76 Estimating the Estrada Index

Domene MC, Fowler PW, Mitchell D, Seifert G, Zerbetto F (1997) J Phys Chem A 101:8339–8344 Energetics of C_{20} and C_{22} fullerene and near-fullerene carbon cages

Došlić T (2005) Chem Phys Lett 412:336–340 Bipartivity of fullerene graphs and fullerene stability

Estrada E (2000) Chem Phys Lett 319:713–718 Characterization of 3D molecular structure

Estrada E (2002) Bioinformatics 18:697–704 Characterization of the folding degree of proteins

Estrada E (2004) Proteins 54:727–737 Characterization of the amino acid contribution to the folding degree of proteins

Estrada E, Hatano N (2007) Chem Phys Lett 439:247–251 Statistical-mechanical approach to subgraph centrality in complex networks

Estrada E, Rodríguez-Velázquez JA (2005a) Phys Rev E 71:056103 (1–9) Subgraph centrality in complex networks

Estrada E, Rodríguez-Velázquez JA (2005b) Phys Rev E 72:046105 (1–6) Spectral measures of bipartivity in complex networks

Estrada E, Rodríguez-Velázquez JA, Randić M (2006) Int J Quantum Chem 106:823–832 Atomic branching in molecules

Fowler PW (2002) Croat Chem Acta 75:401–408 Resistance distances in fullerene graphs

Fowler PW (2003) MATCH Commun Math Comput Chem 48:87–96 Complexity, Spanning Trees and Relative Energies in Fullerene Isomers

Fowler PW, Manolopoulos DE (1995) An atlas of fullerenes. Clarendon Press, Oxford. Reprinted: Dover, New York, NY (2006)

Ginosar YI, Gutman IT, Mansour TM, Schork M (2008) Chem Phys Lett 454:145–147 Estrada Index and Chebyshev polynomials

Grünbaum B, Motzkin TS (1963) Can J Math 15:744–751 The number of hexagons and the simplicity of geodesics on certain polyhedra

Gutman I, Estrada E, Rodríguez-Velázquez JA (2007a) Croat Chem Acta 80:151–154 On a graph-spectrum-based structure descriptor

Gutman I, Furtula B, Glisić B, Marković V, Vesel A (2007b) Indian J Chem A 46:723–728 Estrada Index of acyclic molecules

Gutman I, Graovac A (2007c) Chem Phys Lett 436:294–296 Estrada Index of cycles and paths

Gutman I, Radenković S (2007d) Z f Naturforsch A 62:254–258 Estrada Index of benzenoid hydrocarbons

Gutman I, Radenković S, Furtula B, Mansour T, Schork M (2007e) J Serbian Chem Soc 72:1321–1327 Relating Estrada Index with spectral radius

Gutman I, Radenković S, Graovac A, Plavšić D (2007f) Chem Phys Lett 446:233–236 Monte Carlo approach to Estrada Index

Manolopoulos DE, May JC, Down SE (1991) Chem Phys Lett 181:105–111. Theoretical studies of the fullerenes:C_{34}–C_{70}

Schwenk AJ (1979) Ann NY Acad Sci 328:183–189 Spectral reconstruction problems

Zhang H, Balasubramanian K (1993) J Phys Chem 97:1034110345 Analytical expressions for the moments and characteristic polynomials of fullerenes containing isolated pentagons